TORRANCE PUBLIC LIBRARY
3 2111 00874 1222

W9-AUL-933

Modern Cosmological Observations and Problems

BY

GREG BOTHUN

University of Oregon

UK Taylor & Francis Ltd, 1 Gunpowder Square, London EC4A 3DE
USA Taylor & Francis Inc., 1900 Frost Road, Suite 101, Bristol, PA 19007–1598

British Library Cataloguing-in-Publication Data

A catalogue record for this book is available from the British Library.

Library of Congress Cataloging Publication Data are available

ISBN 0–7484–0332–9 HB
ISBN 0–7484–0645–X PB

Cover design by Amanda Barragry

Typeset in 10/12pt Times by Graphicraft Typesetters Ltd., Hong Kong

Printed by TJ International Ltd, Padstow, UK

Contents

List of Colour Plates

DEDICATION

FOR JOSH AND SAM

(. . . this is sort of what dad does . . .)

Preface

As of December 1996, cosmology is an extremely data-rich field. In the last five years a tremendous amount of cosmologically relevant data has come from the Hubble Space Telescope (HST), the *Cosmic Background Explorer* (*COBE*), the 10 m Keck telescope and other facilities. This book is designed to be an up-to-date summary of much of this data and how it impacts current generations of cosmological models. It is written from the observational point of view and follows a consistent theme of going where the data leads within the overall framework of standard Big Bang cosmology. It is intended for use in advanced undergraduate courses or beginning graduate courses in observational cosmology. Readers should be familiar with partial derivatives and the calculus of several variables. Given the rapid state of flux in this field, it is hoped this book provides a reasonable summary of the current state of cosmological observations.

The cosmology of today is quite a bit different than the kinds of observations and models that existed just 10 years ago. Over the last decade, several new kinds of theories and observations have come into existence and our expectations of a simple and well-behaved large-scale Universe are proving to be naive. Also gone is our perception of a quiet expanding local Universe and a sensible large-scale galaxy distribution. Theory today is bombarded by a vast array of observational data but there remains no clear and preferred model for the origin and evolution of structure in the Universe. This is the principal theme of the book: to outline in understandable terms what the latest observations are and how they are either consistent or in conflict with competing cosmogenic scenarios. An interesting outcome of this exercise is the realization that many of the conflicts between data and theory can be resolved by resurrecting an old idea, the cosmological constant.

To keep with the current state of information delivery, there is a World Wide Web site associated with this book in which updates will be maintained. A great deal of cosmologically relevant data now exists online and the Web site will serve as a pointer and organizer of these other resources. It will also contain some figures that could not appear in this book due to space or budgetary limitations. The Web address is `http://zebu.uoregon.edu/tandf.html`. In this way, I hope that the basic foundation of the book can be kept up to date and that the Web site, together with the subject-oriented resource lists in the book (and on the Web site), will continually provide interested students with a source of material. In addition, suggested student exercises centered around using real data can be found at the Web site.

Collecting and organizing the information that is contained in this book turned out to be a far more daunting challenge than originally envisioned. Perhaps this is why there are rather few textbooks available at the advanced undergraduate level. Significant new results are beginning to come out on a monthly basis and the number of cosmological models which are now in the literature is truly staggering. I have done my best to summarize and analyze the most recent data and models. Had this book been completed in 1995 it would contain substantially less information than it currently does. Hopefully this does not mean that the book is obsolete in 1997! Many elements of basic cosmology are retained in this book. Those elements have been drawn from four excellent sources of material whose overall scope is considerably more advanced than the material contained herein. The present book is designed to be an observational stepping-stone to reading and understanding these more advanced discussions of cosmology:

- *Galactic Dynamics* by James Binney and Scott Tremaine. This is the sourcebook for understanding all possible dynamical features of stellar systems and provides some of the starting points for derivations done here.
- *Principles of Physical Cosmology* by James Peebles. A classic book that is extremely comprehensive in the discussion of cosmological models. This book has a heavy theoretical basis and it is hoped that our discussion of observational data is a complement to this fine book.
- *The Early Universe* by Rocky Kolb and Michael Turner. These authors are among the most eloquent voices that describe the inflationary paradigm and the exotic physics of the early Universe. This book is heavily based on theoretical particle physics and general relativity.
- Michael Strauss and Jeff Willick have written an excellent summary of peculiar velocities and the large-scale mass distribution. This summary appears in *Physics Reports* **261** (1995) 271. It too is a mostly theoretical approach to the general problem of inferring the mass distribution from redshift surveys and or measures of deviation from expansion motion. It is a very fair and realistic appraisal of the difficulties involved and contains a genuinely refreshing discussion about the role of biases in extragalactic samples. It also reinforces a major theme in this book – no adequate structure formation theory matches all the observational constraints.

The organization of this book is as follows:

- *Chapter 1* is a brief historical summary of cosmological model making followed by a discussion of the foundations of our modern cosmological model.
- *Chapter 2* contains a detailed discussion on the extragalactic distance scale and incorporates the very latest results from the HST. It demonstrates that no consensus is yet available on the measured expansion rate of the Universe.

- *Chapter 3* examines the remarkably complex structure that is now evident in the galaxy distribution. Large voids, great linear chains of galaxies, great attractors which perturb the expansion motion of galaxies, all are part of this structure which stands in stark contrast to the remarkable homogeneity of the Cosmic Microwave Background.
- *Chapter 4* details those observations that establish the presence of dark matter, its distribution and its potential contribution to the overall mass density of the Universe. It discusses potential dark matter candidates in the contexts of astrophysics and particle physics.
- *Chapter 5* focuses on how the nature of dark matter gives rise to different scenarios for the formation of the observed structure under the general framework of gravitational instability, which is explored in detail. The two competing theories of the mid 1980s, hot dark matter (HDM) and cold dark matter (CDM), have given way to a more complex mixture of HDM and CDM in an attempt to account for the most recent observations.
- *Chapter 6* contains material that is generally not found in standard cosmological reviews but which is nevertheless important. It focuses on how the baryons are distributed in the Universe and discusses evidence for, or limits on, the diffuse background radiation in the Universe at different wavelengths. Alongside this is the evidence for a new population of very diffuse galaxies that have significantly altered our views of galaxy formation and evolution. These newly discovered low surface brightness (LSB) galaxies may constitute as much as 50% of the general galaxy population. Their discovery plainly shows that we should never be surprised at what observations of the Universe have to offer.

Over the years, the author has learned the most about cosmology in discussions with several colleagues. Many of them have provided invaluable data and/or comments on this book and, in reality, this work is the aggregate sum of many contributions. I would specifically like to thank Margaret Geller for telling me to always go where the data leads, Bob Schommer for showing me that there is nothing like good data, Jeremy Mould for helping to put me in a position where I could actually write this book and Wal Sargent for teaching me what a perspective is.

I would also like to acknowledge the following individuals for helpful, insights and valuable data: Matt Bershady, Mike Disney, Richard Ellis, Harry Ferguson, Wendy Freedman, Craig Hogan, Chris Impey, Christine Jones, Rob Kennicutt, David Koo, Limin Lu, Dave Monet, Rachel Pildis, Joel Primack, Chuck Steidel, Christopher Stubbs, Michael Strauss, Don Vandenberg, Ted von Hippel, Michael West, and Ned Wright.

CHAPTER ONE

Cosmological Model Building

1.1 Overview

A measure of the advancement of a particular civilization may well lie in its ability to provide comprehensible answers to fundamental questions. Two examples of such fundamental questions are (1) What is the nature of the Universe and (2) What is the Universe made of? Seeking the answers to these questions defines the discipline of physical cosmology. While this discipline is a convolution of elementary particle theory, general relativity and astronomical observations, there is still room for elements of mysticism and imagination in our cosmological models. Indeed, history has taught us that humans have an insatiable appetite for grand ideas about the nature of the world. Furthermore, at any given time in history, everyone always thinks their world model is correct. Since the answers to our two posed questions continue to elude us, it is safe to say that no cosmological theory has yet proven to be entirely satisfactory. Hence one should expect, and even demand, continual revisions and challenges to any existing cosmological paradigm.

This book will focus on recent challenges and observations in cosmology. Theory today is bombarded by a vast array of observational data but there remains no clear and preferred model for the origin and evolution of structure in the Universe. This book outlines some of the most outstanding observational problems associated with our modern cosmological model. Although much mathematical formalism is essential in order to convey the context of these problems, this book is not about exploring the physics of the early Universe or about the details of general relativity. Many excellent reviews of those areas are available and some are listed after Chapter 7. Instead it will focus on issues not normally addressed in one collection. In brief, it focuses on the current (1996) status of the following issues:

- Our observational attempts to determine the expansion rate (H_0) and density (Ω) of the Universe.

- Our characterization of large-scale structure and the discovery of deviations from pure expansion motions.
- Observational evidence for the the presence of dark matter, its distribution in the Universe and its overall nature (e.g., baryonic vs. nonbaryonic).
- A detailed investigation of the gravitational instability paradigm for structure formation and the current set of observational constraints that need to be satisfied by structure formation models.
- How many baryons are contained in bright, easy-to-detect galaxies compared to those that might constitute a diffuse background or are contained in very dim, diffuse galaxies that generally escape detection.

The resolution of these issues has profound influences on our cosmological model; they form the basis for intense observational inquiry utilizing the most sophisticated telescopes and detectors. Indeed, with a refurbished Hubble Space Telescope, the working 10 m Keck telescope and the coming of the general 8–10 m class telescope, one can look forward to an extremely data-rich era in cosmology. It is the theme of this book to go where the data leads and to use good data as a strong constraint on various cosmological models. As we shall see, when this is done, no existing model is able to explain the observations in full and this situation is likely to be exacerbated in this new era. This, however, is not disheartening but instead is an indication that we are merely in the preliminary stages of cosmological inquiry as we continue to gather and identify fundamental data. Indeed, this has been the case throughout our history of cosmological model making.

1.1.1 Ancestors of our modern cosmological model

Over the last few thousand years, cosmological models have migrated from the realm of the imagination towards the realm of certainty. Along the way, model making has been plagued by mistakes in reasoning and erroneous observations. Historical memory of cosmological model making and the methodology involved are important, as there are lessons to be learned. For instance, just as we now view the ancient notion of an Earth-centered Universe as absurd, we should be reminded that this model arose from a combination of observation and ignorance. That same state exists today, specifically in the area of dark matter, which is now invoked in a plethora of cosmological contexts, even though we have very little clue as to its real nature and only partial clues as to its very existence. Perhaps in another millennium or two, physicists will have discovered the existence of another long-range attractive force in the Universe, thus moving our modern cosmological model, with its various dark matter dominated Universes, into the same realm as the absurd Earth-centered Universe.

Our probing of the nature of the Universe is guided by observations and the kinds of questions that we pose. Lingering ambiguity is often the main result of this process but such is to be expected. Our inquiries about the cosmos are naturally primitive and naive. We peer through large pieces of polished glass at dim sources in the night, trying to discern the grand architecture of the Universe. It is a noble endeavor and, when combined with generational patience, will lead to understanding. Indeed, this process has been under way for some time now.

We begin 100 000 years ago when interaction with the environment was largely sensory and survival was the key. This was a world awaiting discovery, but for now, it was a world to be divided into animate and inanimate objects. As the millenia passed, human culture took root and began to flourish. With a growing awareness of self and a need to feel connected to the cosmos, humanity projected its personality onto the celestial canopy. Certain patterns of stars were endowed with human or animal characteristics. Mythologies were created to explain the origin and workings of the natural and perhaps supernatural universes. Observations (against some very dark skies) of the cycles of sunrise and sunset, lunar phases, and seasonal motions of the stars, became ingrained in the cosmological models created by various cultures. It is interesting to note that, despite variations in the details of individual cultural mythic cosmologies, most of them share the common concept of the Cosmic Womb. The primal female gives birth to the Universe and there is no distinction between organic and inorganic objects as all are made in Her Womb. The Cosmic Womb represents an original and profound idea in cosmological model making because it embodies the concept that the entire Universe was created in a single event.

1.1.2 The legacy of the Ancient Greeks

Bearing in mind that a rich diversity of cosmological models have been proposed by many different peoples and cultures throughout history, let us focus on the model which has undoubtedly exerted the greatest influence on our modern cosmological theory – the model of the Ancient Greeks. The advent of Pythagorean geometry coupled with a bias towards perfection in the Universe, ultimately led to the Aristotelian (circa 350 BC) system of crystalline spheres, with the Earth (Athens?) at the center. A key feature of this cosmological model is the important concept that the Universe is finite. The earlier Universe of mysticism, wonder and discovery had now evolved to a logical, apparently simple, geometric Universe which could now be drawn on a small piece of paper! Figure 1.1 (see plate section) shows this simple two-sphere Universe.

The physical foundation of this Universe was defined in terms of four primal forces – earth, air, fire and water. All that is observed or created in this Universe could be understood as the interaction between these primal forces. This was a highly deterministic Universe where the role of chance was explicitly dismissed. However, roughly 100 years before Aristotle, Democritus raised a dissenting view. In his cosmology, nothing existed except for "Atoms and the Void." All else, he argued, was "Opinion and Illusion." Furthermore, in this cosmological model, the Universe essentially ran by itself and was, arguably, a random occurrence, resulting from the collisions of atoms.

Interestingly, much of this framework crafted by the early Greeks is still retained in our modern cosmology. In particular, (1) we still use geometry as the defining characteristic of our cosmological model, (2) we still attribute all observations to the interaction of four forces, in this case the strong and weak nuclear forces together with gravity and electromagnetism and (3) the role of chance, as described via quantum mechanics, plays a key feature of the very early Universe. These three ingredients define an evolving Universe which, as the Greeks understood, becomes available to us through observation and measurement.

1.1.3 Towards Renaissance cosmology

The two-sphere Universe of Aristotle was to persist, relatively unaltered, for about 1700 years. Possibly this was a consequence of the loss of distinction between cosmology, philosophy and religion; as all three were now unified under Aristotle's simple geocentric model. Such a model has a powerful allure and may explain why, instead of using unbiased observations to test and refine the model, cosmologists of the time forced the physical world to conform to this simple geocentric model. Chief among these early cosmologists was Ptolemy (circa AD 100–150), who created a remarkably cumbersome model to account for the observed retrograde motion of the planets while keeping the earth the center of all planetary orbits. Figure 1.2 shows that many parameters are required to account for each orbit. Belief in an Earth-centered Universe was so strong that Ptolemy's model went unquestioned for centuries. The lack of challenges brought to the Ptolemaic model is an important historical reminder of the dangers of thinking that any particular cosmology – including ours today – is wholly correct.

In the mid sixteenth century, Copernicus (1473–1543) proposed a profound paradigm shift which swept away the previously accepted cosmology. With one idea, Copernicus removed the uniqueness of the Earth by placing the Sun at the center of the solar system. Although history rightly identifies Copernicus as starting this revolution, many of its basic tenets were written down ≈ 150 years earlier in the essay *On Learned Ignorance* by Cardinal Nicholas de Cusa. It postulated that:

- All motion was relative.
- No matter where one stood in the Universe the same pattern of stars would be "strewn out in front of them."
- The Earth may not be stationary (as demanded in the two-sphere Universe).

Embodied in these postulates are the themes of relativity, homogeneity and the explicit concept that the Universe has no center. The Copernican notion of non-uniqueness would appear to be a subset of these grander ideas. As is usually the case in history, the original source of these grand ideas remains obscure. Cardinal Nicholas may have received much of his inspiration from the Roman poet Lucreutis, who in about 100 BC wrote about the infinite, atomist, nondeterministic Universe

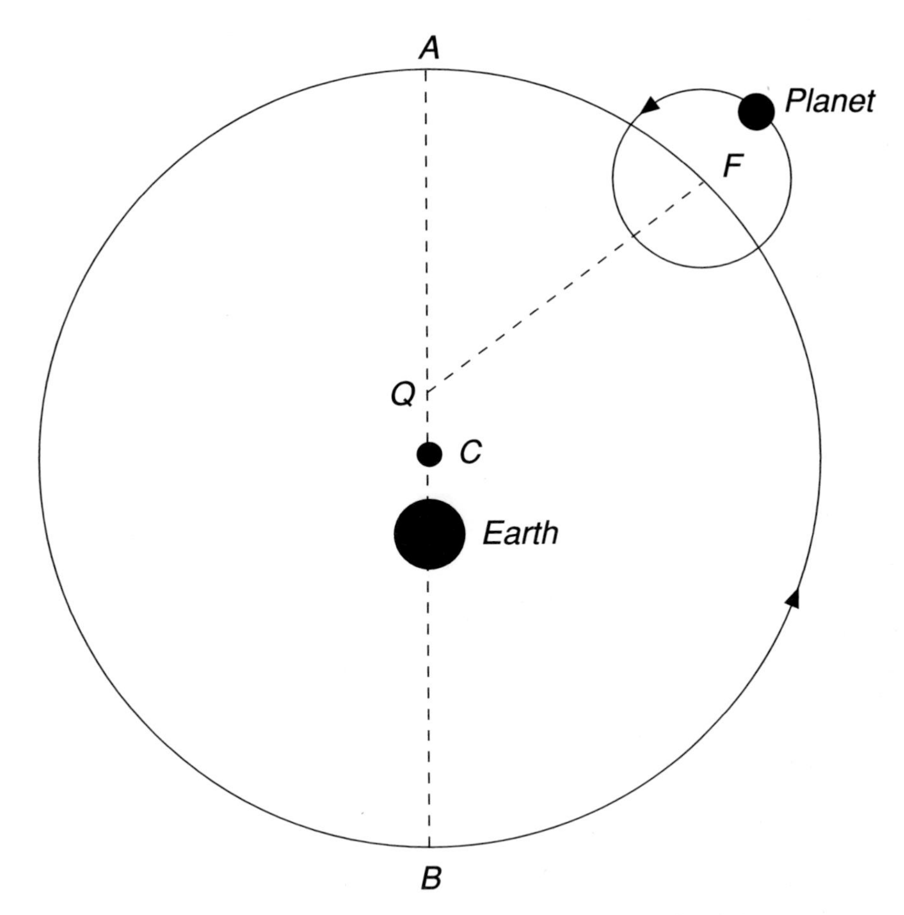

Figure 1.2 Representation of the Ptolemaic system showing the three main elements: the eccentric, the epicycle and the equant. Here the Earth is slightly offset from the center of the circle *C* defined by diameter *AB*. The planet, in orbit about the Earth, moves on an epicycle centered on *F* which is "affixed" to the larger circle defined by diameter *AB*. The planet's movement on the epicycle could qualitatively account for observed retrograde motion. The final element, the equant, represents a fixed point *Q*, about which rotates the epicycle center *F*. Since *Q* is offset from *C*, the distance between *F* and *Q* varies slightly throughout one orbital cycle, like the distance between the Earth and the planet.

in his poem *On the Nature of the Universe*. This poem was lost for centuries until it was discovered in 1417 in an Italian monastery, approximately 100 years before Nicholas's essay.

Support for the Copernican revolution came from other Renaissance scientists. Galileo (1564–1642), upon observing that, in the case of Jupiter, small objects moved about the bigger object, embraced the heliocentric cosmology of Copernicus by analogy with the Jovian–Galilean satellite system. Tycho Brahe (1546–1601)

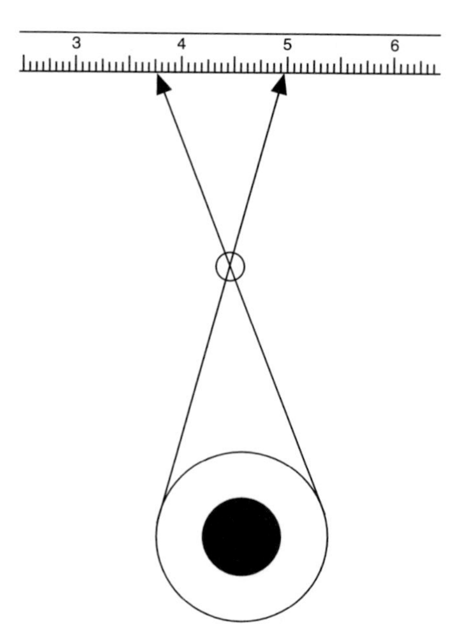

Figure 1.3 Schematic representation of stellar parallax. Distant stars act as a fixed background reference coördinate system (shown here as a ruler). Nearby stars, when observed 6 months apart, will show a small movement with respect to the background of fixed stars.

realized it was possible to prove that the Earth did indeed revolve around the Sun, through direct measurement of stellar parallax. Stellar parallax is the reflection of the annual motion of the Earth around the Sun, which causes a perceptible shift in the positions of nearby stars compared to more distant stars (Figure 1.3). Tycho made very accurate naked-eye measurements but failed to detect any positional shifts. Based on his failure to detect parallax, Tycho concluded erroneously that the heliocentric model must be incorrect. The alternative explanation, that the stars are too far away to allow for a naked-eye determination of stellar parallax, was apparently never considered by Tycho. This is but one of numerous examples throughout the history of cosmology when scientific judgment was clouded by human ego or naiveté; an experimenter who believes in the infallibility of their skills at detecting observational phenomena cannot be considered an impartial observer. In Tycho's case, the most scientifically plausible explanation for his nondetection of stellar parallax was simply that the effect was below the sensitivity of his experiment.

Still, Tycho did one very important thing that accelerated the pace of cosmological model building – he published his data, in particular his data on the position of Mars. These measurements were later used by Kepler (1571–1630) to empirically show that the orbits of the planets could not be perfect circles, as defined by Pythagoras and Aristotle and retained under the Copernican model. Armed with this improved positional data, Kepler was able to geometrically describe the orbits of the planets around the Sun and, for the first time, to set the relative scale of the solar system. Kepler's second law of planetary motion, that planets sweep out equal areas in their elliptical orbits at equal times, demands that orbital speed is a function of distance from the Sun. Hence, Kepler indirectly discovered the presence of a force acting upon the planets to change their velocity. However, Kepler's laws remained a set of empirical rules without a dynamical basis. The link between these laws and the physical world would be established about 50 years later by Isaac Newton (1642–1727).

1.1.4 Towards the physical Universe

Newton's fundamental insight into nature allowed the discovery of the R^{-2} force law of universal gravitation. From this, Kepler's laws follow trivially. We can readily derive Kepler's third law from Newton's equations as follows.

Assume that M_1 is a small mass in a circular orbit about a much larger mass M_2. We can write down the force law on M_1 using Newton's formulations:

$$F_g = -\frac{GM_1M_2}{R^2} \tag{1.1}$$

$$F_g = -M_1 a \tag{1.2}$$

Combining terms yields

$$a = \frac{GM_2}{R^2} \tag{1.3}$$

In an orbit governed by a central force, the centripetal acceleration is given by

$$\frac{-V_c^2}{R} \tag{1.4}$$

For a circular orbit, the circular velocity V_c is just $2\pi R/P$, where P is the orbital period. Equating (1.3) and (1.4) and writing V_c in terms of R and P then yields

$$4\pi^2\left(\frac{R^3}{P^2}\right) = GM_2 \tag{1.5}$$

in which the harmonic law of Kepler is now apparent as the R^3/P^2 term. Measurement of R and P for any lesser body (M_1) in orbit about a larger body (M_2) now leads directly to a physical quantity, mass.

But this derivation assumed some physical insight in the form of equation (1.4). The genius of Newton was his ability to use calculus and coordinate transformations to reveal the underlying physics. For example, equation (1.4) is really derived using the following coordinate transformation appropriate for an object in a circular orbit.

$$x = r\cos\theta, \qquad y = r\sin\theta \tag{1.6}$$

Newton defines the first time derivative of a spatial coordinate as a velocity and the second time derivative as an acceleration. An object in circular orbit has a constant acceleration in the r and θ directions. Differentiation of equation (1.6) yields

$$dx/dt = V_x = \frac{dr}{dt}\cos\theta - r\frac{d\theta}{dt}\sin\theta$$

$$dy/dt = V_y = \frac{dr}{dt}\sin\theta + r\frac{d\theta}{dt}\cos\theta \tag{1.7a}$$

A second differentiation with respect to time now yields

$$d^2x/dt^2 = \left[\frac{d^2r}{dt^2} - r\left(\frac{d\theta}{dt}\right)^2\right]\cos\theta - \left[\frac{rd^2\theta}{dt^2} + 2\frac{d\theta}{dt}\frac{dr}{dt}\right]\sin\theta$$

$$d^2y/dt^2 = \left[\frac{d^2r}{dt^2} - r\left(\frac{d\theta}{dt}\right)^2\right]\sin\theta - \left[\frac{rd^2\theta}{dt^2} + 2\frac{d\theta}{dt}\frac{dr}{dt}\right]\cos\theta \tag{1.7b}$$

and the accelerations in the r and θ directions are just

$$f_r = \frac{d^2r}{dt^2} - r\left(\frac{d\theta}{dt}\right)^2 \tag{1.8a}$$

$$f_\theta = \frac{rd^2\theta}{dt^2} + 2\frac{d\theta}{dt}\frac{dr}{dt} \tag{1.8b}$$

For a circular orbit d^2r/dt^2 is zero, so that

$$f_r = -r\left(\frac{d\theta}{dt}\right)^2 = \frac{-V_\theta^2}{r} \tag{1.9}$$

and equation (1.4) is recovered.

Newton's ability to predict Kepler's laws signifies the arrival of the first physical cosmology. Such a cosmology is defined by the ability of (unbiased) observation to provide information and insight into the physical properties of objects that inhabit the Universe. For Newton, these physical properties were mass, position, velocity, acceleration, energy and momentum. Objects with those properties would interact to define the Universe and the interaction mechanism was gravity.

1.1.5 How far to the nearest star?

For the next two centuries cosmological advances were tied to determining how big the Universe is (note that this effort continues today). During much of the eighteenth century a significant effort was made toward producing stellar catalogs with accurate positions from which determinations of stellar parallax could be made. This was an age of great anticipation, for the first accurate measurement of the stellar parallax of even just one star would provide the first observational determination of the size of the universe. The original stellar catalog of Tycho had a positional accuracy of one arcminute and the catalog of Edmund Halley (1656–1752) did not improve upon this, although Halley was able to detect the proper motion of nearby stars, thus at least identifying good candidates for parallax measurements. These efforts continued largely through the work of James Bradley (1693–1762) and William Herschel (1738–1822). Many attempts were made to use these data to make stellar parallax measurements but the results were not very consistent or accurate. The first accurate stellar parallax measurement of 0.32 second of arc for the star 61 Cygni was published by F.W. Bessel in 1838. This was followed in 1839 by Thomas Henderson's measurement of about one arcsecond for α Cen and the 1840 measurement of α Lyrae (Vega) at 0.26 arcsecond. With these three measurements, it now became clear that the distance to even the closest stars was about one million times larger than the Earth–Sun distance. Such vast distances implied that the energy outputs of stars had to be huge.

The accurate observations of double-star motions by Herschel now took on added meaning, since these motions were accurately described by Newtonian gravity, hence the same force that held over the (by now) small scale of the Solar System also held over very much larger scales. This is perhaps the first verification of an important principle, known as the cosmological principle, which asserts that the Universe must be homogeneous and isotropic. Put another way, the cosmological principle demands that the Universe is not arbitrary, hence any local physics must hold on larger scales. The motions of binary stars therefore showed that the same physics which held the Solar System together also occurred at a location many millions of times farther away.

1.1.6 General and special relativity

The early twentieth century brought a number of profound developments on the observational and theoretical fronts, which once again necessitated a fundamental revision to the accepted cosmological framework. To begin with, a refinement of Newtonian mechanics was required, and this was offered by Einstein in the form of the general theory of relativity. Although the theory is best understood in equation form (e.g., Weinberg 1972; Coles and Lucchin 1995), two of the most important ideas of general relativity theory can be summarized as follows:

- *Energy and mass are equivalent*: Any object which has energy also has mass and is therefore affected by gravity. Mass can be converted into energy, energy into mass. This discovery could now account for the enormous energy source that powered the stars and which could render them visible to us over distances of thousands of light-years.
- *Gravity is the manifestation of the curvature of spacetime*: Space communicates with matter and instructs it how to move; matter communicates with space and instructs it how to curve.

In addition to general relativity, Einstein also developed the case of special relativity, which asserts that the laws of physics are identical in all inertial frames. For this to be true, the speed of light cannot be different in different frames, even if those frames are in relative motion. This condition is met if the speed of light is a universal constant. This finite communication time, which was not a component of Newton's description of gravity, allows for cause-and-effect relations, which lead to the concept of causality. Among many things, special relativity shows that time runs slower for objects moving at high velocity; lengths also become shorter and masses increase. The increase in mass can be understood through relativity as a relation between the increase of kinetic energy in one frame and a mass increase in another. This leads to the important principle of equivalence, $E = mc^2$. The conditions of general relativity follow from special relativity but are applied to an accelerating frame (such as a gravitational potential). Hence inertial and gravitational forces are the same phenomenon, and there is an identity between inertial masses as derived from Newton's laws of motion and gravitational masses which produce the force.

The basic prediction is that Newtonian gravity is very accurate when the gravitational field is weak (meaning that space is locally flat) but breaks down when the gravitational field is very strong (meaning that space is locally curved). This was first manifested observationally when precision measures of the orbit of Mercury disagreed with Newtonian mechanics. The resolution is provided by general relativity. Mercury is sufficiently close to the Sun that it orbits in curved space, hence Newtonian mechanics provides an incomplete specification of its orbital parameters. The precession of the perihelion of Mercury's orbit and the bending of starlight when it passes near the Sun are in excellent agreement with predictions from general relativity and serve as its best verification.

1.1.7 Einstein's biggest blunder?

Einstein also realized that, since gravitation is a universally attractive force, the Universe is unstable to collapse. Newton realized this, too, and postulated that the Universe must be infinite in extent to avoid collapse. Einstein's method to avoid universal collapse was to postulate the existence of another field, which acts as a repulsive force. In the field equations this can be identified with a term that represents vacuum energy and acts as a source of negative pressure in the Universe. This source of pressure prevents the Universe from collapsing through an effective repulsive force, which grows with distance. In this way, gravitational and repulsive forces are balanced. However, this balance is on a point of unstable equilibrium. A slight perturbation in either direction would send the Universe into a state of terminal collapse or expansion. The possibility that the Universe was expanding on its own, without being assisted by vacuum energy fields, was not considered by Einstein. The observational discovery of universal expansion by Edwin Hubble would come a few years later and would prompt Einstein to consider his vacuum energy term his "Greatest Blunder." Ironically, it now seems that reinvoking this vacuum energy field may be the solution to some of our most profound cosmological problems.

1.1.8 The discovery of universal expansion

The last major step in observations which have led to the development of our modern cosmological model was provided by Hubble. Earlier observations by Slipher (1914) had shown that galaxies have spectral features (emission or absorption lines) and that most of them exhibit a Doppler shift towards longer wavelengths, commonly called a redshift. Hubble and Humason systematically measured more galaxy spectra and again found that most galaxies exhibited a redshift, indicating radial motion away from the observer. Hubble noticed that not all galaxies exhibited the same redshift, so the amplitude of the redshift had to depend upon some other property. But before researchers could discover it, they needed to establish whether these nebulae (galaxies) were objects located in our own galaxy or well beyond its boundaries. Resolution of this issue would require measuring the distance to the Andromeda nebulae (M31).

The first credible effort was made in 1922 by E. Opik, who derived a distance to M31 of about 450 kpc or approximately 1.5 million light-years – substantially larger than the estimated size of the Milky Way galaxy (Opik 1922). Opik reasoned that, since galaxies exhibited approximately equal flux ratios in different filters (that is, the colors of galaxies are not greatly dissimilar), then to first order, all galaxies have similar stellar populations, hence similar values of mass-to-luminosity ratio (M/L). If M/L does not vary greatly from one galaxy to another, then a galaxy's rotational velocity, which from Newtonian mechanics must be determined by M, is also an indicator of its intrinsic luminosity, since M/L is assumed to be constant. Under this assumption, the distance to M31 follows trivially from measuring its apparent flux.

Further support for the extragalactic nature of M31 came in 1925 when Hubble discovered variable stars in M31 that had similar properties to those detected earlier in the Large Magellanic Cloud (Hubble 1925). This provided a relative distance scale between the LMC and M31 and indicated that M31 was located at a distance of ≈ 300 kpc. Hubble discovered Cepheid variables in most of the galaxies in the Local Group, but in this environment, redshift is not well correlated with distance, as the Local Group is loosely gravitationally bound. Observations of Local Group galaxies would therefore not reveal universal expansion.

Hubble took spectra of fainter and smaller galaxies as well, and noticed that their observed redshifts were considerably larger than anything in the Local Group. Hubble also noticed that galaxies which were faint and had small apparent angular sizes, tended to have larger redshifts than galaxies which appeared bigger and brighter. Lacking a suitable means for determining distances to these smaller and fainter galaxies, Hubble made some assumptions about their nature. By assuming that galaxies were either of constant brightness or constant physical diameter, Hubble could deduce that the smaller and fainter galaxies with the higher redshifts were farther away than the brighter bigger galaxies with smaller redshifts. In this way, he could make a plot of galaxy distance versus observed galaxy redshift.

This plot, using Hubble's original data, is shown in Figure 1.4. Although it appears quite noisy, there is a general trend for more distant objects to exhibit

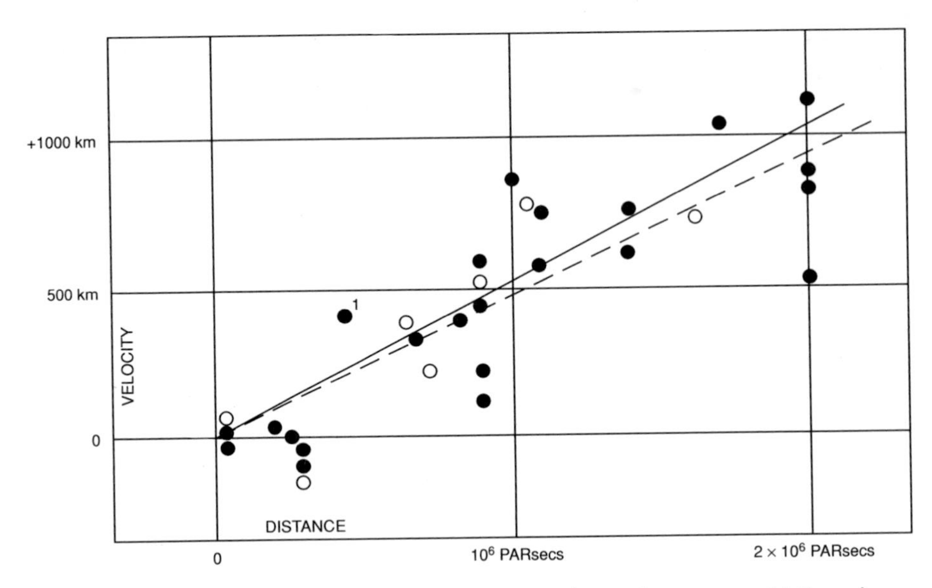

Figure 1.4 Hubble's original plot shows the correlation between redshift and distance for his sample of galaxies. Although the data is quite noisy, the overall trend is adequately represented by a linear law in which redshift is directly proportional to distance. That law is depicted by the solid line. Reproduced from Hubble (1929).

a larger redshift. Hubble's data was sufficient to empirically demonstrate that recessional velocity is proportional to distance, hence the Universe is in a state of uniform expansion. Einstein's general relativity and the static Universe could now be resolved – the Universe itself was expanding. This determination of uniform, linear expansion of the Universe has a clear prediction. If galaxies are moving apart from one another today, then in the past they must have been closer together. Indeed, there must have been a time when all the galaxies in the Universe were together in the same space. At this time, the Universe was very dense and in a physical state well removed from how it appears today.

1.1.9 Models of a dense Universe

In 1929 Hubble presented the observational evidence for an expanding Universe (Hubble 1929). In 1934 Tolman made an analogy between the expanding Universe and a thermodynamic state (Tolman 1934). Equating expansion with the thermodynamic concept of entropy led him to conclude that the expansion must cool any background which has a thermal spectrum. Chandrasekhar and Heinrich (1942) later suggested that if the Universe in its early history achieved thermal equilibrium at conditions of temperature and density near 10^{10} K and 10^7 g cm^{-3}, then equilibrium abundances of lighter elements would have frozen out in ratios that roughly agreed with observations. But this argument is only partially correct, since the high matter density at the time would have required rapid expansion (to avoid early collapse of the Universe), hence equilibrium calculations are not appropriate. Gamow (1942, 1946) used this as the basis for his set of arguments that physical processes in the early Universe were dynamic in nature, as they must have occurred in a rapidly expanding and cooling environment.

Gamow (1948) later reasoned that, at sufficiently high temperatures (kT in excess of the rest mass energy of a neutron), the energy density in the photon field must have greatly exceeded the energy density of the matter field. These conditions allow the radiation field to photodissociate any nuclei, so the early Universe had to consist of photons and free elementary particles. Without going into detail, the ability for a neutron–proton capture to occur (thus creating deuterium and helium) is strongly related to both the number density of neutrons and protons as well as the energy density in the photon field (fixed by the number density of the photons and the temperature of the Universe). The current cosmological abundance of deuterium (a difficult observation to make) and helium thus tells us much about these early physical conditions. Gamow used the estimated abundances of helium to predict that, at the current epoch, the photon field has been redshifted to millimeter wavelengths. Similar conclusions were also reached by Alpher and Herman (1948).

1.1.10 The discovery of the Cosmic Microwave Background

Gamow's prediction would be verified in 1965 with the accidental discovery of the Cosmic Microwave Background (CMB) by Penzias and Wilson of Bell Laboratories (Penzias and Wilson 1965). A curious coincidence also led to the unknown discovery of the CMB as early as 1941. The energy density in the CMB is very similar to the energy density of ambient starlight in our own galaxy. Observations of the excited states of the CN molecule made by McKellar (1941) indicated excitation by a photon background with a characteristic temperature of 2.3 K. But since the CN molecule could only be found where our galaxy was, the excitation was attributed to the ambient energy density of starlight in our galaxy, not to energy outside it (i.e., the CMB). The initial measurements of Penzias and Wilson indicated that the flux density of photons at their millimeter receiver was independent of position in the sky. Failing to see any 24 h modulation of this signal, the remaining logical conclusion was that the CMB was indeed of cosmological origin and therefore everywhere. Further observations of this CMB showed that its spectral signature was consistent with a blackbody, as predicted earlier by Gamow.

Today we have a precise measurement of the CMB due to the *Cosmic Background Explorer* (*COBE*) instrument launched by NASA in 1989. *COBE* made very sensitive measurements of the CMB and any anisotropy that might be present (crucially important to the structure formation models discussed in Chapter 5). The *COBE* measurements indicate that the CMB is a nearly perfect blackbody characterized by a temperature of 2.74 ± 0.02 K. Prior to *COBE*, some balloon-borne experiments indicated slight deviations from a perfect blackbody spectrum, including the now infamous submillimeter excess in the CMB that many theorists

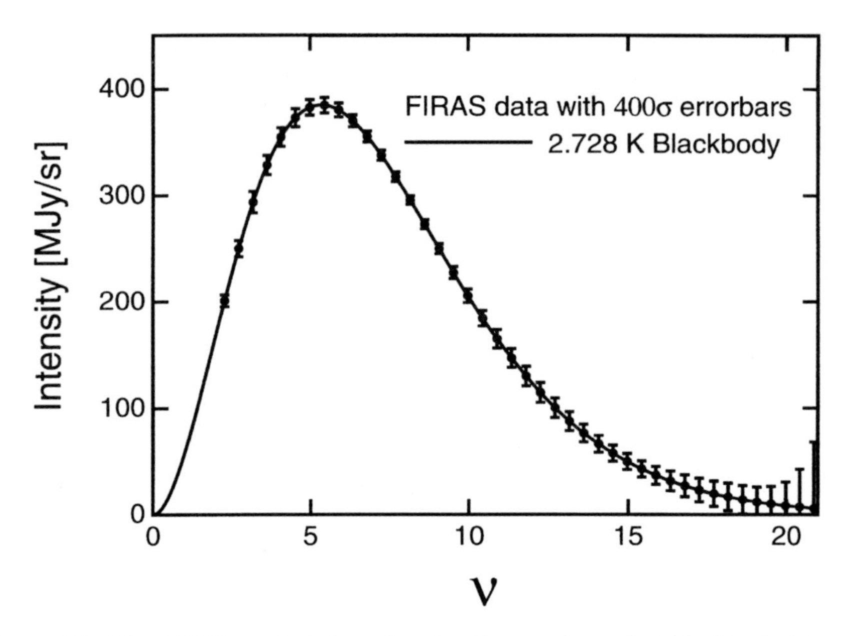

Figure 1.5 The *COBE* spectral data showing the nearly perfect blackbody fit. The plotted error bars are 400 times greater than their actual value, in order that some idea of the errors can be represented. Each data point represents one of the 42 spectral bands of *COBE*. The *x*-axis is the wavenumber in cm^{-1} and the *y*-axis is the intensity of the microwave emission. Courtesy of Edward L. Wright (UCLA).

attributed to the presence of hot dust in the early Universe. However, there is nothing like good data, and the spectrum obtained by the *COBE* mission (Figure 1.5) completely rules out any additional contributions to the CMB.

1.1.11 The Hot Big Bang model

On the basis of two observations – (1) the Universe is currently in a state of uniform expansion and (2) the Universe is filled with photons that come from background radiation at a temperature of 2.74 K – we can construct our generic cosmological model known as the Hot Big Bang model. This model is but a mere 30 years old, so we should not expect it to be a complete description of what we observe, and indeed it is not. But there are some things the model can explain well:

- *The origin of the CMB*: From its initially extremely hot and dense state, the CMB arises due to photon production from matter–antimatter annihilation. Once the photon background is produced, it simply cools with the expansion of the Universe in the thermodynamic manner first specified by Gamow. The observed entropy S of the Universe, measured by the ratio of baryons to photons, is $\approx 5 \times 10^{-10}$. Essentially this means for every five billion autiquarks, there were five billion and one quarks. The residue of this tiny mass asymmetry has produced the observed stars and galaxies in the Universe.

- *The uniform expansion of the Universe*: By simple extrapolation to very small times, the currently observed expansion law, $V = HD$, must produce a situation where all the matter is concentrated into a small volume. At some point in this small volume, unknown physics initiated the expansion. Since the observed expansion today shows no preferred direction, the expansion became isotropic very early on.
- *The abundance of light elements*: The Universe at an expansion age of about 1 second was a hot and dense mixture of electrons, protons, neutrons, neutrinos and photons. The ratio of protons to neutrons at this time was unity as interactions with neutrinos mediated the neutron-to-proton and proton-to-neutron conversion. However, at an expansion age of 2 seconds, the Universe had cooled to the point where it became transparent to neutrinos and this mediation was gone. Since free neutrons decay with half-life ≈ 900 seconds, the proton-to-neutron ratio (p/n) began to increase. As the Universe continued to expand, it cooled to the point where some of the nucleons could fuse into light elements, such as deuterium and helium, through the same series of fusion reactions that are presently occurring in our Sun. At the time of this hydrogen-to-helium fusion, the value of p/n was 7. Thus for every 14 protons there were two neutrons. Since the proton–proton chain ultimately binds two protons to two neutrons, so that ^{4}He is produced, then 12 protons result along with a ^{4}He nucleus. Since the mass of ^{4}He is very nearly equal to the mass of four protons (hydrogen atoms), an initial p/n of 7 makes a definite prediction about the ^{4}He baryonic mass fraction of the Universe. Since the total baryonic mass of the Universe is essentially the mass in hydrogen plus the mass in helium, the helium mass fraction is given by

$$\text{Helium mass fraction} = \frac{\text{mass of helium}}{\text{mass of hydrogen} + \text{mass of helium}}$$

or, in units of the proton mass, $4/(4+12) = 1/4$. The observed abundances of deuterium, ^{3}He and ^{4}He are all quite consistent with simple predictions based on the theoretical conditions of the Universe at time 1–2 seconds (Kernan and Sarkar 1996).

These three observations really form the observational foundation of the Hot Big Bang model. A fourth observation, that the Universe is filled with galaxies arranged in a complex structure, cannot easily be accounted for in this model. Although the general idea that structure formation via gravitational instability should produce observable anisotropies in the CMB is consistent with our observations, the overall complexity of the observed distribution of galaxies is not yet well understood. Later chapters will examine this issue in great detail.

1.2 Overview of relevant cosmological equations

1.2.1 The Robertson–Walker metric

To place the Hot Big Bang model into a physical context necessitates a sensible mathematical formulation. To assist with this formulation we assume that the

Universe on a large scale is both homogeneous and isotropic. This assumption is known as the cosmological principle and the observed isotropy of the expansion and the CMB are strong testaments to its validity. And having accepted its validity, our task is to construct a geometrical model of the Universe that explicitly incorporates large-scale homogeneity and isotropy. Ideally, this model should be described by a relatively small number of parameters, all of which can be observationally determined. Much of this book is devoted to a modern discussion of attempts to determine these parameters from observations. However, before doing that, we must describe the framework that allows observations to be directly connected to our cosmological model.

To begin with, we note that general relativity is a geometrical theory concerning the overall curvature of spacetime. Within that context we seek to specify the coordinate properties of a homogeneous, isotropic, expanding Universe. If we are to fully describe the Universe in geometrical terms, we must derive a metric which describes the coordinate paths that objects are allowed to take. In deriving this metric, we must introduce the concept of an **event**. An event is something which occurs at a certain place at a certain time. Hence all events in the Universe can be thought of as occurring in a four-dimensional **spacetime** continuum, with three spatial dimensions and one dimension of time. To compute the separation between any two events in spacetime, it is necessary to specify the **metric**. As a simple example, consider the surface of a sphere, which can be thought of as a two-dimensional analog to the four-dimensional spacetime. Using simple spherical trigonometry, the metric of a sphere can be written as

$$ds^2 = R^2[(d\phi)^2 + \cos^2\phi(d\theta)^2] \tag{1.10}$$

where ds denotes the distance between two points on the surface of the sphere, R is the radius of the sphere, and $d\phi$ and $d\theta$ are the difference in latitude and longitude between the two points (measured in radians). With this expression it is possible to compute the separation between any two points along the surface of the sphere. Hence the geometry of the sphere and the physical specification of events is completely described by its metric.

The geometry of four-dimensional spacetime is described by an analogous metric. However, instead of computing the distance between two points on the surface of a sphere, we wish to compute the separation between two events, which involves both space and time. Special relativity allows one to show that the spacetime interval ds between two events which occur near each other in flat space is given by

$$ds^2 = dt^2 - \frac{1}{c^2}(dx^2 + dy^2 + dz^2) \tag{1.11}$$

where dt is the time interval between the two events (as determined by an inertial observer), c is the speed of light, and dx, dy, dz correspond to the separation between the two events in each of the three spatial dimensions. Note that, unlike the metric for an ordinary sphere, the spacetime metric need not always be positive. The geometry of spacetime is completely specified by equation (1.11). A **geodesic** is the shortest interval between any two points in spacetime.

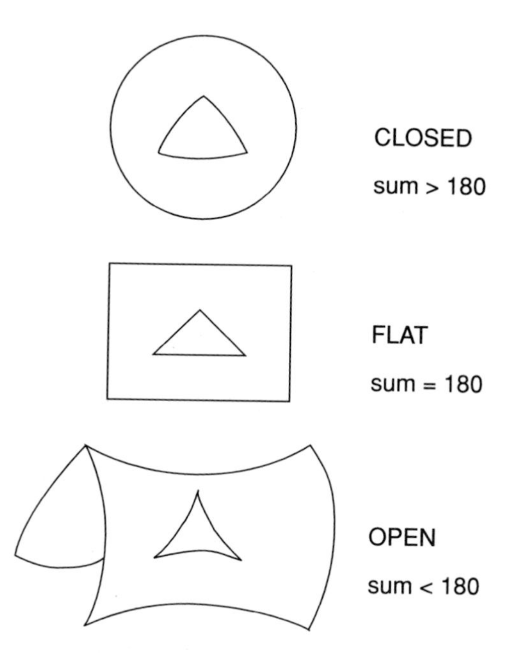

Figure 1.6 Schematic representations of the possible geometries of spacetime. These are two-dimensional analogs of curved space. In positively curved space the sum of the angles of a projected triangle on that surface is greater than 180°; in negatively curved space it is less than 180°.

Equation (1.11) assumes a flat Euclidean geometry, in which initially parallel lines always remain parallel. However, according to Einstein's theory of general relativity, spacetime is curved by gravity, which is a manifestation of the energy density of matter. The separation between two events will therefore depend on the curvature of spacetime. This is schematically shown in Figure 1.6 for three specific curvatures.

For a homogeneous and isotropic universe, the most general metric in curved spacetime is the **Robertson–Walker metric**, first proposed in 1934. Expressed in spherical polar coordinates (r, θ, ϕ), this metric takes the form

$$ds^2 = dt^2 - \frac{R^2(t)}{c^2}\left[\frac{dr^2}{1-kr^2} + r^2 d\theta^2 + r^2 \sin^2\theta\, d\phi^2\right] \tag{1.12}$$

Where $R(t)$ is the **universal scale factor** which describes the time evolution of the universal expansion, k denotes the curvature of spacetime ($k = -1,0,1$ for negative, zero, or positive curvature), and the coordinate r is **comoving** with the universal expansion. If we imagine a particle at rest with a given set of coordinates r, θ, ϕ, then as long as no external forces operate on this particle, the particle remains at those coordinates. These coordinates are called **comoving** coordinates. They are related to physical coordinates through the scale factor:

$$\text{Physical distance} = R(t) \times \text{comoving distance} \tag{1.13}$$

The Robertson–Walker metric is independent of any particular gravitational theory. Gravity enters through the scale factor $R(t)$ and the curvature constant k; the distance between any two spacetime events therefore depends on the choice of cosmological model. This specific cosmological model is determined by the values of $R(t)$ and k, which can be observationally determined.

1.2.2 The dynamics of an expanding Universe

Now we have specified the metric that holds in a homogeneous and isotropic Universe, the next step is to consider the dynamics of an expanding Universe. Since the Universe has mass, then at all times the expansion must compete against the combined (attractive) gravitational acceleration of that matter. In Newtonian mechanics this acceleration is given by

$$\nabla^2\Phi = -4\pi G\rho \tag{1.14}$$

where ρ is the matter density and Φ is the gravitational potential. This equation is historically known as Poisson's equation.

This simple equation for gravitational acceleration does not apply in the very early Universe, due to the presence of very high energy photons. The majority of the mass–energy in the early Universe is in the form of radiation moving at c. The high radiation pressure drags the matter along with it and effectively counters the tendency for the matter to collapse. In this sense, the Universe acts as a relativistic fluid with a pressure term whose behavior is not adequately described by Newtonian mechanics. The details about the stress–energy tensor in Einstein's field equations are beyond the scope of this book (see Weinberg 1972; Peebles 1993) but they lead to the following generalization of equation (1.14):

$$\nabla^2\Phi = -4\pi G\left(\rho + \frac{3p}{c^2}\right) \tag{1.15}$$

where p/c^2 is the pressure (which we subsequently set to p; $c = 1$) and the combined term $\rho + 3p$ effectively becomes the gravitational mass density ρ_g, which produces the net gravitational acceleration of material that decreases the expansion rate.

If we now consider a sphere of radius r_s and volume V which has some mean gravitational mass density within it, the total mass of that sphere is given by

$$M = \rho_g V = \tfrac{4}{3}\pi(\rho + 3p)r_s^3 \tag{1.16}$$

The acceleration at the surface of the sphere is given by Newton's law of gravitation as $-GM/R_s^2$. Multiplying equation (1.13) by the term $-G/R_s^2$ then yields

$$\text{Acceleration} = \ddot{r}_s = -\tfrac{4}{3}\pi G(\rho + 3p)r_s \tag{1.17}$$

where $\ddot{r}_s^2$ refers to the second time derivative of the spatial coordinate r_s (the first time derivative $\dot{r}_s$ is a velocity). Equation (1.17) is a standard equation in general relativity and it describes the evolution (e.g., expansion or contraction) of a homogeneous and isotropic mass distribution. Within this sphere there is some net energy E_n. This energy is ρV. If the material enclosed in r_s moves so that it changes r_s, then E_n changes in accordance with how much work is done by the pressure of the fluid on the surface of the sphere. By conservation of energy we then have

$$dE_n = \rho\, dV + V d\rho = -p dV \tag{1.18}$$

Equation (1.18) states that the change in net energy is exactly equal to the change in volume multiplied by the pressure. Rearranging the terms involving dV and defining the volume as $V = 4\pi r_s^3/3(\dot{V} = 3\dot{r}_s r_s^2)$ we obtain

$$\dot{\rho} = -(\rho + p)\frac{dV}{V} = -3(\rho + p)\frac{\dot{r}_s r_s^2}{r_s^3} = -3(\rho + p)\frac{\dot{r}_s}{r_s} \tag{1.19}$$

Solving for p in equation (1.19) and plugging that solution in equation (1.17) yields the following differential equation:

$$\ddot{r}_s = \tfrac{8}{3}\pi G \rho r_s + \tfrac{4}{3}\pi G \dot{\rho}\left(\frac{r_s^2}{\dot{r}_s}\right) \tag{1.20}$$

This is a messy differential equation. If we multiply both sides by the term $\dot{r}_s$, choose units such that the quantity $(4\pi/3)G = 1$ and let r_s be x, then we arrive at the following functional form:

$$\dot{x}\ddot{x} = 2\rho x \dot{x} + \dot{\rho} x^2 \tag{1.21}$$

Now the right-hand side is just

$$\frac{d}{dt}(\rho x^2) \tag{1.22a}$$

and the left-hand side is just

$$\frac{1}{2}\frac{d}{dt}(\dot{x}^2) \tag{1.22b}$$

Integrating both sides over time then yields

$$\frac{1}{2}\int \frac{d}{dt}(\dot{x}^2)dt = \int \frac{d}{dt}(\rho x^2)dt \tag{1.22c}$$

Switching back to normal units yields the first integral of equation (1.20):

$$\dot{r}_s^2 = \tfrac{8}{3}\pi G \rho r_s^2 + K \tag{1.23}$$

where K is a constant of integration which we can identify with the curvature term in the Robertson–Walker metric. Equations (1.17) and (1.23) are the main equations of this cosmological model.

If we consider the case of a static Universe, where r_s is constant by definition, then all derivatives are zero, and equations (1.17) and (1.23) become

$$0 = \tfrac{4}{3}\pi G(\rho + 3p), \qquad \tfrac{8}{3}\pi G\rho + K = 0 \tag{1.24}$$

Since the mass density ρ must be positive, then to satisfy the constraint of a static Universe, p must be negative. Since normal matter cannot have negative pressure, Einstein introduced the cosmological constant Λ into the field equations to serve as the source of negative pressure. In the static Universe, Λ balances the net gravitational acceleration. But the Universe is not static, it is expanding according to the expansion scale factor $R(t)$ given in equation (1.13). Our hypothetical sphere radius r_s will then be different at some later time, t, such that

$$r_s(t) = r_s(t = 0) \times R(t) \tag{1.25}$$

Equation (1.17) now becomes

$$\frac{\ddot{R}(t)}{R} = -\frac{4}{3}\pi G(\rho + 3p) + \frac{\Lambda}{3} \tag{1.26}$$

where the first of the two terms on the right-hand side is for the pressure and density of ordinary matter (e.g., stars and galaxies) and the second term includes the contribution of the cosmological constant. Equation (1.26) then describes the relativistic acceleration of the expansion, which in principle can be dominated by Λ at late times, as ρ and p decrease with time but Λ stays nearly constant. This in fact is the physical manifestation of nonzero Λ. In this case the Universe evolves from being radiation dominated, to being matter dominated, to being vacuum energy dominated.

The second of the two cosmological equations (1.23) can now be expressed as

$$\left(\frac{\dot{R}(t)}{R(t)}\right)^2 = \frac{8}{3}\pi G\rho + K/R(t)^2 + \frac{\Lambda}{3} \tag{1.27}$$

Equation (1.27) demonstrates that the rate of change of the scale factor $R(t)$ is affected by three things: the net gravitational force acting to decelerate the Universe, which is determined by the matter density (ρ); a curvature term related to the geometry of the Universe ($K = 0$ is flat; $K = +1$ is positive curvature, $K = -1$ is negative curvature); and a term related to the vacuum energy, which acts as a long-range repulsive force.

The quantity $\dot{R}(t)/R(t)$ is the rate of change of the scale factor and is parameterized as H, the Hubble constant; H is a measurable quantity. For $\Lambda = 0$, if we can measure H and ρ (the present-day mass density of the Universe) then we will have solved for $K/R(t)^2$, hence under the Robertson–Walker metric we can completely specify the geometry of spacetime. In this way, we have formulated a mechanism where observations can fully determine the cosmological model – there are no hidden variables. In the special case where the curvature of the Universe is zero ($K = 0$) and $\Lambda = 0$, we have

$$H^2 = \tfrac{8}{3}\pi G\rho \tag{1.28}$$

The rate of change of ρ with time is given by equation (1.19), which can be rewritten in terms of the scale factor $R(t)$ as

$$\dot{\rho} = -3(\rho + p)\frac{\dot{R}(t)}{R(t)} \tag{1.29}$$

With zero pressure ($p = 0$) we have

$$\frac{\dot{\rho}}{\rho} = -3\frac{\dot{R}(t)}{R(t)} \tag{1.30}$$

The only solution is that $\rho \approx R(t)^{-3}$ ($\dot{\rho} = -3R(t)^{-4}\dot{R}(t)$). Note how the expression $\rho \approx R(t)^{-3}$ is also the one that satisfies the condition of maintaining the mass $4\pi/3\rho(t)R(t)^3$ within a sphere constant as a function of time. From equation (1.28) we then have

$$H^2 = \left(\frac{\dot{R}(t)}{R(t)}\right)^2 \sim R(t)^{-3}$$

which is only satisfied if the scale factor $R(t)$ goes as $t^{2/3}$. The quantity $\dot{R}(t)/R(t) = H$ is now

$$H = \frac{(2/3)t^{-1/3}}{t^{2/3}} = \frac{2}{3t}$$

Thus, in the case of zero curvature, the expansion age of the Universe is

$$T_{exp} = \frac{2}{3H} \tag{1.31}$$

If space is devoid of mass (hence negatively curved), we can set $\rho = 0$ to yield

$$\left(\frac{\dot{R}(t)}{R(t)}\right)^2 = \frac{K}{R(t)^2}$$

which has a solution of the form $R(t)$ goes at t, and the expansion age tends to H^{-1}. In either case, observations which determine H also then reveal the approximate age of the Universe. The fact that the ages of the oldest stars in the Universe are in the range $0.67H^{-1}$ to $1.0H^{-1}$ has led some (e.g., Liddle 1996) to claim this forms another observational pillar for the Hot Big Bang theory. However, it is important to realize the expansion age of the Universe is only $\leq H^{-1}$ in the special case where $\Lambda = 0$. For $\Lambda \geq 0$, the relationship with H^{-1} is considerably more complicated, as Λ acts to make the expansion rate decrease more slowly because it is a repulsive force. In fact, it can be shown that

$$H_0 t_0 = \frac{2}{3}\Lambda^{-1/2}\ln\left(\frac{1 + \Lambda^{1/2}}{(1 - \Lambda)^{1/2}}\right)$$

(Kolb and Turner 1991; Krauss and Turner 1994) in the case of inflation (Chapter 4) in which $\Omega + \Lambda = 1$ (where $\Lambda = 0$ is the usual case). $H_0 t_0 > 1$ is satisfied for $\Lambda > 0.74$. Hence, to reconcile the possible age problem by invoking nonzero Λ, this requires a value of Λ much bigger than Ω, meaning that at the present epoch the Universe is dominated by vacuum energy.

The final useful relation to derive is the expression for the critical density of the Universe. This is defined as the density required to bring the expansion of the Universe to an eventual halt by mutual gravitational contraction. Like the previous derivations, this one can be done in terms of energy conservation, assuming the Universe is a sphere of uniform density. In this case the total mass is given by equation (1.16) with $p = 0$. Consider now a galaxy trying to escape the surface of this sphere. Its potential energy is given by

$$\mathrm{PE} = -GMm/R = \frac{-4\pi m r_s^2 \rho G}{3} \tag{1.32}$$

and its kinetic energy is given by

$$\mathrm{KE} = \tfrac{1}{2} m v^2 \qquad (v = Hr_s) \tag{1.33}$$

The velocity v of this galaxy is determined by the expansion rate expressed by the Hubble law. Thus $v = Hr_s$. To escape this sphere of radius r_s, KE must exceed PE, so we have the critical condition

$$\frac{1}{2} m H^2 r_s^2 + \frac{4\pi m r_s^2 \rho G}{3} \geq 0 \tag{1.34}$$

We can readily eliminate m and r_s^2 to arrive at the expression for critical density:

$$\rho_c = \frac{3H^2}{8\pi G} \tag{1.35}$$

The critical density is not dependent upon the size and mass of the Universe but only on its expansion rate. If its real density exceeds ρ_c, the Universe is destined to collapse. The Universe cannot collapse at early times due to entropy production and the associated high radiation pressure. Equation (1.35) strictly only applies in the matter-dominated era.

1.3 Modern cosmological puzzles

We close this first chapter by describing the topics we will pursue in detail throughout the rest of this book. These topics are all relevant to the determination of H, ρ and Λ, which are needed to fully specify our cosmology. Our approach will be to show how observations have been used to constrain these cosmological parameters. In addition to this topic, we will also focus attention on the dark matter content of the Universe and the nature of that dark matter as well as the formation of structure (galaxies and clusters) in the Universe. In general, galaxies and clusters of galaxies are used as test particles or probes to determine the cosmological parameters. However, most of these determinations are complicated by the unknown distribution and nature of the dark matter. Recognition of the importance of dark matter is the single biggest difference between current cosmological models and those that were popular a mere 20 years ago.

We devote a chapter to each of the following issues and use the most modern observations available to characterize our current state of knowledge:

1. What is the age of the Universe as determined from the observed expansion rate and cosmological distance scale? Is there a need to invoke the cosmological constant to reconcile the ages of the oldest stars with the value of the Hubble constant?
2. What is the nature of the large-scale distribution of matter in the Universe as traced by the three-dimensional galaxy distribution, and do we have a sample that accurately characterizes it?
3. What is the evidence for the existence of dark matter and what is its overall contribution to the total mass density of the Universe? Do we really live in a Universe dominated by an exotic form of matter?
4. How did structure form in the Universe and what formation scenarios are consistent with the current observational data?
5. Where do the baryons reside? Are they predominately inside or outside of galaxies? How efficient was the process of galaxy formation?
6. What is the true nature of the galaxy population? Do we have a representative survey of galaxies in the nearby Universe from which coherent arguments about galaxy formation and evolution can be made?

Our theme in this book will be to go where the data leads, even if the data leads us towards scenarios that challenge our physical understanding. Along the way, we will encounter acrimonious debates over the distance scale and new but controversial methods of measuring distances. We will show how our knowledge of galaxy clustering is still rudimentary, since we have not yet sampled a "fair" volume of the Universe, and how this impacts on our ability to measure H. We will see how theoretical prejudice shapes our thinking of what the dark matter content of the Universe is, and how this thinking is in apparent conflict with the observations. This conflict means that ρ remains rather unconstrained as its value is uncertain by a factor of ≈ 10. We will examine how most structure formation models fail to match the observed power spectrum of the small-scale galaxy distribution with the large-scale anisotropy determined by the *COBE* data. Finally, we will examine where the normal matter in the Universe resides. That is, how much of the baryonic content of the Universe is conveniently located in bright, easy-to-discover galaxies compared with the amount hidden in very diffuse galaxies that are difficult to discover or in still more diffuse backgrounds that faintly glow at different wavelengths. So let's ride the data and see where we arrive.

References

ALPHER, R. and HERMAN, R. 1948 *Nature* **162**, 774
CHANDRASEKHAR, S. and HEINRICH, L. 1942 *Astrophysical Journal* **95**, 288
COLES, P. and LUCCHIN, F. *Cosmology: the Origin and Evolution of Cosmic Structure*, New York: Wiley
GAMOW, G. 1942 *Journal of the Washington Academy of Sciences* **32**, 353
GAMOW, G. 1946 *Physical Review* **70**, 572
GAMOW, G. 1948 *Physical Review* **74**, 505
HUBBLE, E. 1925 *Observatory* **48**, 139
HUBBLE, E. 1929 *Proceedings of the National Academy of Sciences* **15**, 168
KERNAN, P. and SARKAR, S. 1996 *Physical Review D* **54**, 3681
KOLB, R. and TURNER, M. 1991 *The Early Universe*, Reading MA: Addison Wesley
KRAUSS, L. and TURNER, M. 1995 *General Relativity and Gravity* **27**, 1137
LIDDLE, A. 1996 The Early Universe, Lectures given at the Winter School "From Quantum Fluctuations to Cosmological Structures", Morocco, December 1996
MCKELLAR, A. 1941 *Publications of the Dominion Astrophysical Observatory* **7**, 251
OPIK, E. 1922 *Astrophysical Journal* **55**, 406
PEEBLES, P.J.E. 1993 *Principles of Physical Cosmology*, Princeton NJ: Princeton University Press
PENZIAS, A. and WILSON, R. 1965 *Astrophysical Journal* **142**, 419
SLIPHER, V.M. 1914 Spectrographic Observations of Nebulae, Paper delivered at the 17th Meeting of the American Astronomical Society
TOLMAN, R. 1934 *Relativity, Thermodynamics and Cosmology*, Oxford: Clarendon Press
WEINBERG, S. 1972 *Gravitation and Cosmology*, New York: Wiley

CHAPTER TWO

The Extragalactic Distance Scale

2.1 Overview

In Chapter 1 we were able to parameterize the important equations by the observational parameter $H = \dot{R}(t)/R(t)$, which specifies the rate of change of the scale factor $R(t)$. This rate of change today is known as the Hubble constant (H_0) which is a measure of the present-day expansion rate of the Universe. Since the Universe has mass, the expansion rate is slowing down due to gravitational attraction, so the value we measure for H today is different (smaller) than its value in the past. The importance of determining an accurate value for H_0 with respect to our cosmological model is threefold:

- It specifies the expansion age of the Universe and therefore provides a consistency check with other, independent ways of dating the Universe. As we will see, current values of H_0 indicate an expansion age less than the ages of the oldest stars, which is clearly physically impossible.
- It specifies the value for the critical density of the Universe which, when compared to the observed value (a difficult parameter to measure), determines whether the Universe is open (expands forever) or closed (expands to a maximum radius then contracts)
- It sets the overall scale of the Universe.

This chapter summarizes the many observational techniques which have been employed by various groups using the best available telescopes, including the Hubble Space Telescope (HST). We will adopt a rigorous, step-by-step approach to show how the extragalactic distance scale is determined. Although some of these steps are tedious, it is necessary to understand them in some detail in order to properly sort out the current controversy over the value of H_0. The goal is to measure H_0 to an accuracy of $\pm 10\%$ in order to distinguish between various cosmological models. This goal is not yet attainable, and may never be attainable using current methods of measuring distances to external galaxies.

Besides H_0, there are other parameters to be measured. Rigorous discussion comes in later chapters, but here is a list to introduce them:

- *The actual density, ρ, of the Universe*: If ρ exceeds the critical density, the Universe is said to be closed. If ρ is less than the critical density, the Universe is said to be open, meaning it will expand forever. In the cosmological equations of Chapter 1, ρ is an important parameter. As will be seen in Chapters 3 and 4, estimates of ρ vary by a factor of 10, whereas estimates of H_0 vary by a factor of 1.8. Thus, the dominant observational uncertainty in our determination of the spatial curvature of the Universe is the measurement of ρ.

- *Whether or not Λ is zero*: In the cosmological equations of Chapter 1, Λ appears as a long-range repulsive term and physically acts like a source of negative pressure. As will be discussed in Chapter 4, Λ represents the remaining vacuum energy after inflation has ended. Although convenient, Λ does not necessarily have to be zero. Indeed, a positive value of Λ contributes vacuum energy to the Universe, which allows the expansion rate to increase with time. In practice this means the Universe has an expansion age $\approx$ 10–20% larger than H_0^{-1}.

- *Age estimations of the Universe*: A consistent model is one in which the expansion age, as inferred by measuring H_0 (and ρ), agrees with an independently estimated age for the Universe. The most readily available independent estimate comes from the age of the oldest stars, as found in globular clusters. This is the subject of much research and much uncertainty associated with how stars evolve and how much convective mixing of material is important in later stages of evolution. In general, mixing of material down into a stellar core increases stellar lifetimes. The broadest range of estimates of globular cluster ages is $t_{glob} = 10$–18 Gyr (Flannery and Johnson 1982; Stetson, Vandenberg, and Bolte 1996) but in recent years age estimates have converged to the 15–18 Gyr range. A recent estimate is that of Bolte and Hogan (1995), who derive $t_{glob} = 15.8 \pm 2.1$ Gyr. Jiminez and Padoan (1996) use a more refined, but controversial technique, to fix the ages at $t_{glob} = 16.0 \pm 0.5$ Gyr. Chaboyer *et al.* (1996) concludes that the best age is 14.6 ± 1.7 Gyr with a 95% confidence lower limit of 12.2 Gyr.

However, a new generation of stellar models that take into account helium mixing have just been published by Sweigart (1997). They suggest that these ages of 15–17 Gyr should be revised downward. It is well beyond the scope of this book to delve into stellar interior models, but suffice it to say that globular cluster ages remain highly model dependent and we still don't understand the physics of the interiors of low mass, metal-deficient stars, certainly not well enough to do accurate age dating. Although Sweigart's recent results are likely to be controversial (e.g., they favor a steeper relation between [Fe/H] and M_v for horizontal branch stars than is actually observed), they should serve mostly as a reminder that globular cluster ages are not cast in stone. Furthermore, as pointed out in Chaboyer, Demarque, and Sarajedini (1996), it is likely there is a range of formation times, and hence a range of ages for galactic globular clusters. So it will be necessary to age date most of the halo population of globular clusters in order to find the oldest ones.

Another means to age date our galaxy, and therefore the Universe by adding a reasonable time for galaxy formation, is to make use of nuclear cosmochronometers based on the decay of long-lived radioactive elements. Age dating via this technique is also quite model dependent since nuclear cosmochronology is quite sensitive to the history of very heavy element creation and the amount of infall or mixing of enriched gas with primordial gas in our galaxy. Current models give age estimates in the range 13–21 billion years (Cowan, Thielmann, and Truran 1991), which is consistent with the age estimates from globular cluster stars. A more recent treatment of the problem by Chamcham and Hendry (1996) suggests that a minimum age for our galactic disk is 12 billion years. Their preferred model gives an age of 13.5 Gyr, but this number is subject to systematic uncertainties.

Perhaps the most important aspect of alternate ways of age dating our galaxy is the establishment of a minimum age. From nuclear cosmochronology this minimum age is 12 Gyr. Another means of establishing a minimum age is to identify the coolest white dwarfs and to use white dwarf cooling theory (Winget *et al.* 1987) to determine ages. It is difficult to observationally identify cool white dwarfs, so finding the coolest or oldest in our galaxy is therefore an elusive goal. Furthermore, due to their intrinsic faintness, searches for cool white dwarfs are generally restricted to our galactic disk, although HST observations of halo fields are now being used to find halo white dwarfs (von Hippel, Gilmore, and Jones 1996). So far, the basic result from these searches is the rather firm minimum age for our galactic disk of 9–10 Gyr (Winget *et al.* 1987). Since the disk is the last galactic component to form, the age of our galactic halo could be considerably larger.

Finally, there is a potentially very interesting constraint that can be put on the age of the Universe from observations of high redshift galaxies. If a spectrum of a $z \approx 3$ galaxy reveals the presence of strong stellar absorption lines, indicative of a stellar population of galaxies a few billion years old, then the Universe must necessarily be a few billion years old by this redshift. The age as a function of redshift is given by

$$\mathrm{Age}(z) = \frac{t_0}{(1+z)^{3/2}}$$

where $t_0 \sim t_{glob}$ is the age of the Universe. At $z = 3$ the age is then 1/8 the age of the Universe. The discovery of a galaxy at $z = 3$ with a stellar population of ≈ 2 billion years then constrains t_0 to be at least 16 Gyr. Dunlop *et al.* (1996) have discovered a galaxy with $z = 1.55$ that does have spectral features indicative of a stellar population of age a few billion years. Their preferred age is 3.5 Gyr. At this redshift, the Universe is 1/4 of its present age, which again suggests a minimum age of 16 Gyr, assuming this galaxy took a few hundred million years to form.

Even stronger evidence for relatively old galaxies at high redshift is provided by the work of Steidel *et al.* (1996). For galaxies dominated by ultraviolet (UV) emission there should be a pronounced break in the energy distribution at 912 Å as emission at wavelengths shorter than this is completely absorbed by neutral hydrogen. This absorption edge is known as the Lyman limit. At redshifts of 3–3.5 the Lyman limit is redshifted into a ground-based U (ultraviolet) filter at wavelength $\approx$ 3500–4000 Å. Hence, a faint galaxy which “disappears” when observed through this filter is a candidate for a high redshift, star-forming galaxy. Besides having found a number of actively star-forming galaxies at this redshift, Steidel *et al.* have also discovered, through spectroscopic follow-up, absorption lines indicative of an older stellar population perhaps with mean age as great as 2 billion years. At the very least, the Steidel *et al.* result shows that some galaxies at redshifts 3–3.5 have already produced a significant number of stars.

2.2 Possible kinds of cosmological model

In terms of the parameters H, ρ, t_{glob} and Λ, five different combinations are consistent with extant data. The role of observations is to collapse these five combinations down to a single one that fits all the data. By convention, instead of using ρ, cosmologists use another parameter, $\Omega = \rho/\rho_c$. A spatially flat Universe has $\Omega = 1$. Although many combinations of these parameters are allowed, a cosmology in which H_0 is 100, $\Omega = 1$ and $\Lambda = 0$ produces an expansion age of only 6.7 Gyr; it can therefore be ruled out. As will be discussed later on, there is strong theoretical prejudice for $\Omega = 1$, which then necessarily pushes estimates of H_0 to lower values. Here then are the current cosmological options which are more or less consistent with observations:

- *Option 1*: H_0 is 50, $\Lambda = 0$, the ages of globular clusters are well understood with $t_{glob} \approx 16 \pm 1$ Gyr, and $\Omega \approx 0.1$ (the Universe is open); most of the matter is baryonic. Arguments for this cosmology were first made in Sandage (1972) and again most recently in Sandage and Tammann (1997). If correct, it is a nice, self-consistent cosmology in which the expansion age and the globular cluster ages agree. Of course, it's difficult to account for inflation in this model.
- *Option 2*: H_0 is 100, $\Lambda = 0$, the ages of globular clusters are at their lowest possible limit, and the Universe must be very open. The age of the Universe in this cosmology is 10 Gyr. This option dates back to the work of Gerard De Vaucouleurs in the early 1960s and was strongly revived in the early 1980s by Aaronson *et al.* (1980, 1986) Recent improvements in stellar evolutionary code, however, would now seem to preclude such young globular cluster ages, hence they rule out this cosmology. In addition, other age indicators strongly suggest that our galactic disk is at least 10 Gyr old. Like option 1, this option is also inconsistent with the inflationary scenario.
- *Option 3*: H_0 is 75, $\Omega = 0.1$–0.3, and the ages of globular clusters are marginally understood with t_{glob} in the range 12–15 Gyr. This is the compromise option, in which overlapping error bars accommodate most observations and prejudices, although $\Omega = 1$ produces an age of only 9 Gyr with this value for H_0.
- *Option 4*: H_0 is 50, the Universe is critical ($\Omega = 1$), and the ages of globular clusters are marginally understood with t_{glob} in the range 12–13 Gyr. This option is preferred by inflationary cosmologies which predict a critical Universe. This cosmology was strongly favored in the 1980s to the point that observations which suggested $H_0 \geq 80$ were essentially ignored by a large segment of the theoretical community, since $H_0 \geq 80$ would conflict with the $\Omega = 1$ constraint coupled with the ages of globular clusters.

- *Option 5*: We understand the ages of globular clusters with current models putting them at 13–17 Gyr. By the consensus of most of the last 10 years' worth of data, H_0 lies in the range 70–90. Going where this data now leads then suggests (a) the Universe is open or (b) the inflationary constraint of flat space has to be satisfied by a combination of Ω and Λ. This option is a significant departure from the cosmological models of the 1970s and the 1980s, and as we will see later, it has a fair bit of observational support.

The current situation is clearly unsatisfying as a very large range of cosmological models remain consistent with the data. As we will see in Chapter 4, although the inflationary paradigm is elegant, it has very little observational support. At the same time, there seems to be real conflict between the ages of the oldest stars and the inferred expansion age based on measurements of H_0. We are thus left with either a positive value for Λ or a low value for H_0 as the alternatives.

2.3 Establishing the extragalactic distance scale

H_0 is determined by measuring redshifts and distances to galaxies. Redshifts are trivial to measure but distances are difficult to measure. When determining extragalactic distances, there are two routes to take. We focus on the conventional route, determination by means of a distance scale ladder. The astronomical distance scale ladder works in a bootstrap manner, where certain techniques are used over short distance scales to calibrate other techniques to use over longer distance scales. The main problem with this approach is that all errors are cumulative. Thus, the final technique used to derive distances to galaxies for purposes of determining H_0 carries with it all systematic errors that occurred in the previous steps. Each step of the ladder must be free from systematic errors, and this requires forming a representative sample of objects for each distance measuring technique. Much of the disagreement about the value of H_0 derived in this manner stems from issues of sample selection. It is simply impossible to summarize all the various objections that distance scale practitioners raise about each other's data sets. Most of these objections center around the argument that conspiratorial sample selection effects combine to produce erroneous results (i.e., the wrong value for H_0).

Besides, it is not clear that modern determinations of H_0 have ever been made without preconceived ideas about the age of the Universe As early as 1963, the theory of stellar evolution, combined with color–magnitude (CM) diagrams of globular clusters, suggested ages as high as 17 billion years (Sandage 1970; Smith *et al.* 1963). This naturally leads to values of $H_0 \leq 55$ km s^{-1} Mpc^{-1} in order to avoid a potential conflict between the expansion age of the Universe and the age of its oldest stars. Indeed, it is clear from history that the age of the Universe has been used as a constraint on H_0. Hubble's original determination of H_0, later modified by Baade, led to an age for the Universe that was younger than the estimated age of the Earth. Since the age of the Earth is known with precision, this is a valid constraint. However, the ages of the oldest stars in our galaxy are not known with precision, so they should never be used a priori to exclude possible values for H_0.

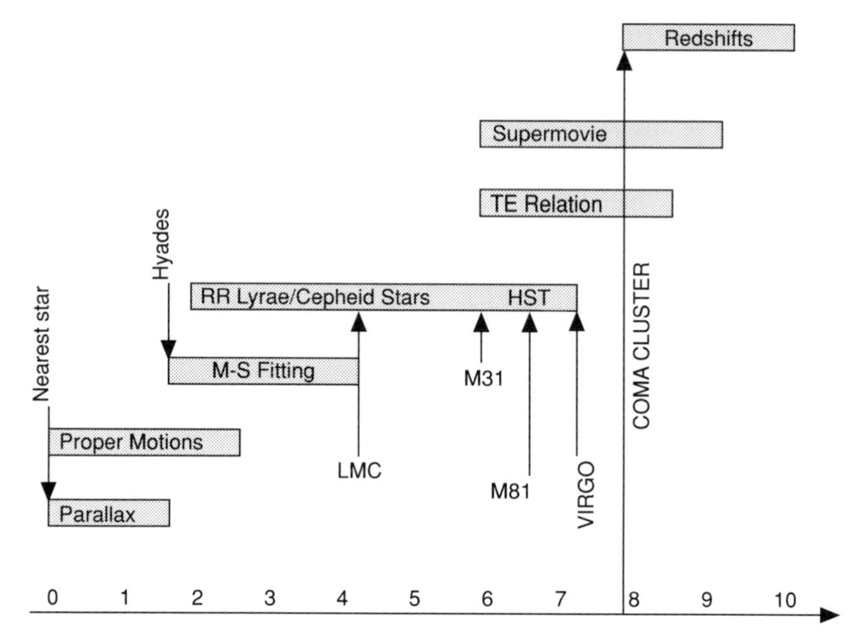

Figure 2.1 The distance scale ladder: different techniques are used over certain distance ranges. To get to larger distances, one technique is used to calibrate another. Individual galaxies and/or clusters are set at their nominal distances. The use of the Hubble Space Telescope has extended the range of Cepheid-based distances to galaxies considerably, further away. The numbers along the arrow represent the log of the distance in parsecs.

Moreover, the age of the Universe is only directly related to H_0^{-1} in a specific cosmology, one in which $\Lambda = 0$. This means that pursuit of H_0 in the absence of prejudice concerning the age of the Universe can potentially provide evidence that Λ is not zero. Such a finding would alter our cosmology in a far more profound way than determining the actual value of H_0.

Perhaps the best way to resolve the controversy over H_0 is to develop some physical technique for measuring distances directly. This would have the effect of bypassing the cumbersome distance scale ladder altogether. In theory these direct methods are rooted in understood physics and are immune to issues associated with sample selection. These techniques are beginning to surface but are not yet sufficiently robust to employ in a credible manner. Nevertheless, we discuss some promising techniques (e.g., timing delays associated with gravitational lensing, the Sunyaev–Zeldovich effect) at the end of this chapter. At present, these alternative distance measuring techniques produce results that are fairly model dependent but which are not inconsistent with the results obtained from the distance scale ladder.

Figure 2.1 summarizes the ideology and techniques used in constructing the distance scale ladder. One technique is illustrated calibrating another technique, and so on. In the subsequent discussion, Population I refers to stars in our galactic disk, which are generally younger and more metal-rich than Population II stars, which occur in our galactic halo and in globular clusters. The goal is to achieve a representative sample of objects at each step; this ensures the final measure of H_0 is free from systematic bias. As we shall see, this is difficult to achieve.

2.3.1 The first rung: calibrating the main sequence luminosities of stars via stellar parallax

Galaxies emit optical radiation by the combined luminosity of their stellar populations. Therefore we must know the absolute brightnesses of different types of stars if we are to make any progress in determining the extragalactic distance scale. Hence the first rung of the distance scale ladder is firmly grounded in trigonometric parallaxes of nearby stars that anchor the lower end of the main sequence. The phenomenon of stellar parallax is a consequence of the Earth's revolution around the Sun. Relative to the baseline defined by the Earth–Sun distance, a nearby star traces out a small ellipse in its apparent position as measured with respect to more distant stars. To accurately determine the distance to that nearby star requires two precise measurements:

- The Earth–Sun distance in physical units
- The semimajor axis of the ellipse in angular units

The Earth–Sun distance can now be measured with high accuracy by first determining the distance to Venus via radar reflection timing. That distance, combined with the Sun–Venus–Earth angle, directly determines the distance to the Sun. The second measurement is more problematic. In general, atmospheric aberration limits the angular resolution of a single ground-based image (photographic or digital) to 0.5–1.0 arcsecond. The distance a star must have in order to have a parallax of 1 arcsecond is 3.26 light-years (3.086×10^{18} cm). This distance is known as a parsec. The closest star to the Sun has a distance of ≈ 4 light-years. A star at a distance of 20 parsecs from the Sun would exhibit a parallactic angle of only 0.05 arcsecond, approximately a factor of 10 below the best angular resolution that can be achieved in a single image. Although the centroid of a stellar image can be determined to a positional accuracy that is many times smaller than the seeing disk, the limiting factor for parallax measurements is systematic error associated with the determination of absolute position in the sky. These errors are best eliminated or reduced by repeated observation of the same star. Thus, determining the absolute luminosity scale of stars, based on distances to nearby stars, is a time-consuming process that requires many *years* of measurement in order to reduce the random error in the distance determination. In general, random errors as low as ± 0.01 arcsecond have been obtained using photographic plates as the imaging system. Still, at a distance of 20 parsecs, this represents a 20% error in distance and, worse still, a 45% error in intrinsic stellar luminosity. Although there are some 8000 stars whose parallax has been determined, most of them are relatively low quality and have formal errors that are not well determined. We thus regard 20 parsecs as a limiting volume out to which stellar parallax can be directly determined with reasonable accuracy from photographic measurements.

If we consider the disk of our galaxy as having a radius of 10 kiloparsecs (kpc), and a thickness of 1 kpc, then the volume that we sample with trigonometric parallax is only 10^{-7} of the total volume. How can we be sure this tiny volume contains a representative sample of stars? Fortunately, a theoretical argument involving stellar lifetimes serves as a consistency check. In general, star formation events in our galaxy make many more low mass stars than high mass stars. Since high mass stars have very short lifetimes, we would not expect to find many in a volume-limited sample, nor would we expect to find many stars that were in short-lived evolutionary phases. Hence, a representative sample of stars should be heavily weighted towards low mass, main sequence stars (plus a few white dwarfs) and this is generally what is observed in the trigonometric parallax sample.

A long-term program under way at the United States Naval Observatory employs CCD measurements to improve the accuracy of stellar parallax determinations. To date, measurements of approximately 200 stars indicate a median error of ±1.1 milliarcseconds, an order of magnitude improvement over what is available in the current Yale Parallax Catalog. In principle then, distance measurements for stars with distances up to 200 parsecs can now be made. In addition, the *Hipparcos* astrometric satellite made measurements (Perryman *et al.* 1997) of about 100 000 stars down to 8 magnitude with a median error of 1.8 milliarcseconds. These new databases should greatly improve the precision for which we know lower main sequence luminosities. More important, the larger volume accessed by these two surveys should allow for direct determinations of distances to Population II subdwarfs. As will be seen later, this can provide a very important and independent check of the distance scale(s), which are ultimately derived from the positions of stars in the Hertzsprung–Russell (HR) diagram.

The *Hipparcos* data and the CCD data of the United States Naval Observatory allow a check to be made on the reliability of the old photographic parallax scale. Any systematic error in the distances to the nearest stars changes the luminosity of the main sequence; hence it affects all distances that are subsequently derived. A preliminary analysis of the *Hipparcos* parallax data by Perryman *et al.* (1995) shows the following:

- The median standard error for all parallax measurements is 1.5 milliarcseconds.
- Any systematic errors are less than 0.1 milliarcsecond.
- There is no significant difference in the position of the lower main sequence as defined by previous parallax measurements of stars located within 10–15 pc.
- There are some serious disagreements for individual stars with distances nominally within 25 pc of the Sun between the *Hipparcos* data and specific ground-based observations.

The last item will cause a reevaluation of the masses and luminosities of these stars, but overall it will not have a profound effect on the placement of the main sequence (many of the discrepant stars are not main sequence stars but red or blue giants).

In sum, the parallax measurements do allow for a reliable determination of stellar luminosity for main sequence stars. This provides a means to calibrate the distance to a stellar cluster in which the main sequence can be identified (see below).

2.3.2 The second rung: distances to stellar clusters

Stellar clusters are important empirical astrophysical laboratories since they represent a group of stars at a common distance which were born at a common time. Differences in stellar evolutionary rates then allow the HR diagram to be filled out after a few million years of stellar evolution. Let's suppose that intermediate age stellar clusters, which contain a few thousands of stars, include a certain type of star that can serve as a standard candle, but this certain type of star is not contained in the currently available trigonometric parallax samples. Let's further suppose this star is intrinsically bright, so it can be seen at large distances. If we can determine the distance to the cluster containing that star, we can calibrate its absolute brightness. We can then use that star to probe larger distances.

Three basic types of star found in stellar clusters can be used to determine distances in our own galaxy as well as in external galaxies:

- *RR Lyrae variable stars*: These variable stars typically have pulsational periods of a few days and there is no correlation between pulsational period and luminosity. They are evolved stars and are found in the oldest clusters, like globular clusters. Although there is still some disagreement over their absolute magnitude (more fully discussed below), RR Lyrae variables have $M_v \approx +0.5$, hence they can be used as a distance indicator out to ≈ 1 Mpc, where they have an apparent magnitude fainter than 25.0 mag, the limit of ground-based telescopes.

- *Cepheid variable stars*: These variable stars show a strong relationship between intrinsic luminosity and pulsational period. In practice this relationship is empirically defined by Cepheids in the Large Magellanic Cloud (LMC), hence an accurate distance to the LMC would calibrate the relationship. However, there is some concern that the Cepheid period–luminosity (PL) relationship depends on metallicity, hence the LMC relationship may not be universal. Cepheids are also found in young, open clusters in our galaxy but, as we shall see, calibrating their intrinsic luminosity in those clusters is quite difficult. For the longer-period Cepheids (periods of a month or so) the intrinsic luminosity is quite large, $M_v \approx -7$. Hence, ground-based measurements can detect this population out to a distance of ≈ 4 Mpc. However, the improved angular resolution available with the HST has allowed individual Cepheid variables to be detected out to distances of ≈ 15 Mpc.

- *The brightest red supergiants*: These are young massive stars which are in a short-lived evolutionary phase at the tip of the red giant branch. There luminosities can approach $M_v = -9$, hence ground-based measurements can detect them in Virgo cluster galaxies and beyond.

In January 1997 a conference was held in which some of the first *Hipparcos* results were made public and discussed. The most relevant of these new results comes from Feast and Catchpole (1997), who discuss a parallax sample of 26 Cepheids. These stars are at the very limit of the useful range of accurate parallax measurement of *Hipparcos*, so the measurements are potentially subject to systematic error. Notwithstanding this, Feast and Catchpole derive a zero point for the Cepheid PL relation, which is approximately 0.2 mag brighter than previous measurements indicate. This has significant consequences for the Cepheid distance scale that we describe later in this chapter. However, there is still much uncertainty in this new zeropoint estimate as (1) there may be systematic errors in the parallax measurements themselves for these large distances, (2) Feast and Catchpole derive reddening estimates to the Cepheids based only on blue (B) and visual (V) photometry, (3) they assume a metallicity correction of −0.04 mag, where the correction comes from the Laney and Stobie (1992) metallicity calibration and the metallicity of the individual Cepheids is inferred from the B – V color. Unfortunately, the reddening and metallicity corrections are degenerate when only B – V is used. Hence, apparent change in the zero point of the Cepheid PL relation requires additional confirmation.

To take advantage of these stellar distance indicators, one has to accurately calibrate their absolute magnitudes by measuring good distances to nearby clusters which contain these objects. In the case of RR Lyraes, this means globular clusters and essentially there are no nearby globulars. For Cepheids and M-supergiants, young, open clusters contain a handful of these objects. These clusters are generally located in the plane of our galaxy; hence they are reddened by interstellar dust.

Main sequence fitting

This method relies on using the intrinsic luminosity of the main sequence (in practice the lower main sequence). Since stars in a cluster are at a common distance, differences in apparent magnitude reflect differences in intrinsic luminosity. If enough of the main sequence can be detected, it is possible to fit that sequence in apparent magnitude space with the calibrating sequence obtained from trigonometric parallax. This is schematically shown in Figure 2.2. In principle this method should be very good as there are many stars used in the fit, hence the zeropoint errors are formally small. A hidden assumption in this fitting procedure is that the shape of the main sequence is the same as the shape of the calibrating sample of stars. To first order, this is a safe assumption. To second order, differences in metal abundance and age slightly affect the shape and placement of the main sequence, depending upon the filter system used to measure the effective temperature.

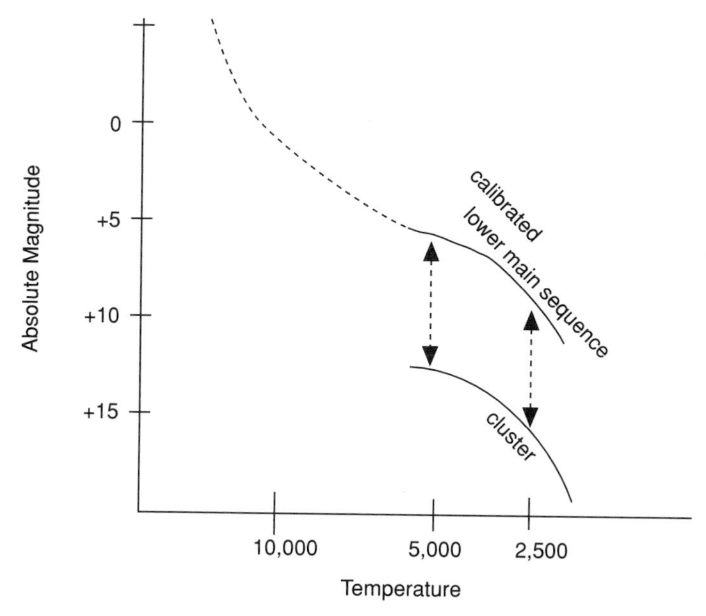

Figure 2.2 Main sequence fitting: the calibrated main sequence is plotted in absolute magnitude space, where the absolute magnitudes of the lower main sequence stars (solid) have been calibrated from trigonometric parallax samples. The rest of the main sequence (dashed) indicates that very few of these stars are in the available parallax samples. The hypothetical lower main sequence of a nearby cluster is shown. The offset with respect to the calibrating relationship is 7 mag, a distance of 250 pc. The lower main sequence in this representation is fairly vertical, meaning that small errors when determining the spectral type of a star translate into a significant error for its distance.

As shown in Figure 2.2, in the traditional B – V vs. V CM diagram, the lower main sequence is quite steep; it is the vertical offset that determines the distance. This means the spectral type or surface temperature of the stars that comprise the lower main sequence in the stellar cluster must be quite well known, otherwise large random errors can be introduced. For instance, a change of just one subspectral class type (e.g., G8 vs. G9) introduces an error of 12% in the distance. It is therefore much better to use a color system in which the main sequence is relatively flat, so that errors in spectral classification are not as severe. This can be done if a long-baseline color index, such as V – K is used. Until recently, this was not practical, but now with relatively large format IR arrays coming in line, these important measurements can be made.

There are two important limitations to deriving distances to nearby stellar clusters via main sequence fitting:

- The calibration of main sequence luminosities, from the parallax samples, is most reliable for stars with masses less than the mass of the Sun. These so-called red dwarfs have absolute magnitudes in the range M_v +6 to +10. Hence, even at a distance of 1 kpc, the bottom end corresponds to an apparent magnitude of +20. At this faint flux level it is difficult to accurately measure an apparent magnitude.

- Star formation in our galaxy generally occurs within the confines of dusty molecular clouds. Newly formed clusters are often surrounded by a cocoon of dust. After a million years or so, the young cluster has migrated out of this dusty cocoon but is still located in the plane of our galaxy. Nearby clusters have relatively large angular sizes on the sky (e.g., the radius of the Pleiades cluster is 4°). Over this angular extent, the amount of foreground reddening will vary across the face of the cluster; hence systematic errors will result if a single reddening determination is made. And because of the large angular size, the only feasible manner of doing stellar photometry is via photographic plates; wide-field CCD imaging systems are a rather recent development. Thus, the available photometry is mostly photographic, with reddenings to individual stars that are not well determined.

Convergent point method

Distances to some nearby clusters can also be determined using a second method which takes advantage of the fact that open clusters are only weakly gravitationally bound, so the individual stars are gradually escaping and the cluster is essentially expanding. If one can determine the true space motion of the individual stars, by measuring the radial velocity and proper motion components, the individual vectors will point back towards a region of small radius. This point is known as the convergent point, and a distance estimate to the cluster can be made as follows.

For sufficiently nearby stars, the transverse component of their velocity can be measured over periods of decades. This motion is usually measured in seconds of arc per year. If we denote the proper motion by μ, the true transverse velocity by T and the parallax of the star (measured in seconds of arc) by Π_p, then the quantity μ/Π_p is equal to T in astronomical units per year. One astronomical unit per year is equivalent to 4.74 km s^{-1}. Therefore

$$T = 4.74\mu/\Pi_p \tag{2.1}$$

and the true space velocity of any star is

$$v_{space} = (V_r^2 + T^2)^{1/2} \tag{2.2}$$

We can then apply the convergent point method (also known as the moving-cluster method) to the Hyades cluster. Referring to Figure 2.3, V_h is the true space velocity of the Hyades cluster, which is deconvolved into T and V components. Each individual star has some angle θ between the Sun and the extrapolated convergent point. This geometry then permits the following:

$$V = V_h \cos\theta, \qquad T = V_h \sin\theta = 4.74\mu/\Pi_p \tag{2.3}$$

Solving for Π_p yields the distance

$$\Pi_p = 4.74\mu/V \tan\theta \tag{2.4}$$

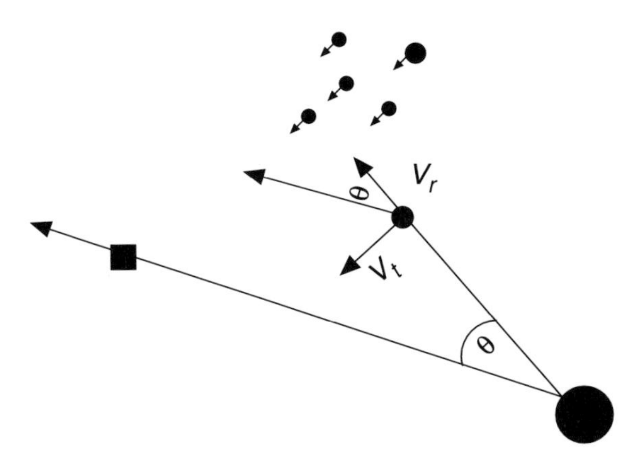

Figure 2.3 Convergent point method: the individual stars (small filled circles) are moving on the plane of the sky in a roughly parallel trajectory towards a common point (filled square). If it is possible to measure the angle between this common point, called the convergent point, and an individual star, then determining the radial and tangential components to the velocity will yield a distance, as derived in the text.

The Hyades and Pleiades clusters

The two nearest young clusters to the Earth are the Hyades and Pleiades clusters. Each contains a population of luminous main sequence stars. Hence an accurate distance to one or both of these clusters will serve to calibrate the upper main sequence, where the Cepheids are located. Unfortunately, neither the Hyades nor the Pleiades contains a Cepheid. In general, upper main sequence stars are missing from the ground-based parallax sample because they are intrinsically rare in a volume-limited sample. The Hyades is the closer of the two; hence measurements of the individual proper motions and radial velocities of its stars can yield its distance using the convergent point method. This has been done by many groups. The most reliable distances rely on proper motion samples of stars that can be used to define Hyades members. However, stellar luminosities as a function of main sequence mass have a small dependence on the opacity in the atmosphere, which in turn depends upon the overall metal abundance of the star. This is especially true in those stars where H^- opacity dominates (usually red giant atmospheres) because the principle donors of electrons are metals. Because the Hyades is a relatively metal-rich cluster, its observed main sequence may be systematically offset compared to the more metal-normal main sequence found in the more distant clusters that contain Cepheids.

The Pleiades is approximately twice as far away as the Hyades and is a more metal-normal cluster. This provides a better calibration of main sequence luminosities. Distances to the Pleiades can be determined by main sequence fitting to the few A and F stars that are in the trigonometric parallax sample. Unfortunately, the resulting distance modulus depends upon which parallax sample is used and whether or not Lutz–Kelker corrections should be applied. Lutz–Kelker (Lutz and Kelker 1975) corrections are rather important and occur for the following reason. If there is a uniform distribution of stars (a reasonable assumption for nearby stars) then the number of stars varies as distance cubed. Each determination of trigonometric parallax has an associated error. Thus, if you define a distance-limited sample (say all stars with measured trigonometric parallax 0.05 arcsecond or greater), then some stars with larger distances will scatter into your sample (because of measuring errors) and some stars with smaller distances will scatter out. Because there are more stars at larger distance, the net effect is that your sample will be contaminated by more distant stars and your estimation of intrinsic luminosity will be systematically low. For extragalactic samples, the same problem is present and is commonly known as the Malmquist effect.

The Lutz–Kelker method offers a statistical correction for this effect and should always be applied. It is rather unclear why some authors do not apply it. Due to these effects, determinations of the distance modulus ($m - M = 5\log(r/10)$ where r is measured in parsecs) of the Pleiades varies from 5.57 ± 0.08 to 5.68 ± 0.04. Proper motion studies of the Pleiades members are more difficult but yield distance moduli that are consistent with these estimates. Nonetheless, this is a 10% range in the distance to the Pleiades, which ultimately becomes part of the error in determining H_0.

Using the distance to the Hyades or Pleiades, combined with the ground-based trigonometric parallax sample, now gives us in principle the entire range of main sequence luminosities, which we can use to derive distances to clusters that actually contain Cepheid variable stars. The typical distance to a cluster like this is a few kiloparsecs. If we consider a young cluster at a distance of 3 kpc ($m - M = 12.4$), the Sun would have an apparent V-magnitude of 17.2. Using a CCD device, 1% photometry at this brightness level is easy, but with photographic photometry it is not. This is why it is essential to involve A and F stars in the main sequence fitting, as they fall in the magnitude range 12–15, where precision photographic magnitudes have already been obtained. In general, accurate photometry of lower main sequence stars, where the calibration is most secure, does not exist in distant, young galactic clusters that contain Cepheids.

2.3.3 The third rung: determining the zero points of standard candles

The Cepheid variable period–luminosity relation

The relationship between the period of pulsation and the intrinsic luminosity of Cepheids has been known for 75 years. In general, the calibration of the zero point of the Cepheid PL relation based on main sequence fitting to young galactic clusters is not reliable. This is due to poorly determined photographic magnitudes and especially to poorly determined foreground reddenings. In the early 1980s a vigorous program of Stromgren and $H\beta$ photometry of eight open clusters which contain Cepheid variable stars was carried out by E. Schmidt. The choice of photometric system here was designed to give maximum sensitivity to reddening variations across the cluster as well as the mean metallicity of the cluster. Schmidt's (1984) redetermination of the distance moduli of these eight clusters with well-established Cepheid members yielded an astounding result: the zero point of the Cepheid PL relationship was too bright by 0.5 mag. Hence, any galaxy whose distance was determined by measuring the periods of Cepheids located in that galaxy would have a distance which was too large by 0.5 mag (28%). That better photometry of stars in open clusters would yield a systematic error so large is worrisome testimony to the poor quality of the previously obtained photographic magnitudes and reddening estimates. In recent years there has been some question about the accuracy of the $H\beta$ photometry obtained by Schmidt, but a better recalibration of these open clusters has not yet appeared. Because of these discrepant results, and the generally poor quality of photographic photometry, most practitioners of the extragalactic distance scale ironically *do not* use the zero point of the Cepheid PL relationship, as defined by open clusters in our galaxy; instead they rely on the zero point obtained by deriving the distance to the LMC, a fundamentally different galaxy than the Milky Way. We will explicitly consider how the distance to the LMC is determined a bit later on in this chapter.

Possible problems with Cepheids

There are several nearby galaxies within the 4 Mpc Cepheid horizon, and in the 1970s and early 1980s there were vigorous photographic campaigns aimed at detecting and measuring Cepheid variables in these galaxies. Cepheid magnitudes were generally determined at either blue or visual wavelengths, both of which are sensitive to small amounts of reddening in the host galaxy itself. This is a vexing problem for observations of Cepheids in that they are usually located in or near dusty spiral arms. A breakthrough in the credibility of the Cepheid magnitude measurements occurred in the mid 1980s with the work of Barry Madore and Wendy Freedman. Thanks to improvement in detector technology, it became possible to make measurements of Cepheid variables in the near-IR part of the spectrum, where the sensitivity to host galaxy reddening was down by an order of magnitude. Furthermore, the amplitude of the variability increases as you go to longer wavelengths, thus Cepheids are easier to detect in time series observations of galaxies at these longer wavelengths. Not surprisingly, when IR measurements of Cepheids were compared to B- or V-band measurements, different distances to the *same* galaxy were found. Unfortunately, there is a paucity of data available to calibrate the near-IR Cepheid PL relationship. For instance, the most practical band in which to make the measurements is the I band, and to date, there are only 22 Cepheids in the LMC with measured I-band magnitudes (although the MACHO observations, described later, will add significantly to this database) compared to the hundreds that have been measured (photographically) in the B and V bands.

The use of the Cepheid PL relationship as a distance indicator is the best example of the Population I distance ladder, as ultimately the zero point of the relationship is derived through main sequence fitting of young clusters. However, there has been some concern that the Cepheid PL relationship may depend on metallicity. If this is the case, the PL relationship would have a different zero point for the LMC than say for M31, as the LMC has a significantly lower metallicity. Theoretically, a small dependence would be expected as the atmospheric opacity of these stars can be dominated by electron scattering. This would also affect the observed color of the Cepheids, which suggests there should be an observable period–luminosity–color relationship. Observationally, it is quite difficult to test for the dependence of the Cepheid PL relationship on metal abundance. Currently the best test is to examine the radial dependence of the PL relationship in the Andromeda galaxy, where there is a detectable abundance gradient. Although observations have failed to find any significant difference in the PL relationship as a function of radius in M31, its overall abundance gradient is not very steep. Furthermore, the observations of Freedman and Madore (1990) have recently been challenged by Gould (1994a), who suggests that the data does indeed reveal a metallicity dependence on Cepheid luminosity. In particular, Gould suggests that metal-poor Cepheids are less luminous than metal-rich Cepheids at a given period, although his argument is not very convincing as the sample size is small. The issue remains somewhat unresolved and could be important at the 10% level.

RR Lyrae absolute magnitude scale

By analogy with the Cepheid PL scale, it is possible but much more difficult to use a Population II distance scale ladder, based on the Population II main sequence, for calibrating the absolute magnitude of RR Lyrae variable stars. These stars are found in globular clusters and in the general galactic halo field star population. However, their absolute magnitudes are around +0.5, which limits their applicability to distances of 0.5–1.0 Mpc. But since they are found in galactic globular clusters, RR Lyrae stars can be used to determine distances to these globular clusters, which leads to the globular cluster luminosity function (GCLF) in our galaxy. If we assume the GCLF is invariant from galaxy to galaxy (explored in detail later in this chapter), the Population II distance scale can be extended out to distances as large as 20 Mpc due to the large intrinsic brightness of globular clusters.

Seeking the galactic calibration of the absolute magnitudes of RR Lyrae stars has been an ongoing endeavor for 30 years. Throughout that time there have been various lines of evidence which suggest the absolute magnitude has a small dependence on metal abundance, although this evidence has never been entirely clear. Two recent determinations are from Carney (1992):

$$M_v = 0.15[\mathrm{Fe/H}] + 1.01(\pm 0.08) \tag{2.5}$$

and from Simon and Clement (1993):

$$M_v = 0.30[\mathrm{Fe/H}] + 0.94 \tag{2.5a}$$

The other preferred calibration is to assume no metallicity dependence and a zero point of +0.60 mag, that is

$$M_v = +0.6 \tag{2.6}$$

It would be desirable to reach resolution between equations (2.5), (2.5a) and (2.6). Note that equation (2.6) is recovered in the case where [Fe/H] = −2.6 for equation (2.5) and [Fe/H] = −1.1 for equation (2.5a). Most galactic RR Lyraes come from a stellar population with [Fe/H] ≈ −1.5. A better calibration of M_v for RR Lyrae stars is potentially achieved by determining distances to the host globular clusters via main sequence fitting. Unfortunately, nearby Population II main sequence stars are quite rare, and up to now only one such star has an accurate parallax measurement. At the moment, it is unknown how many stars in the *Hipparcos* database are Population II main sequence stars, but one expects several dozen good candidates out of their sample of 100 000 nearby stars. Over the next five years we should see a more reliable calibration of the Population II main sequence, and hence improved distance determinations to globular clusters from main sequence fitting.

The Baade–Wesselink technique

We note finally that, for any pulsationally driven variable star, there is a dynamical way for determining the luminosity of the star. This technique, called the Baade–Wesselink technique, works on the principle that the pulsational period is driven by the dynamical timescale of the star, which depends on $\rho^{-1/2}$. This means that luminosity variations are directly proportional to variations in stellar radius. An accurate radial velocity curve as a function of phase is required for determination of the radius, and this is the observationally difficult part of the technique. Once the radius is known, the luminosity of the star can be determined from the well-known relation between luminosity, effective temperature and radius:

$$L = 4\pi\sigma R^2 T_{eff}^4 \tag{2.7}$$

The effective temperature of the star is determined from its color or absorption line spectrum. Up to now the method has been applied to a small sample of RR Lyrae and Cepheid variable stars with mixed results. In particular, application of the method to RR Lyrae stars by Storm *et al.* (1994) yields a value +0.2 mag fainter than the zero point of equation (2.6).

Brightest red supergiants

Calibration of M-supergiant luminosities in open clusters in our galaxy suggests that they reach a maximum luminosity of $M_v \approx -9$. At these luminosities, M-supergiants can be detected as individual stars out to distances of at least 10 Mpc. The overall population of these stars in a galaxy depends strongly on its star formation rate and, for instance, an actively star-forming galaxy like M101 (Figure 2.4) would be expected to be rich in M-supergiants. But there are several practical difficulties associated with their use as distance indicators:

- Like Cepheids, their calibration rests generally on poor photographic photometry and main sequence fitting to open clusters which have reddening variations across them.

Figure 2.4 CCD Image of M101 taken by the author in blue light. M101 is a typical high surface brightness, actively star-forming spiral galaxy which is rich in Cepheids located in and near the spiral arms.

- It is completely unclear whether there is a threshold number of stars that a galaxy must contain in order for it to host at least one brightest red supergiant. Hence, statistical population effects may come into play; a galaxy like the Milky Way with some 10^{11} stars may have a few brightest red supergiants, but dwarf galaxies like IC 1613 and NGC 6822 with only 10^{8} stars, while having supergiants, may not have the brightest red supergiant.

- These stars are located in active regions of star formation in external galaxies, hence they will be somewhat reddened.
- Using only UBVR photometry, it is difficult to unambiguously distinguish between a foreground M-dwarf in our galaxy and a distant M-supergiant in another galaxy. Indeed, there have been cases where foreground M-dwarfs were mistakenly assumed to be M-supergiants in another galaxy (in this case M101; see Humphreys *et al.* 1986). Infrared photometry is needed to clearly distinguish between an M-dwarf and M-supergiant.

2.3.4 Distances to Local Group galaxies

The foundation of the extragalactic distance scale lies in obtaining distances to a sample of nearby galaxies that have (hopefully) the same range of properties as more distant galaxies. For instance, to pick an absurdly simple situation, suppose all spiral galaxies had the same exponential scale length. If this were true, a measurement of the distance to the Andromeda galaxy would then directly secure distance measurements to all other spiral galaxies. Alas, things are not so simple and distances to a variety of different galaxies must be obtained. Unfortunately, the Local Group of galaxies, which necessarily acts as the only available nearby sample of galaxies, is noticeably deficient in two areas:

- There are no massive elliptical galaxies.
- There is only one large spiral galaxy (M31).

Most of the Local Group galaxies are low mass, dwarf galaxies whose heterogeneous properties do not lend themselves as distance indicators. Figure 2.5 shows images of the Local Group dwarf irregular galaxies IC 1613 and NGC 6822.

We summarize the distances to several key Local Group galaxies below using the Cepheid and RR Lyrae distance methods together with some others.

The Large Magellanic Cloud

Because of the large number of empirical observations of its Cepheid and RR Lyrae content, obtaining an accurate distance to the LMC is absolutely crucial. In principle this distance could define the zero point of both the Cepheid PL relation and the RR absolute magnitudes. When measuring the distance to the LMC there are two important sources of systematic uncertainty:

- *Foreground reddening*: The reddening along the line of site to the LMC is not well known. This is primarily because the reddening is low, hence difficult to measure accurately. The canonical value is $E(\mathrm{B}-\mathrm{V})=0.07$ (or $A_v \approx 0.23$). But values of $E(\mathrm{B}-\mathrm{V})=0.11$ up to $E(\mathrm{B}-\mathrm{V})=0.18$ (Walker 1991) have been advocated for certain lines of sight. Although the value of $E(\mathrm{B}-\mathrm{V})=0.18$ seems excessive, values in the range $0.06 \leq E(\mathrm{B}-\mathrm{V}) \leq 0.11$ are consistent with the data. This translates directly into a 0.17 mag range for A_v, which is 8% in distance. Hence, from this limitation alone, it does not seem possible to derive a distance to the LMC that is more accurate than $\approx 10\%$.

(a) (b)

Figure 2.5 Images of the Local Group irregular galaxies IC 1613 and NGC 6822. Both galaxies are forming stars, hence they have a small population of Cepheids. Both have some Population II stars and some RR Lyraes, so they can serve as consistency checks on the RR Lyrae/Cepheid distance scale, as discussed in the text. (a) A blue band CCD image of IC 1613 taken by the author. (b) A photograph of NGC 6822 taken by David Malin with the AAT telescope and reprinted with permission.

Figure 2.6 Unique CCD red band image of the Large Magellanic Cloud (LMC). The image was taken with a parking-lot camera arrangement (Bothun and Thompson 1988) and the image scale is 36 arcseconds per pixel. The LMC is the major galaxy on which is based the Cepheid distance to the Virgo cluster.

- *Depth effects*: The true orientation of the LMC is not known and indeed is quite confusing. This can be seen by directly comparing R-band and Hα images of the LMC (Figures 2.6 and 2.7) and noting the different appearance of the LMC. The bar area is highly flattened and has an axial ratio of 0.25. The main body of the LMC (area surrounding the bar) has an axial ratio of 0.70 ± 0.03, corresponding to an inclination to the line of sight of 48°. Including the detached 30 Doradus and Constellation III areas, however, raises the axial ratio considerably and lowers the derived inclination to 27°. Hence, the front-to-back distance of the physical structure that defines the LMC is not well known but may be as large as 10% (e.g., 5 kpc) of its distance from the Milky Way.

In the following discussion all measurements have been placed on a reddening scale of $E(\mathrm{B}-\mathrm{V}) = 0.07$. Distances to the LMC determined from Cepheids show remarkable agreement. Such distances, however, are based on calibration using Cepheids in open clusters in our galaxy, whose distances have been determined by

Figure 2.7 CCD image of the LMC taken through a narrowband filter that highlights the emission from H II regions. Note the different appearance of the LMC compared to the R-band image of Figure 2.6.

main sequence fitting using either the Hyades or Pleiades fiducial sequences. Hence, some small systematic error could be present. The various distance moduli which have been obtained are 18.52 from BVI photometry (Feast 1988), 18.52 from JHK photometry (Laney and Stobie 1988), 18.50 from I-band photometry (Madore and Freedman 1991), and 18.42 from 1.05 μm photometry (Visvanathan 1989). Although the numbers of stars in these samples are relatively small (22 in the I-band sample), the agreement is very good. Moreover, there is good agreement between the optical and IR measurements, indicating that our adopted foreground reddening value is sensible. Madore and Freedman (1991) use multicolor photometry toward each Cepheid to determine the reddening. This has the advantage that both the foreground reddening of our galaxy and the internal reddening of the LMC are determined (though not in a separable manner). There is very little evidence to date for either significant amounts of internal reddening or variations in internal reddening for the LMC. Modal averaging of these distance estimates yields a distance modulus to the LMC of 18.50.

RR Lyrae derived distances, however, are not completely consistent with the Cepheid distances. Walker (1992) presents comprehensive measurements of RR Lyrae stars in seven LMC clusters. Excluding NGC 1841, which has a large reddening, the average V-magnitude is 18.97 ± 0.03 at a mean metallicity of $[\mathrm{Fe/H}] = -1.9$. At this value of [Fe/H], equation (2.5) yields $M_v = 0.73$; hence $(m - M) = 18.24$, which is in conflict with the Cepheid value of 18.50. If we use the alternative calibration (2.5a) then the distance is increased by +0.36 in $(m - M)$ to 18.60. If we use the alternative metallicity-independent calibration (2.6) then the distance modulus increases to 18.36. If we instead believe that the Cepheid calibration is correct, and therefore solve for the zero point in equation (2.5) using the mean V-magnitude of 18.97 for the RR Lyrae sample, we derive a zero point of 0.76 or 0.25 magnitude brighter. Finally, if we assume that equation (2.5) is the correct calibration, then a distance modulus of 18.50 can be recovered from the RR Lyrae data if we assume $E(\mathrm{B} - \mathrm{V}) = 0.0$.

On the theme of going where the data leads, we conclude that the galactic calibration of the Cepheid and RR Lyrae distance scales cannot be simultaneously correct. Indeed, the LMC is not the only Local Group galaxy to exhibit significantly different distance moduli between Cepheids and RR Lyraes. The same magnitude of difference is seen is the case of IC 1613, where the Cepheids give $(m - M) = 24.42$ (Freedman 1988) and the RR Lyraes give $(m - M) = 24.10$ (Saha, Hoessel, and Krist 1992). The discrepancy between these two distance measuring techniques has been explored in more detail by van den Bergh (1995a), who eventually decides that the zeropoint discrepancy indeed remains unresolvable. However, van den Bergh (1995a) points out that most of the evidence favors equation (2.5) as the correct calibration in that (1) it can be theoretically reproduced using horizontal branch models (Dorman 1992); and (2) for a typical galactic RR Lyrae ($[\mathrm{Fe/H}] \approx -1.5$) equation (2.5) predicts $M_v = +0.78$, which is in excellent agreement with the statistical parallax value of $M_v = +0.76 \pm 0.14$ derived by Hawley *et al.* (1986). Equation (2.5a), on the other hand, suggests $M_v = +0.49$ for a typical galactic RR Lyrae star, significantly brighter than actually observed.

A possible way to resolve the RR Lyrae–Cepheid disagreement on the distance to the LMC is to find another set of distance indicators that agree either with the Cepheid or the RR Lyrae distance. Unfortunately, these other distance determinations to the LMC give yet different values. Here are some of these distance determination techniques:

- Classical novae in our galaxy show a relation between maximum amplitude and the rate of luminosity decline. As most of these novae are located in young clusters in our galactic disk, the calibration of their intrinsic luminosities suffers from the same problems that plague the calibration of the galactic Cepheid PL relation. Hence, the galactic calibration of nova intrinsic luminosity is somewhat uncertain. This is clearly seen in the application of the nova method to the LMC by Capaccioli *et al.* (1990), which yields $(m - M) = 18.71 \pm 0.30$.
- Certain long-period variables, known as Miras, can also serve as standard candles, particularly in the infrared. K-band photometry of long-period variable stars

Figure 2.8 Blue band CCD image of M33 taken by the author. M33 is a low mass spiral member of the Local Group and serves as one of the fundamental calibrators of the Tully–Fisher relation.

by Hughes and Wood (1990) yields $(m-M) = 18.64 \pm 0.05$. This distance rests on the galactic distance to the globular cluster 47 Tucane, which contains these kinds of stars. Hence, this is a Population II distance indicator.

- As will be discussed later, the luminosity of planetary nebulae can be used as a distance indicator. Measurements of the mean magnitude of planetary nebulae detected in the LMC by Ciardullo and Jacoby (1992) yield $(m-M) = 18.44 \pm 0.18$. This distance, however, is calibrated against the luminosity function of planetary nebulae in M31, hence it requires knowing the distance to M31 and we haven't gotten there yet.
- In 1984 Schommer *et al.* made a bold attempt to determine direct distances to LMC clusters via the technique of main sequence fitting (Schommer, Olszewski, and Aaronson 1984). Using CCD photometry, Schommer *et al.* were able to track the main sequence down to an apparent magnitude of $v \approx 23$ ($M_v \approx +4.5$) in three intermediate age LMC clusters. Fits to these intermediate age clusters initially produced values of $(m-M) = 18.2$, which are consistent with the galactic RR Lyrae calibration but inconsistent with the galactic Cepheid calibration. Although the measurements were difficult, the fitting procedure was sound, provided the theoretical zero age main sequence (ZAMS) for young stars of the same metallicity was sound. Adjustments in that scale by van den Berg and Poll (1989) showed that the previous ZAMS was too bright by ≈ 0.15 mag, thus adjusting $(m-M)$ upwards to 18.35 (± 0.15).
- Distance determinations to the LMC received a serendipitous boost due to SN1987A. Analysis of its expanding photosphere (Eastman and Kirshner 1989) and its light echo as manifested by the circumstellar ring (Panagia *et al.* 1991) both yield $(m-M) = 18.50 \pm 0.15$. But these results are model dependent and different models can give slightly different results (Gould 1994b). Indeed, a recent reanalysis of the ring data by Gould (1995) gives a distance to SN1987A of $(m-M) = 18.37 \pm 0.04$.

Based on all these data, we feel the true distance modulus to the LMC lies in the range 18.35–18.55 and, within this range, there is no resolution towards any particular value. There certainly doesn't seem to be a single best distance indicator to the LMC. For instance, the measurements which are least sensitive to uncertainties in foreground reddening are the H-band observations of Cepheids, which give $(m-M) = 18.5$, and the ring data, which give $(m-M) = 18.35$. It seems unlikely that depth effects have conspired to produce this range in distances (e.g., SN1987A is on the very near side of the LMC). Finally, the metallicity-dependent galactic calibration of the RR Lyrae stars gives $(m-M) = 18.25$ based on equation (2.5), or $(m-M) = 18.60$ based on equation (2.5a). Hence, uncertainty in the distance to the LMC remains one of the larger sources of systematic uncertainty in the determination of H_0.

The Andromeda galaxy – M31

M31 is the only large spiral in the local group, so it is a fundamental calibration point for virtually all extragalactic distance indicators that are applicable beyond

5 Mpc. There have been several recent determinations of distances to M31 and, in general, these indicators show convergence to a specific value. However, most of these distance estimates depend on knowing the distance to the LMC. In addition, M31 has an inclination of $\approx 80°$, hence internal reddening variations across its disk are important. Here are the most recent and credible determinations of the distance to M31, assuming the total V-band absorption to M31 (foreground reddening + M31 average internal extinction) is 0.4 mag:

- Welch *et al.* (1986) measured H-band magnitudes of seven Cepheids, and using 18.45 of the LMC modulus, they obtained $(m - M) = 24.26 \pm 0.16$. Although these observations have the advantage of being mostly unaffected by reddening, the Cepheids themselves are rather faint in the H band. Moreover, the observations were not done with an imaging detector, but with a single-channel photometer. Imaging observations are preferred, as a more precise value of local sky can be determined from them and they are freer from the crowding problems that can strongly affect single-channel aperture photometry.
- Freedman, Madore, and Wilson (1991) measured BVRI magnitudes of a few Cepheids and obtained $(m - M) = 24.38 \pm 0.13$. The difference from the H-band data may be due to an underestimate of the internal reddening in M31 or systematic errors in the H-band magnitudes. In any event, the difference is small and within the respective error bars.
- M31 has some bright, extended globular clusters. Albeit with some, noise color–magnitude diagrams of stars in those clusters can reveal the giant branch. In principle this giant branch can be fitted to the giant branches of similar galactic globular clusters of "known" distance. Christian and Heasley (1991) obtain $(m - M) \approx 24.3$ with this method, though the errors are somewhat large and difficult to estimate properly.
- A similar procedure was performed by Mould and Kristian (1986), who fitted metal-poor giant branches to red giants measured in the halo of M31. This has the advantage that the halo fields are relatively unreddened. They obtain $(m - M) = 24.23 \pm 0.15$ but this calibration rests on distances to galactic globular clusters.
- RR Lyrae stars have been detected anu measured in the halo of M31 by Pritchet and van den Bergh (1987) with a resultant modulus of $(m - M) = 24.33 \pm 0.15$. This assumes $M_v = +0.6$ for RR Lyraes.
- Two other indicators are available: novae (Capaccioli *et al.* 1989) give $(m - M) = 24.27 \pm 0.20$ and carbon stars, again using the LMC as the calibrator, give $(m - M) = 24.45 \pm 0.18$ (Richer, Crabtree, and Pritchet 1990).

In general, all these distance indicators yield a pretty uniform set of distances to M31 but all of them, except for novae and RR Lyraes, depend upon the distance to the LMC. For an LMC modulus of 18.45, the most probable distance modulus to M31 is in the range 24.25–24.40. In the case of M31 there is very good agreement between the Cepheid and RR Lyrae distance scales. This further strengthens the case for a slightly different zero point for the RR Lyraes in the LMC compared to our galaxy and M31, showing that we still do not understand the dependence of absolute magnitude on metallicity for RR Lyraes.

The other spiral – M33

The last local group calibrator of the extragalactic distance scale is M33 (Figure 2.8). M33 has an inclination of 57°, hence internal reddening problems are not as severe as in the case of M31. We use a reddening of $E(\mathrm{B-V}) = 0.07$ (van den Bergh 1991) and adjust all quoted distances below to this reddening value. Techniques similar to those for obtaining the distance to M31 have been applied to M33 and assume the same distance modulus to the LMC. Notably, M33 also shows a discrepancy in distance determinations between the Population I indicators and the Population II indicators. We begin with the Population I distance scale:

- H-band photometry of a handful of Cepheids, again obtained using a single-channel aperture device, yields $(m-M) = 24.20 \pm 0.14$ (Madore *et al.* 1985); a distance actually less than the distance of M31. In contrast, BVRI photometry by Freedman, Madore, and Wilson (1991) yields $(m-M) = 24.58 \pm 0.13$. It seems unlikely this discrepancy is the result of underestimates of the reddening in the Freedman *et al.* sample, so it probably reflects systematic errors in the H-band photometry of faint Cepheids. We emphasize that, since the H-band sky background is ≈ 14.0 mag arcsec^{-2}, accurate determinations of H-band magnitudes of stars as faint as $h = 19.0$ require a very accurate determination of the sky background. This is nontrivial with a single aperture as the detector. Ultimately, JHK imaging observations of Cepheids in M33 will be required to sort out the discrepancy noted above.
- Mould *et al.* (1990) used photometry of long-period variables in M33, compared with those in our galaxy, to derive $(m-M) = 24.52 \pm 0.17$.
- Studies of the rates of decline of classical novae in M33 give $(m-M) = 24.38 \pm 0.20$ (Della Valle 1988).

An average of these Population I indicators gives a distance to M33 of $(m-M) = 24.42 \pm 0.17$. Removal of the suspect H-band observations from this aggregate yields $(m-M) = 24.49 \pm 0.10$, which we adopt as the Population I distance scale to M33.

The Population II distance indicators are the following:

- Again assuming $M_v = +0.6$ for RR Lyrae stars, the observations of Pritchet (1988) yield 24.60 ± 0.23. This distance is only discrepant by 0.1 mag with respect to the Population I result. The real problem lies in the determination made by Mould and Kristian (1986) using the halo giant branch of M33. Recall that the same procedure applied to M31 gave results quite consistent with the Cepheid scale. In the case of M33, Mould and Kristian derive $(m-M) = 24.80 \pm 0.3$. Analysis of another sample of halo red giants by Wilson, Freedman, and Madore (1990) yields $(m-M) = 24.60 \pm 0.3$. The relatively large error bars on these measurements are an accurate indicator of the difficulty of this distance determination method as (1) accurate photometry of the halo stars must be done, (2) very good background subtraction must be done in order to properly isolate the halo giant branch in the CM diagram of the general field, (3) there must be a good calibrating giant branch available for a globular cluster in our galaxy. Since the halo of M33 is quite sparsely populated, an extra degree of difficulty comes into play compared to M31. Taking the average of these three indicators (there is nothing in the data that suggests the MK result is any less valid than the results of Wilson, Freedman, and Madore) yields $(m-M)=24.67\pm0.11$ as the Population II distance scale to M33. This is consistent with a recent review by van den Bergh (1995b), who derives $(m-M)=24.67\pm0.07$ for the Population II distance to M33. This error bar is unrealistically low and gives the erroneous implication that the Population II scale is more precisely determined than the Population I scale.

We regard the distance to M33 as unresolved, but note that if we just take the Cepheid BVRI data and the RR Lyrae data, then we obtain a consistent distance of $(m-M)=24.6$. This would imply that the galactic zero points of both the Cepheid and RR Lyrae scales are good – in sharp contrast with what is found in the LMC. On the basis of all the data, we would place M33 at a distance of $(m-M)=24.4$–24.8, leaning towards a probable value of 24.6.

2.3.5 Extragalactic distance indicators

To reach distances beyond 4 Mpc, a variety of extragalactic distance indicators have come into existence. Most of them are calibrated against the distance to the LMC, M31 and M33. The five most reliable and well-used extragalactic distance indicators are as follows.

The Tully–Fisher relation

This was originally used by Opik in 1922 and is basically a dynamical method for determining intrinsic luminosity. The argument goes like this.

Circular velocities in galaxies scale as

$$M \propto V_c^2 R \tag{2.8}$$

Using the mass-to-light (M/L) ratio, mass can be parameterized as

$$M = L(M/L) \tag{2.9}$$

If we assume the disk galaxy surface brightness (SB = galaxy luminosity per unit area) is the same for all spirals, we can write L in terms of R and SB as follows:

$$\mathrm{SB} = L/\mathrm{area} = L/R^2, \qquad L = \mathrm{SB} \times R^2 \rightarrow L \propto R^2 \tag{2.10}$$

So now equation (2.10) becomes

$$L(M/L) \propto V_c^2 L^{1/2} \tag{2.11}$$

If we now further assume that M/L is constant, we have the simple expression

$$L^{1/2} \propto V_c^2 \rightarrow L \propto V_c^4 \tag{2.12}$$

Converting to magnitude units and remembering that $M = -2.5 \log L$, we arrive at the predicted slope in the relation involving M and V_c:

$$M \propto -10 \log V_c \tag{2.13}$$

The methodology used to derive equation (2.13) has the implicit assumption that the luminous matter dominates the mass of galaxies and the effects of dark matter on the gravitational potential, hence on V_c, are explicitly not considered, even though it is likely that the masses of galaxies are dominated by dark matter (Chapter 4). It is thus quite paradoxical that equation (2.13) has been empirically verified numerous times. This verification indicates that rotating-disk galaxies scale their rotation velocities by their luminosities, despite being dark matter dominated. Hence, there must be some fine-tuned coupling between the amounts of dark and light matter in galaxies that we simply do not understand. Because of this possible fine tuning, it is difficult to argue that the Tully–Fisher (TF) relation has a real basis in physics. Rather, it is an empirical relation that seems to work remarkably well.

For the TF relation to work, three things must be adhered to by all rotating-disk galaxies and each of these three things represents a potential failure point (or at least a point of systematic error) in the derivation of extragalactic distances.

1. Galaxies must be circularly symmetric in order to properly correct the observed circular velocity to the plane of the galaxy to derive V_c. To first order this is true; to second order, galaxies could be embedded in dark matter halos which are flattened, and this would cause systematic deviations from circular motion, producing dispersion in the TF relation. A systematic dependence of halo flattening on galaxy environment would produce systematic zeropoint shifts in the TF relation (Franx and De Zeeuw 1992; Rix and Zaritsky 1995).
2. Galaxies must have the same M/L ratio. In reality this is absurd as the stellar population differences in galaxies clearly produce differences in M/L. However, if the stellar mass is a very small percentage of the total mass, then M/L variations associated with stellar population differences are relatively minor. Still, for the TF relation to work at all, these dark matter dominated galaxies must somehow be programmed to produce a certain amount of luminous matter in order that they obey the TF relation! This brings up the serious possibility that small but systematic differences in dark matter content produce systematically different zero points in the TF relation. The lack of any identified third parameter in the TF relation, despite a wealth of observational attempts to find one, argues that these systematic effects are subtle (e.g., Bernstein *et al.* 1994; Feast 1994; Persic and Salucci 1991; Biviano *et al.* 1990; Mould, Han, and Bothun 1989; Giraud 1987).
3. Galaxies must have the same surface brightness. This is clearly even more absurd than the M/L requirement. In fact, an entire chapter in this book is devoted to the subject of low surface brightness (LSB) galaxies, whose very

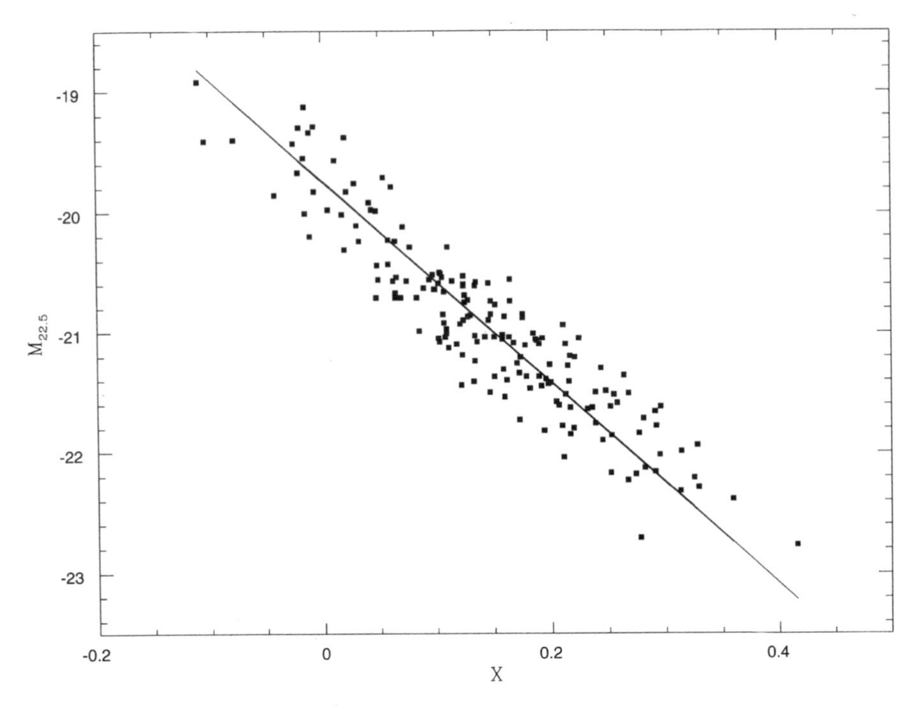

Figure 2.9 Example relation between rotational velocity and I-band luminosity for a large sample of spiral galaxies. The slope of the line is –8.4, which is close to the expected slope –10 relation derived from the virial theorem. The *X*-axis is the log of the line width where zero corresponds to log $\Delta V = 2.5$ (~300 km s^{-1}.) Reproduced from Dell'Antonio, Bothun, and Geller (1996).

existence straightforwardly shows that disk galaxy SB can not be constant. Interestingly, LSB galaxies do define a TF relation like that of normal galaxies, although the scatter is much larger (Sprayberry *et al.* 1995). Moreover, surface brightness issues are only important if the calibration sample has a different range in surface brightness (over the same range in V_c) as the distant sample; only then will a systematic offset occur (Bothun and Mould 1987). This is relevant to the case of M33 as a fundamental calibrator of the TF relation in that it is generally of higher SB than other galaxies of similar V_c.

Despite the fact that galaxies do not all have the same SB or *M*/*L*, there remains a good correlation between V_c and luminosity in almost any sample of inclined disk galaxies. Figure 2.9 shows a recent example of this good correlation; the data come from Dell'Antonio, Bothun, and Geller (1996). Although of much the scatter around this relation is due to measurement errors, some of the scatter is intrinsic. Unfortunately, the intrinsic scatter of the TF relation is not well known and estimates lie in the range 0.1–0.7 mag. In principle, if this intrinsic scatter is low, calibration of the TF relation requires only two points. This requirement can be met by M33 and M31. It is desirable to increase the calibration sample in case the intrinsic scatter is moderately large. Some investigators choose to increase the calibration sample by including Local Group galaxies which are less massive than M33. I believe this is not good practice, since the circular symmetry of low mass galaxies is seriously in question. Ideally, one wishes to add spiral galaxies with a V_c value between those of M33 and M31. There are five such nearby spiral galaxies now available (images of each galaxy are available at the Web site). Each of these galaxies is in the HST Key Project Program to determine Cepheid distances.

- *M81*: The best determination of its distance comes from Freedman *et al.* (1994a) using wide-field planetary camera (WFPC) data from the HST to identify and measure Cepheid variables. This distance is $(m-M)=27.80\pm0.20$ (which assumes an LMC modulus of 18.5).
- *NGC 2403*: This galaxy has a slightly higher V_c than M33 but overall is very much like M33. HST observations of NGC 2403 have not yet been concluded at the time of writing. Ground-based measurements of its Cepheids yield a distance modulus of 27.46 ± 0.24.
- *NGC 300*: This is a very low mass, but rather high SB spiral galaxy. It has V_c less than M33 and its not clear if it should be used as a TF calibrator. A Cepheid distance of $(m-M)=26.0\pm0.2$ has been derived from observations of a few Cepheids.
- *NGC 925*: This is a late type spiral galaxy which is somewhat barred and shows multiple spiral structure like NGC 2403. It is a member of the NGC 1023 group of galaxies which also contains NGCs 891, 949, 1003, 1023 and 1058. All of these galaxies have had recent determinations of their distance (Ciardullo, Jacoby, and Harris 1991; Pierce 1994; Tonry 1991; Schmidt, Kirshner, and Eastman 1992). The mean distance modulus to the NGC 1023 group that results is $(m-M)=29.98\pm0.12$ (mean error). Using an LMC distance modulus of 18.5, Silbermann *et al.* (1996) derive $(m-M)=29.84\pm0.16$ from WFPC observations of its Cepheids. Interestingly, Silbermann *et al.* (1996) derive a moderate level of mean reddening for these Cepheids, which again underscores the need to estimate individual reddening from multicolor photometry.
- *NGC 3351*: This is a barred spiral which is a member of the Leo I group of galaxies. Leo I is an important group because it contains the two nearest bright elliptical galaxies (NGCs 3377 and 3379). As will be clear later, it is important to have a collection of spirals and ellipticals in the same group, to check for the consistency of different kinds of distance estimators as applied to different kinds of galaxies. HST observations of its Cepheids by Graham *et al.* (1997) yield a distance modulus of $(m-M)=30.01\pm0.19$.

In principle these five spiral galaxies, as well as M31 and M33, provide a better calibration of the TF relation. This new calibration is shown in Figure 2.10, where it can be seen that NGC 925 and NGC 3351 lie significantly off the relation defined by just M31 and M33. Thus, the addition of more calibrating galaxies, while in principle a good idea, in practice is likely to reveal the intrinsic scatter and subtle galaxy–galaxy differences in the TF relation.

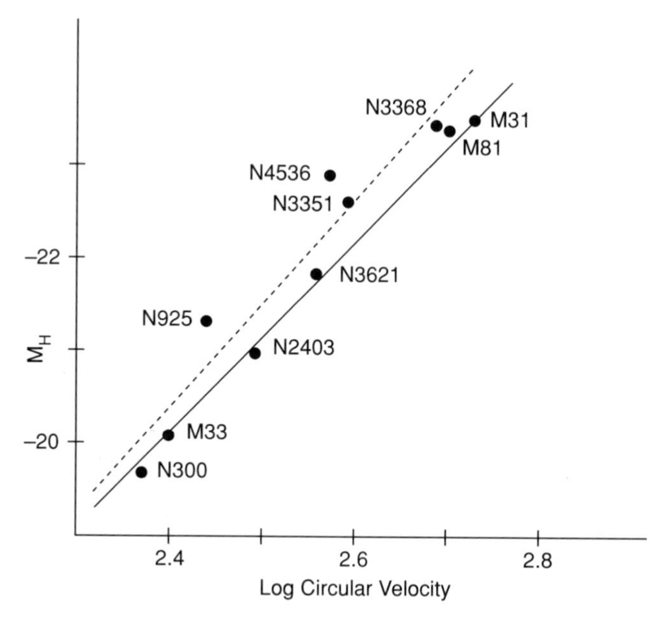

Figure 2.10 Available sample of Local Group galaxies for the calibration of the Tully–Fisher relation. Most of these galaxies have had new distance determinations from the Hubble Space Telescope Key Project (Rawson *et al.* 1997). The solid line represents the relation defined only by M33 and M31; the dashed line is the best fit to all the calibrating galaxies.

The luminosity function of planetary nebulae

Planetary nebulae are evolved stars that are ejecting their outer envelopes. An example is shown in Figure 2.11 (see plate section). The physics of this phenomenon is universal and the evolutionary rates are so rapid in this phase that virtually all currently detectable planetary nebulae in a given galaxy occupy a very small mass range. As the ionizing flux depends upon this mass, the luminosity of a planetary nebulae, as measured in the [O III] emission line, should be nearly constant from nebula to nebula. Slight differences in mass (and perhaps metallicity), which arise from differences in stellar population ages within a galaxy, produce a range in luminosity, which therefore defines the luminosity function of planetary nebulae (the PNLF). Figure 2.12 shows the PNLF for M31. There is a very sharp cutoff in the maximum luminosity which can be used as a standard candle for distance estimates. The main limitations of the PNLF method are as follows:

- It is best applied to elliptical galaxies which have less dispersion in the ages of the stellar populations that will produce the planetary nebulae. Applying it to spiral galaxies is slightly more problematical as a range of stellar population ages contributes to the PNLF.

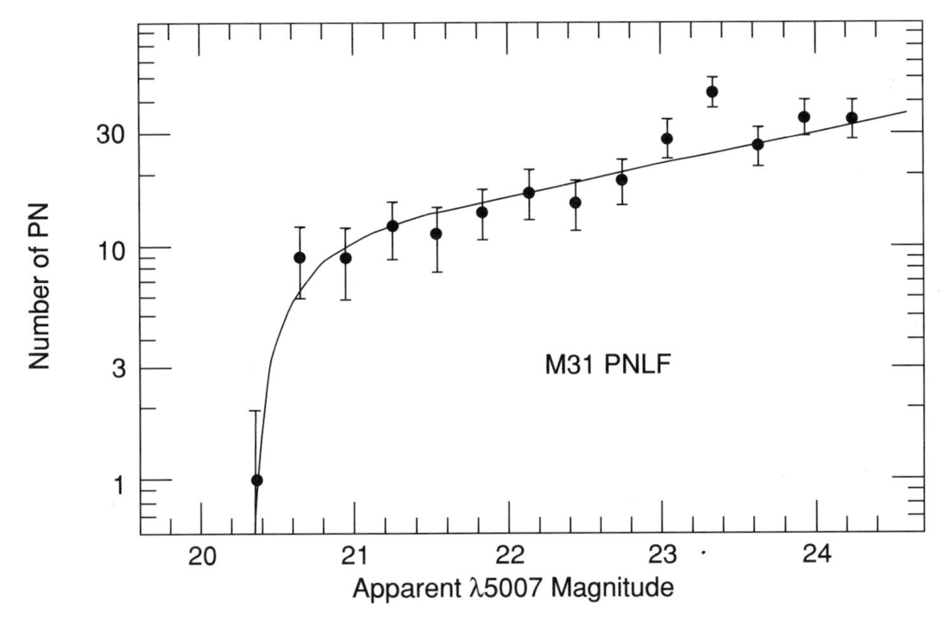

Figure 2.12 PNLF for M31: the sharp cutoff in maximum luminosity, if universal, can be used as the reference luminosity for distance determinations to other galaxies. In this case the distance to M31 serves as the fundamental calibration. Reproduced courtesy of Robin Ciardullo, Department of Physics and Astronomy, Penn State University.

- There is no good way to determine distances to galactic PN, so the PNLF does not have a zero point which can be based on any galactic calibration. Currently, it is the distance to M31 that sets the zero point for the PNLF.
- The method is limited to distances out to about the Virgo cluster. Beyond that, PN become pretty faint for ground-based telescopes.
- As in the case of red supergiants, the PNLF may depend on the total number of stars in the galaxy. Indeed, Bottinelli *et al.* (1991) do find a small dependence of distance modulus on galaxy luminosity, in the sense that the more luminous galaxies have smaller distance moduli. Sandage (1994a) has taken the extreme position that this population bias is far more severe than the proponents of the PNLF method (Jacoby *et al.* 1992) realize. But, examining the actual data (e.g., Figure 2.12) reveals that the PNLF does have a relatively sharp cutoff at some maximum luminosity, though there is insufficient data to reliably measure the intrinsic scatter at the bright end of the PNLF. This likely introduces a small systematic error.
- Recent observations of the UV light of elliptical galaxies by Brown, Ferguson, and Davidsen (1995) indicate that the UV light is dominated by the kinds of stars, e.g., postasymptotic branch (PAGB) stars, that are the ionizing stars in PN. They find small differences in mean mass of these PAGB stars from one elliptical

to another. These mass differences will cause differences in the luminosity of the [O III] line in PN, so they will produce some scatter in the PNLF. The mass differences are generally around 0.01–0.03 solar masses and it is unclear how much scatter in the PNLF is actually introduced by mass differences this small.

On the positive side, PNLF distances, calibrated from the PNLF for M31, for four galaxies (two ellipticals and two early type spirals) in the Leo I group yield a set of distances that are self-consistent to an accuracy of 5%. Furthermore, these distances agree well with the Cepheid distance scale for NGC 3351 cited earlier.

Surface brightness and luminosity fluctuations

This is a statistical method which is based on the precepts of Baum (1955). It is meant to provide information in the stellar population, specifically the giant branch, for a stellar cluster or galaxy which is at known distance. An inversion of this process as applied to the extragalactic distance scale has been championed by John Tonry and collaborators in various publications (e.g., Tonry *et al.* 1997). Rather than explain the theory, it is better to give a practical example of how the method is applied.

Suppose we image an elliptical with a detector with scale 1 arcsecond = 3 pixels. We determine that the intrinsic RMS pixel-to-pixel fluctuation signal in the galaxy image is 10% (we have already subtracted the sky fluctuation signal and the intrinsic luminosity gradient within the galaxy). We then assume that the light from the galaxy is dominated by one kind of star. In this example we assume it is a K0 star, which has an intrinsic $M_v = +0.7$ and B – V = 0.85. We do the observations in the blue. The 10% fluctuation signal comes from the Poisson noise in the distribution of K0 stars per pixel. Thus the fluctuation signal is $N^{1/2}/N$. For the case of 0.1, $N = 100$. The combined luminosity of 100 K0 giants is $M_v = -3.5$. We also measure the mean surface brightness of the galaxy over which we measure the fluctuations, and we find its value is $b = 20.0$ mag arcsec^{-2}. At our pixel scale, 1 pixel would then correspond to a magnitude of 22.5 and each pixel would have an absolute magnitude of –3.5. The distance modulus to this galaxy is therefore $(m - M) = 22.5 - (-3.5) = 26.0$.

Tonry *et al.* (1997) have refined this basic technique using multicolor photometry. They conclude that fluctuation magnitudes in the I band are the most practical to measure. As this fluctuation signal is dominated by giants, differences in individual temperatures and luminosities between the giant stars produce a color dependence on the fluctuation magnitude. This is easily understood as cooler giant stars tend to be more luminous. From direct observations as well as theoretical estimates from stellar population models, Tonry *et al.* (1997) derive a relation between the average absolute fluctuation magnitude and the color of the stellar population:

$$\overline{M}_I = -1.74 + 4.5[\,(\mathrm{V} - \mathrm{I})_0 - 1.15] \tag{2.14}$$

Distances can be derived by measuring $\overline{M}_I$ and V – I, correcting for reddening then using equation (2.14).

The limitations in the application of the surface brightness fluctuation (SBF) method to determining extragalactic distances are as follows:

- The method is really best applied to single-age stellar populations such as those that hypothetically exist in elliptical galaxies or the bulges of spiral galaxies. Furthermore, it is sensitive to small-scale features such as globular clusters and/or structure in the dust distribution. To first order these effects can be identified and removed but they may still leave some residual signal that contributes to the observed value of $\overline{M}_I$.
- In reality it is not one single kind of star that dominates the fluctuation signal, but the entire stellar luminosity function, which must be estimated by some a priori criteria or some observational constraints (such as broadband colors). In the case of giant-dominated luminosity fluctuations there should be correlated color fluctuations. In practice the variation in *K*/*M* giant ratio per pixel can be determined via this correlation. Very good data is required to see these color fluctuations, as the sky fluctuations at wavelengths greater than 7400 Å become severe. To a large extent, the recent work that has resulted in equation (2.14) overcomes this limitation.
- Determining the zero point of the SBF method is extremely difficult. The only galaxy in the Local Group that the method works well for is M32, but it seems unlikely the giant branch in M32 is very similar to that in more luminous ellipticals (e.g., Elston and Silva 1992). One can attempt to use the bulge of M31 but severe complicating effects, such as small-scale dust, possible small-scale stellar clusters and cool luminous giants in a metal-rich population, can all conspire to give a fluctuation signal unique to M31. It is also possible to determine the zero point directly from theoretical considerations (e.g., Worthey 1993). In fact, the zero point of this method used by Tonry (1991) is 0.35 mag brighter than the zero point derived by Tonry *et al.* (1997). In general, this is a signature of a still evolving method. The 1991 zero point, however, was only based on M31 and M32, and used a different color term than the one found by Tonry *et al.* (1997); which is based on a much larger data sample and a better set of theoretical models.

The luminosity of supernovae

If the maximum luminosity of supernovae of type Ia or type II is a constant, they will provide the fundamental determination of the extragalactic distance scale, as their high intrinsic luminosity allows detection at very large distances. Indeed, some particularly optimistic groups hope to use this as a means for estimating the deceleration parameter (1.17) of the Universe by measuring the rate of change of H_0 over look-back times of a few billion years (Perlmutter *et al.* 1995). This is possible because the high intrinsic luminosities of supernovae allow them to be detected at cosmological distances.

Supernovae of type II (which come from the detonation of very massive stars) are known to have considerable variation in their luminosity. However, for these

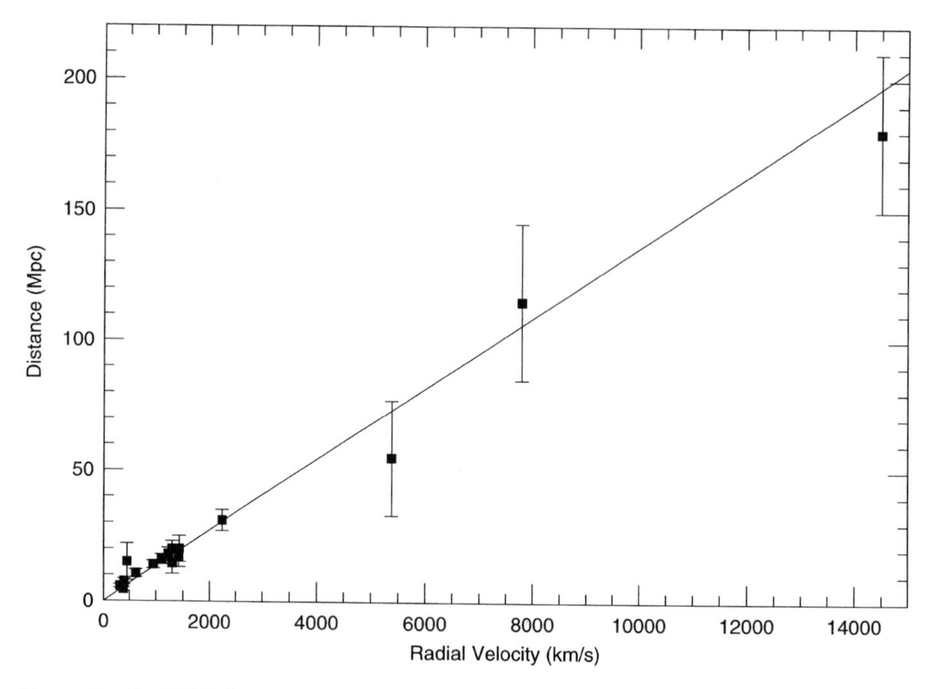

Figure 2.13 EPM distances to 16 type II supernovae vs. redshift of the host galaxy, corrected for Virgo infall. The solid line shows the best fit for the sample. Although there is significant scatter about the relation, the technique does show promise. The data is from Schmidt *et al.* (1994) and their initial calibration yields $H_0 = 73$. But notice that most of the available sample is local, hence it is subject to significant peculiar velocities.

supernovae, one can apply a kinematic model to the expanding photosphere to relate changes in radius to an absolute distance, once an accurate radial velocity curve is measured. This method, called the expanding photosphere method (EPM) is similar to the Baade–Wesselink method as it basically compares the angular size of the photosphere with its measured expansion velocity. It requires a good observational determination of the SN radial velocity curve and a good estimate of the reddening. Application of this method to the type II SN1987A in the LMC can provide its zero point but differences in metallicity and SN environment likely preclude that zero point from being universal. Other possible zero points include type II SN that have occurred in M81 and M101. The recent Cepheid distance to M101 based on HST observations by Kelson *et al.* (1996) gives $(m-M) = 29.34$, in excellent agreement with the EPM distance of $(m-M) = 29.35$. The EPM distance to M81, however, is 25% smaller than the HST Cepheid distance found by Freedman *et al.* (1994a).

The most recent use of EPM is by Schmidt *et al.* (1994) for a small sample of type II SN out to distances of 14 000 km s^{-1}. The method is promising but still prone to large error bars (Figure 2.13) and requires a very good model atmosphere to understand the evolution of effective temperature, which controls the overall spectral shape. In addition, EPM does not yet have a secure zero point. Full details of EPM are given in Eastman, Schmidt, and Kirshner (1996).

Supernovae of type Ia are thermonuclear explosions of degenerate white dwarfs (WDs) near the Chandrasekhar limit (cf. Wheeler and Harkness 1990), or perhaps mergers of WDs in a binary system (Paczynski 1985). Detonation or deflagration models (Nomoto, Sugimoto, and Neo 1976; Arnett 1969) then produce the visible energy release that characterize the supernova light and velocity curves. Detailed models show considerable differences in these scenarios (Khokhlov, Muller, and Hoflich 1993), but their large intrinsic luminosity coupled with the assumed universal physics involved in the Chandrasekhar mass limit have led to strong statements concerning the use of type Ia supernovae as distance indicators (Branch and Tammann 1992) as they are thought to be standard candles. However, this point remains controversial as there is a demonstrable spread in the real luminosities of type Ia supernovae (see immediately below).

In addition, Phillips (1993) first pointed out a correlation between the supernova luminosity and its decay with time (usually over a 2–3 week period). Hence, one can also use the form of the supernova light curve as a second parameter in estimating supernova luminosity and therefore distance. Application of the luminosity decline rate relation to SN Ia can reduce their luminosity scatter down to ± 0.3 mag, making them competitive with other distance indicators (see below). Particularly optimistic uses of this relation, combined with analysis of multicolor supernova light curve shapes, have yielded an apparent dispersion as low as 0.12 mag (Riess, Press, and Kirshner 1996). This is shown in Figure 2.14 and it certainly paints an optimistic picture for the use of SN Ia objects as distance indicators, although von Hippel, Bothun, and Schommer (1997) show why one should consider this treatment with some skepticism.

The complications to deriving distances from SN luminosity or evolution of its expanding photosphere are as follows:

- The supernova must be detected and measured at the time of maximum light. The rise to maximum can be very sudden (timescales of hours). Early detection of the supernova is therefore critical and this is extremely difficult to do observationally. Hence, the available sample of SN II for calibration is necessarily limited.
- Reddening estimates towards supernovae are notoriously difficult to determine. Spectroscopy of the supernova itself can give some constraints if the intrinsic spectrum is understood. Broadband color evolution also has constraining power but requires good observations at the time of maximum light.
- The physical mechanisms which produce supernovae are not well enough understood to be comfortable with predicting that their maximum luminosity should be a constant. Indeed, there is now a very good indication that SN Ia have a significant spread in peak luminosity. Phillips (1993) has compiled data on nine SN Ia to galaxies which have had distances determined via other techniques (e.g., TF relation) and finds a dispersion in peak magnitude of 0.6 mag with a

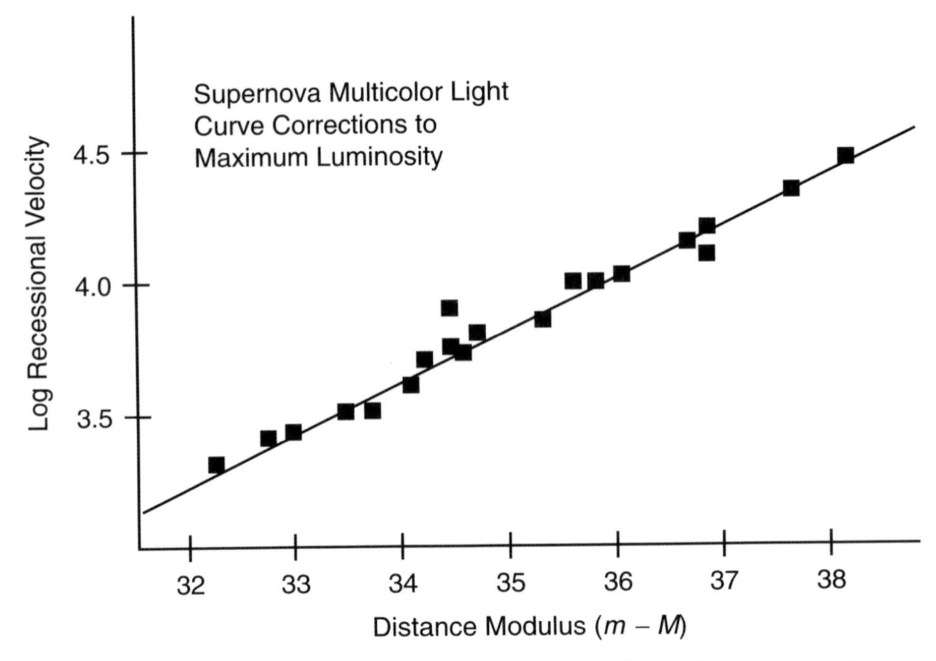

Figure 2.14 Hubble diagram for type Ia supernovae whose distances have been derived using multicolor light profiles and corrections as described in Riess, Press, and Kirshner (1996) and whose data is reproduced here. The scatter in their data is 0.12 mag, making it potentially the best available extragalactic distance indicator. And their initial calibration yields $H_0 = 64$.

total spread of 1.6 mag. This spread is also seen in the larger Calan–Tolo survey (Figure 2.15). However, this large range or dispersion in SN Ia luminosity does not preclude their usefulness as a distance indicator because there is a very tight relation between the maximum SN Ia luminosity and the rate at which the luminosity declines (Hamuy *et al.* 1995).

- An accurate calibration of the supernova luminosity scale would require a supernova occurring in M31 or M33. Eventually one will occur, but astronomers are sufficiently impatient to find H_0 that they use secondary calibrating galaxies, in which an SN Ia does occur.

Secondary galaxies such as NGC 5253, NGC 5128 or IC 4182 have distances of 3–8 Mpc, which have been determined mainly from Cepheids. Two of these calibrating galaxies are shown in Figures 2.16 and 2.17 (see plate section) and all three have problems with respect to being a good calibrator:

- *NGC 5128*, host to SN1986G, is an extremely peculiar and very dusty galaxy. It is therefore difficult to properly deredden the SN light curve. For this reason, NGC 5128 is not used as a calibrator even though it is the nearest galaxy ($d \approx 3$ Mpc) to host an SN Ia.

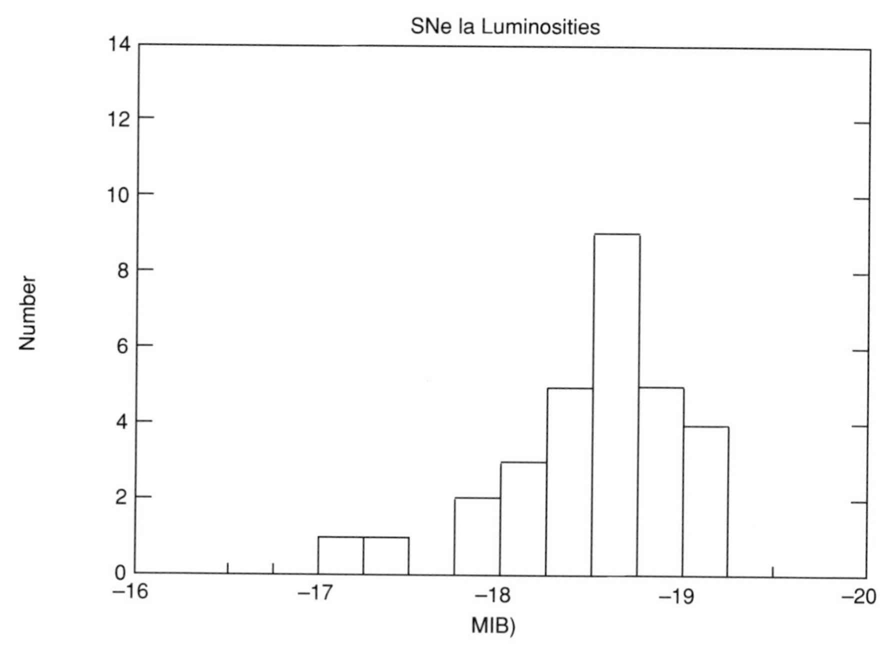

Figure 2.15 Distribution of intrinsic type Ia luminosities for the Calan–Tololo sample (Hamuy *et al.* 1995; von Hippel, Bothun, and Schommer 1997). To effectively use type Ia luminosity as a distance indicator requires significant compression of this intrinsic luminosity range by using luminosity corrections of the kind advocated by Riess, Press, and Kirshner (1996) and Hamuy *et al.* (1995). Reproduced from von Hippel, Bothun, and Schommer (1997).

- *NGC 5253* is a low mass, amorphous galaxy that has hosted two SN Ia over the last 100 years (SN1895B and SN1972E; see Schaefer 1995 for a recent analysis). The occurrence of two SN Ia in a galaxy of this low mass over the last 100 years is odd. A distance modulus of $(m - M) = 28.06 \pm 0.06$ mag has been determined from HST measurements of 11 Cepheids by Sandage *et al.* (1994).
- *IC 4182* is another low mass, amorphous galaxy which was the host of SN1937C. Sandage *et al.* (1992) derive a distance based on HST measurements of Cepheids. These measurements, however, were made with the first cycle of HST time when the optics were not yet corrected for spherical aberration. Concern has also been expressed about the reliability of the photographic photometry of SN1937C (Pierce and Jacoby 1995).

The current calibration of the SN Ia distance scale is directly tied to the Cepheid distance scale. This allows for a consistency check if a supernova occurs in a nearby galaxy which has a Cepheid distance. This check is now available for the case of M100, which we discuss later in this chapter. The SN Ia luminosity calibration is uncertain, due to errors in distance to these galaxies plus systematic error

Figure 2.18 True-color CCD image obtained with the Hubble Space Telescope of NGC 4639. This spiral is located in the Virgo cluster and has hosted a type Ia supernova. Using the HST a Cepheid distance to this galaxy has been determined by Sandage *et al.* (1996). This image was created with support to Space Telescope Science Institute, operated by the Association of Universities for Research in Astronomy, Inc., from NASA contract NAS5–26555, and is reproduced with permission from AURA/STScI. Digital renditions of images produced by AURA/STScI are obtainable royalty-free.

from the adopted distances to the LMC or M31. Overall, however, the paucity of nearby calibrating SN Ia host galaxies is of real concern and raises the strong possibility of systematic bias, especially because the two calibrators are low mass, irregular galaxies, and hence rather dissimilar to the host galaxies of more distant SN Ia. Very recently, Sandage *et al.* (1996) have added to the list of calibrating galaxies by including three in Virgo with Cepheid distances, and the galaxy NGC 3627 in the Leo group. We note that the distance moduli determined from Cepheids for the three Virgo spirals range from $(m - M) = 31.10 \pm 0.31$ for NGC 4536 (Saha *et al.* 1996) to $(m - M) = 32.0$ for NGC 4639 (Figure 2.18) (Sandage *et al.* 1996). We further note that Kennicutt, Freedman, and Mould (1995) have identified 16 different papers in the recent literature that use the SN Ia distance scale to derive H_0. These 16 papers yield a factor of 2 range in the value of H_0, which almost certainly reflects (1) an insecure zero point for the method and (2) uncertainty in the correct calibration for the luminosity decline relation.

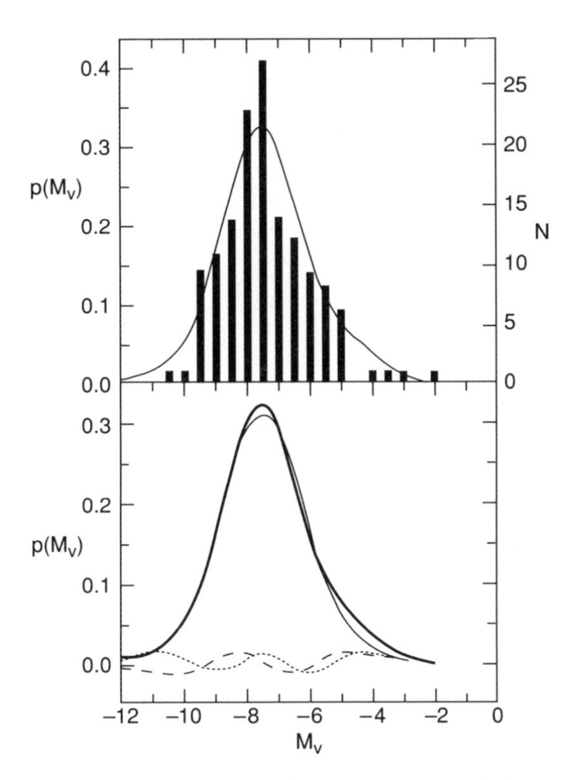

Figure 2.19 Representation of the GCLF for 136 galactic globular clusters: the data has been used to derive an absolute mean visual magnitude of −7.41 with a dispersion of 1.25 mag. Reproduced, with permission, from Abraham and van den Bergh (1995).

The globular cluster luminosity function

This is a Population II distance indicator which rests on the assumption, with very little a priori justification, that the globular cluster luminosity function (GCLF) is universal. For our galaxy, the GCLF is Gaussian with a mean $M_v = -7.5 \pm 0.2$ and a dispersion of ≈ 1.2 mag. This GCLF is shown in Figure 2.19. Determining the distance to an external galaxy involves measuring the peak apparent magnitude of the GCLF. The method has the added advantage of being relatively unaffected by reddening. The complications to this method are as follows:

- The zero point of the galactic GCLF floats around by 0.3 mag, depending upon which sample of globular clusters with "good" distances is used. For instance, Sandage and Tammann (1995) use $M_v = -7.6 \pm 0.1$ for our galaxy and $M_v = -7.7 \pm 0.2$ for M31, using an M31 Cepheid distance of $(m - M) = 24.44$. Abraham and van den Bergh (1995) derive $M_v = -7.41 \pm 0.11$ for our galaxy.
- In apparent magnitude space, different galaxies exhibit different Gaussian widths to their distribution and slightly different behavior at the bright end of the LF.

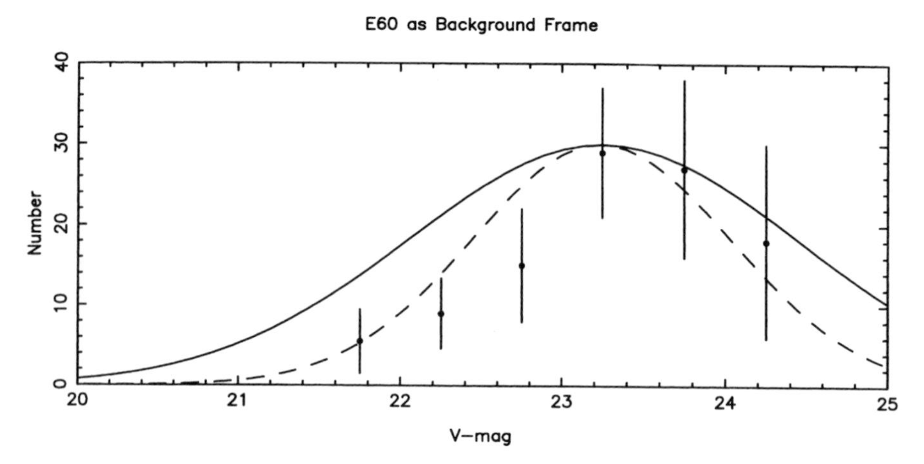

Figure 2.20 The GCLF as measured for NGC 7814 by Bothun, Harris, and Hesser (1992). The dashed line is the best fit to the data and has a dispersion of 0.8 mag. The solid line represents a dispersion of 1.2 mag.

The different dispersions in the GCLFs may be a population statistics effect and strongly indicate that the GCLF does not have a uniform dispersion as is commonly assumed. Figure 2.20 shows the GCLF as determined for the galaxy NGC 7814 (Bothun, Harris, and Hesser 1992), In this case the observed dispersion is significantly lower than for our galaxy.

- Beyond a distance of 5 Mpc, globular clusters cannot be detected individually, hence one relies on a statistical detection of excess faint objects clustered around a galaxy to define the globular cluster system (GCS) and therefore the GCLF. Most measured GCLFs are statistically determined.
- In general, elliptical galaxies have a richer GCS than spirals, yet it is the spiral GCLF which is used as the zero point. It is not clear whether the GCS of a spiral is a good analog to the GCS in an elliptical, especially if some ellipticals were formed by mergers and the globular clusters themselves formed in this process (Zepf, Ashman, and Geisler 1995). However, there is some preliminary evidence that the GCLF does yield consistent distances. For the ellipticals NGC 3379 and NGC 3377 (both members of the Leo group of galaxies), application of the GCLF method gives $(m - M) = 30.05 \pm 0.40$ (note the large error). Application of the PNLF method gives 29.9 and the SBF method gives 29.7 for these two galaxies.

This has summarized the five primary extragalactic distance indicators currently in use. We do not include the Faber–Jackson method in this discussion. This method, now modified to be the D_n–σ relation (Dressler *et al.* 1987), is a dynamical method for determining relative distances between elliptical galaxies, and is used in a similar manner to the TF relation for spiral galaxies. It is not included in the present discussion because the relation has an unknown zero point. There are no elliptical galaxies in the Local Group to calibrate it. The same set of remarks apply to using the brightest cluster member (BCM) as a standard candle or using the "knee" of the galaxy luminosity function (Chapter 5). Perhaps they provide reasonable estimates of relative distances between clusters, but they cannot be calibrated locally; hence they have no absolute zero point.

2.3.6 The distance to the Virgo cluster

Within the limitations discussed above, each of these five distance indicators has been used to determine the distance to the Virgo cluster. The Virgo cluster is the nearest cluster of galaxies and contains about 200 members which are brighter than M33. The core of the Virgo cluster is dominated by the very bright elliptical galaxy M87. This is shown in Figure 2.21. As will be discussed in Chapter 3, knowing the distance to Virgo does not automatically allow for a determination of H_0, but it is one of the last steps to be taken. However, a good distance to the Virgo cluster might be quite useful in establishing the calibration for SN Ia or BCM based indicators, which can be applied over much larger distances.

The TF distance

There are approximately 30 spiral galaxies that are contained in the nominal 6° angular diameter region of the sky that defines the Virgo cluster and which have favorable inclinations. The main limitation is the depth of the sample along the line of sight. If this depth is 5 Mpc (± 2.5 Mpc from the center), then this will introduce ≈ 0.4 mag of range in the TF relation (Fukugita, Okamura, and Yasuda 1993). There is also the possibility that the spiral population of Virgo is actually in the form of a shell of galaxies currently falling into the core. In either case, the front-to-back distance of the spiral sample does seem to be appreciable. There have been

Figure 2.21 The central part of the Virgo cluster defined by the two bright ellipticals M84 and M87, near the right-hand side. Constructed from the Palomar Digital Sky Survey by the author.

many estimates of the distance to Virgo from the TF relation. The most complete sample is that of Pierce and Tully (1988), who derive $(m-M)=30.88\pm0.20$, where the error bar is dominated by the depth of the sample and the zero point is set by their adopted distances to M31 and M33. A comprehensive recent treatment by Yasuda, Fukugita, and Okamura (1997) uses two samples of galaxies for calibration. The first sample uses Cepheid distances for M331, M33, M81, NGC 300, NGC 2403, NGC 3109 (an irregular galaxy, hence a questionable calibrator), and NGC 3368 to yield a calibrating blue band TF relation which has a scatter of 0.29 mag. This sample is combined with distance estimates of 13 galaxies in nearby groups to yield a TF relation for Virgo spirals, a relation which also has a scatter of 0.29 mag. This indicates that the larger sample has not improved the precision of the calibrating relationship but has just made the statistical determination of the zero point more reliable. The analysis of Yasuda, Fukugita, and Okamura (1997) clearly shows that Virgo has a high degree of substructure as well as depth along the line of sight. From a study of 43 spiral galaxies, they derive a mean Virgo distance of $(m-M)=31.18\pm0.40$.

The PNLF distance

Jacoby, Ciardullo, and Ford (1990) apply this method to six elliptical and S0 galaxies in the core of the Virgo cluster to derive $(m-M)=30.84\pm0.05$. The extremely small dispersion reflects the lack of depth in this sample and the fact that the uncertainty in the distance of M31 is not included. A more realistic error bar is closer to 0.10 mag for this sample, which is still rather good.

Surface brightness fluctuation distance

Tonry, Ajhar, and Luppino (1990, 1991) measured V- and I-band fluctuations of 10 elliptical and S0 galaxies to derive $(m-M)=30.86\pm0.13$. The new Tonry *et al.* (1997) calibration of the method coupled with augmentations to the Virgo sample of galaxies has increased this distance to $(m-M)=31.03\pm0.06$.

Supernova Ia distance

Sandage and Tammann (1993) used a sample of 10 SN Ia which have occurred in Virgo cluster galaxies from 1919–1990. Six of these host galaxies are elliptical and the other four are spirals. They derive a mean value of $<M_v>=12.16\pm0.14$. If only the Cepheid distance to IC 4182 is used as the zero point, the Virgo cluster modulus is $(m-M)=31.89\pm0.21$. Using both the IC 4182 and NGC 5253 zero points reduces this estimate by ≈ 0.2 mag to $(m-M)=31.70\pm0.21$. These distance estimates are significantly larger than those obtained by the previously discussed methods. But there are several potential problems with this estimate:

- There is evidence of a correlation between the maximum SN Ia luminosity and B–V color (Hamuy *et al.* 1995). As shown in Tammann and Sandage (1995),

SN1937C and SN1972E are marginally bluer than the supernova observed in Virgo cluster galaxies.

- The most recent SN Ia to occur in Virgo was SN1994D in the host galaxy NGC 4526 (Richmond *et al.* 1995). If $(m - M) = 31.7$ for Virgo, then the photometry of SN1994D shows it to be significantly underluminous.
- Seven of the ten SN Ia used in Sandage and Tammann (1993) occurred in 1961 or before, so they only have photographic photometry.
- The host galaxies to the Virgo SN Ia are all luminous ellipticals and spirals, yet the two calibrating galaxies, NGC 5253 and IC 4182, are low mass, amorphous, irregular galaxies. According to von Hippel, Bothun, and Schommer (1997) there is a small dependence of SN Ia luminosity on galaxy type.
- The Cepheid distance to M100, in Virgo, host of three SN Ia over the last 100 years, is $(m - M) = 31.16$ (see below). M100 is not used in the Virgo cluster SN sample of Sandage and Tammann.

GCLF distance

Based on observations of the statistically determined GCLF of Virgo ellipticals, Harris *et al.* (1991) determine $(m - M) = 31.47 \pm 0.25$. Sandage and Tammann (1995) use that same data set, but apply a different galactic GCLF zero point to derive $(m - M) = 31.64 \pm 0.25$. They arrive at this distance estimate without any discussion of the kind of population sampling biases they previously used to discount the TF and PNLF distances. Such sampling biases are most severe in the case of the GCLF precisely because the GCLF itself is a statistical determination. Interestingly, Nielsen *et al.* (1997) derive a distance of $(m - M) = 31.05 \pm 0.07$ to the elliptical galaxy NGC 4478 using luminosity fluctuations as measured with the HST. These observations also permitted the identification of the GCS of NGC 4478 from which was derived a GCLF distance of $(m - M) = 31.25 \pm 0.2$. This is rather inconsistent with the earlier results for the GCLF distance to Virgo.

2.3.7 Correction for bias in extragalactic samples

These five independent distance estimates to Virgo do not converge to a single value; instead there is a strong dichotomy. The TF, PNLF and SBF methods (which we regard as the most reliable based on the ability to calibrate their zero points using local galaxies) actually give an amazingly consistent distance to Virgo for the very first time. These distance estimates involve samples of ellipticals and spirals. An unweighted mean of these three estimates gives $(m - M) = 30.86 \pm 0.15$ (14.8 Mpc) if we use the Tonry (1991) and Pierce and Tully (1988) data sets. Using the Tonry *et al.* (1997) and Yasuda, Fukugita, and Okamura (1997) data sets yields $(m - M) = 31.01 \pm 0.17$ (17.02 Mpc) for the the distance modulus to the Virgo cluster. These estimates are significantly lower than the $(m - M) = 31.4 - 31.8$ distance that is derived via the other techniques. The lower of the two Virgo distance estimates is difficult to reconcile with $H_0 = 50$. Is there a resolution to these discrepant measures of the Virgo cluster distance modulus or are some of the techniques just wrong?

These discrepant distance estimates to Virgo are very difficult to sort out. Errors in distance can arise from (a) poor observations, (b) systematic errors in zeropoint calibration or (c) lack of a fair sample. Sandage and collaborators in an extensive series of papers (Sandage 1996a,b; Sandage, Tammann, and Federspiel 1995; Federspiel, Sandage, and Tammann 1994; Sandage 1994a,b) vigorously argue that the TF, PNLF and SBF methods all suffer from the lack of construction of a fair and equitable sample, and that sample population biases have not been adequately treated. The general theme of the criticism is that Virgo samples have been constructed in such a way as to yield samples which have the maximum brightness at fixed circular velocity (TF method), or fixed galaxy mass (PNLF method) or metallicity (SBF method). I regard this as remarkable coordination among different observing teams – to produce the same level of conspiracy and systematic error in the distance to Virgo. Moreover, if the criticism is true, then all three distance indicators are worthless because they can never be applied in an unbiased manner to extragalactic distances. This position seems rather extreme and implies that we don't understand very much about the kinds of corrections that should be applied to extragalactic samples. Misapplication of these corrections can certainly lead to erroneous answers, so one must be careful to fully understand their nature.

There are basically three kinds of biases in extragalactic samples that can affect distance determinations:

1. *Sample incompleteness*: At a fixed flux limit, sources disappear beyond some distance. If we know the rough shape of the distribution in distance, we can correct for the missing faint objects. In general, this form of bias is not present in the Virgo cluster samples of galaxies because Virgo is so nearby that even its faintest members can be detected (subject to the surface brightness arguments in Chapter 6).
2. *The Scott effect*: This is frequently confused with the Malmquist bias (item 3). It comes about if some variable, call it S, correlates with intrinsic luminosity (perhaps S could be V_c). At fixed S there is some distribution in luminosity. The tendency to pick only the brightest objects in the distribution at fixed S is then a bias. This is frequently cited as a bias in the TF relation (at fixed line width, only the brightest objects are selected). One test of the existence of the bias is to see whether the dispersion in the TF relation depends upon V_c. In the Virgo cluster samples it does not.
3. *The Malmquist bias*: This occurs when distance errors are involved. In a homogeneous distribution, more objects with intrinsically large distances will be scattered into the sample than will be scattered out of the sample (similar to the Lutz–Kelker corrections). A correction for this bias can be made if the intrinsic scatter of the distance indicator is known. However, the correction only works well in the case where the sample is homogeneously distributed. As we will discuss in detail in Chapter 3, the galaxy distribution is anything but homogeneous,

and application of the homogeneous correction is systematically wrong. For instance, if there is a large void at the back end of our sample, there are simply no distant galaxies to scatter into our sample and a homogeneous correction produces spurious results. The severity of the Malmquist effect is directly proportional to the intrinsic dispersion of the distance indicator (in the case of a homogeneous Malmquist bias this is $1.38\sigma^2$ mag). For the case of the TF relation, if the intrinsic dispersions is ≈ 0.4 mag then the correction is 20% in luminosity or 10% in distance. Estimates of the intrinsic dispersion in the TF relation are difficult to make because they depend on the range of M/L and SB in the sample under consideration. Estimates for the intrinsic scatter have been in the range 0.1–0.7 mag. However, most investigators find this dispersion is ≤ 0.3 mag, meaning that the homogeneous Malmquist correction is small. However, as there is a void behind Virgo, the inhomogeneous Malmquist correction scheme, as first proposed by Landy and Szalay (1992), is more appropriate but very difficult to apply in practice. This correction uses the actual distances estimated along the line of sight to formulate the volume sampling correction, hence it needs a representative sample from which to form the distribution. For the case of Virgo, this correction is substantially smaller than the already small homogeneous correction.

The various bias correction schemes which have been applied to the TF, PNLF and SBF samples by Sandage and coworkers, schemes that produce a 50% larger distance, on the whole I consider them too large. For instance, Gonzalez and Faber (1997) have recently shown that Malmquist bias has only a 3.5–6.5% effect on the derived distance to Virgo – significantly smaller than has been used by Sandage and coworkers (Sandage 1996c) in criticizing and correcting the short distance scale derived for Virgo. In general, these corrections are justified and necessary, but the chosen intrinsic scatter, from which the correction is applied, does seem rather large. This is the key point. If, in fact, the scatter were as large as claimed for the TF, PNLF and SBF methods (≥ 0.5 mag), it is doubtful that the technique would have been used in the first place to estimate the distance to Virgo, let alone produce a consistent set of distances. Nevertheless, Sandage and Tammann (1995) are able to reconcile the TF, PNLF, and SBF distances to Virgo with their estimate of $(m - M) \approx 31.7$ based on the GCLF or SN Ia luminosities. However, if the distance to the Virgo cluster is really 22 Mpc then there are two immediate problems: (1) the measured [O III] luminosity of the PN in the elliptical galaxies reaches a physically implausible value and (2) M31 is significantly underluminous with respect to galaxies in the Virgo cluster with the same V_c.

2.3.8 Hubble Space Telescope distance estimates to Virgo

The most recent determination of the distance to the Virgo cluster has made use of the HST. Determination of the extragalactic distance scale, hence H_0, is one of the HST's key projects. Unfortunately, the initially aberrated optics of the HST prevented

Figure 2.22 Blue band CCD image of NGC 4321 (M100) obtained by the author. M100 is a prominent spiral galaxy in Virgo and was the first galaxy in Virgo searched with the HST for Cepheids. M100 has also hosted four type Ia supernovae since 1900.

work from being done. Corrective optics were installed in December 1993 and observations of the galaxy M100 (Figure 2.22) in the Virgo cluster were soon carried out in order to detect individual Cepheid variables. This galaxy was chosen for three reasons:

1. It is relatively face on, so internal extinction is minimized.
2. It has a high current star formation rate, hence a large population of Cepheids.
3. It is the host galaxy of a well-measured SN Ia (SN1979C), and a good distance to M100 will better calibrate the SN Ia luminosity scale.

The detection and measurement of Cepheids in M100 is empirical proof that the HST now performs up to its original expectations. Freedman *et al.* (1994b) have detected about a dozen Cepheids in M100 and derive the relative M100–LMC distance. Fixing the LMC distance at $(m - M) = 18.5$ yields a distance to M100 of $(m - M) = 31.16 \pm 0.2$. Freedman *et al.* performed observations in the V and I bands to gain a color baseline which could be used to estimate the reddening towards the individual Cepheids in M100. Despite this, the ± 0.2 mag uncertainty in distance remains dominated by the uncertainty in reddening towards the Cepheids. This distance agrees well with the most recent estimates from others using the HST Cepheid data for Virgo galaxies. Ferrarese *et al.* (1996) derive $(m - M) = 31.0 \pm 0.2$ for their sample of HST Cepheids, whereas Saha *et al.* (1996) derive 31.10 ± 0.15 for NGC 4496 and 31.05 ± 0.15 for NGC for NGC 4536.

This distance of M100 is significantly lower than had been derived previously using the SN Ia luminosity calibration based on the distance to NGC 5253 and IC 4182 (whose distance has also been determined from HST observations of Cepheids). One cannot reconcile this difference by changing the distance to the LMC as $(m - M) = 18.5$ is likely an upper limit. Hence, either the distances to NGC 5253 and IC 4182 are systematically high, resulting in a systematically high SN Ia calibration, or the photometry of Freedman *et al.* contains systematic errors, or the reddening has been systematically overestimated. The actual light curves of the Cepheids are very well determined from the HST data, and even though the WFPC detector has a relatively large random error in its photometric zero point (0.04–0.05 mag), there is no evidence for systematic errors in photometry. So it seems likely this improved distance to M100 is also revealing a systematic error in the SN Ia luminosity scale. One possible source would be systematic errors in the SN1937C photographic photometry itself (Pierce and Jacoby 1995). In fact, it remains somewhat disconcerting that the TF, SBF and PNLF distance scales to Virgo can all be calibrated via Cepheid distances to nearby galaxies and all of them give a consistent distance to Virgo, yet that same calibration applied to SN Ia gives a significantly larger distance (Riess, Press, and Kirshner 1996). Clearly there is systematic error somewhere.

Deriving the distance to a large cluster of galaxies like Virgo, on the basis of just one galaxy is problematical at best, as there is no guarantee that the galaxy is in the cluster center. M100 is a gas-rich spiral galaxy, which strongly suggests it cannot be near the center of Virgo because there are various gas-removal mechanisms operative in cluster centers. The observed redshift of M100 is about 400 km s^{-1} higher than the mean Virgo redshift and is located about 4° away from the nominal cluster center (defined by M87). At the mean distance of Virgo, this is a projected separation of ≈ 1 Mpc. Thus, although it is likely that M100 is a member of Virgo, its position and redshift suggest it lies about 10% further away than the cluster center. Interestingly, at this distance, M100 would have a linear diameter of ≈ 35 kpc, making it one of the larger known late type spirals. M100 has been the host of four supernovae since 1901. This is an unusually high frequency of supernovae, which supports our claim that M100 is an intrinsically large spiral. If M100 is 10% farther away than the Virgo cluster center, the distance modulus to Virgo will be $(m - M) = 31.05 \pm 0.20$, in excellent agreement with the TF, PNLF and SBF distance estimates. Using a distance to the LMC of $(m - M) = 18.35$ lowers this distance to 30.90 ± 0.20.

2.3.9 Summary of Virgo distance estimates

An objective summary of these six independent distance measurements to Virgo immediately suggests that one or more of them has a systematic error in the zeropoint calibration. Deciding on which of the methods has the most serious systematic error is an arduous chore which often leads to circular reasoning. Leaving the finger-pointing to others, let us go where the data leads. In this case the data leads to the conclusion that the distance to the Virgo cluster is still unknown at the 20% level. Our more reliable indicators converge on a value of $(m - M) = 30.9$–31.2, but a distance as large as $(m - M) = 31.5$ cannot be rigorously excluded.

Unfortunately, even though we now have a distance determination to the Virgo cluster, we are not yet in a position to determine H_0. To derive H_0 from the distance to Virgo requires a determination of the proper cosmic velocity of Virgo. We derive this cosmic velocity in Chapter 3 after discussing the nature of galaxy clustering in the observable Universe. The large-scale clustering of galaxies causes gravitationally induced deviations from pure expansion (cosmic) velocity. These velocity deviations seriously impact the determination of H_0 at the 10–20% level. But the situation is worse than 10–20% for Virgo, as its velocity must be corrected for both the infall of the Local Group and the possible large-scale motion of Virgo itself toward other mass concentrations. Moreover, it is likely that the Virgo cluster has substructure, so its mean velocity may depend on the choice of Virgo members in the sample. Because of these complications it seems unlikely that even a precise measurement of the distance to Virgo would yield H_0. As we will see, a better way to determine H_0 is to use the relative distance between the Virgo and Coma clusters of galaxies in combination with the cosmic velocity of Coma. This is done at the end of Chapter 3.

2.3.10 Direct astrophysical distance measurements

The cumbersome and somewhat tedious discussion of the previous sections should indicate how the determination of a precise value for H_0 requires a distance measuring technique which completely bypasses the distance scale ladder. There are many steps on the ladder where systematic error comes into play and the entire technique is always open to criticism on the basis of bias associated with sample selection. Direct astrophysical distance indicators do not rely on distance estimates to intermediate galaxies, hence they have the potential to provide definitive distances. Furthermore, these techniques can be applied over large distances where any perturbation from expansion velocity becomes negligible. However, these direct indicators are extremely model dependent; they rely on the astrophysics of some process to provide the distance, and that astrophysics must be modeled. Still, this is a major improvement, as various protagonists can now argue over the physics of the situation, rather than sample selection techniques. Here are some of the most promising kinds of direct astrophysical distance measurements devised up to now.

The Sunyaev–Zeldovich (SZ) effect

The CMB radiation is nearly a perfect blackbody with a temperature of 2.74 K. When this radiation passes through a cluster of galaxies that has a hot intracluster medium (ICM), hot electrons in that plasma scatter the CMB photons to higher energies and thus distort the original blackbody spectrum (Chapter 3). The overall effect is to increase the CMB photon energies preferentially at the shorter wavelengths. At longer wavelengths, this effect causes a net reduction in the temperature of the background radiation that passes through the cluster plasma. The measured decrease in temperature $\delta T/T$ depends on the total amount of scattering, which in turn depends on the electron temperature T_e, the electron density n_e and path length through the plasma dl. The temperature decrement can then be formulated as

$$\delta T/T \propto -\int n_e K T_e dl \tag{2.15}$$

It is convenient to replace the integral in equation (2.15) with average mean values for n_e and T_e, thus

$$\Delta T \propto \langle n_e \rangle \langle T_e \rangle L \tag{2.16}$$

where L is now the physical path length. To estimate the mean values of n_e and T_e we can make use of the observed X-ray flux S_x:

$$S_x \propto \frac{(n_e n_p) T_e^{1/2} V}{D^2} \tag{2.17}$$

where V is the volume of the cluster, n_p is the ion density and D is the distance from the observer to the cluster. In thermal equilibrium $n_p \approx n_e$. For a spherical cluster $V \propto L^3$. More generally, $V \propto L(\theta_x \theta_y)(D^2)$ where θ_{xy} is the angular size of the cluster in projected coordinates x and y. Using equation (2.15) to solve for n_e^2 and inserting the result into equation (2.17) yields

$$S_x \propto \frac{\Delta T^2 T_e^{-3/2} L(\theta_x \theta_y)(D^2)}{L^2 D^2} \tag{2.18}$$

which becomes

$$L \propto \Delta T^2 T_e^{-3/2} S_x^{-1} (\theta_x \theta_y) \tag{2.19}$$

Everything on the right-hand side of equation (2.19) is a measurable quantity from which the physical path length can be obtained. Once the physical size of the cluster is known, its angular size on the sky can be used to determine D and therefore H_0. This method has been applied to a few clusters and the derived values of H_0 are generally in the range 25–75. The most recent determination is based on observations of the Coma cluster, where Herbig *et al.* (1995) derive $H_0 = 74 \pm 29$. The large error bar reflects the difficulty of this method as there are several complications:

- Any small-scale structure in the X-ray gas is fatal; this is because the method uses smooth, average values of n_e and T_e. Small-scale clumping of the gas in which there are differences in n_e and T_e will produce spurious results.
- T_e must be determined, and it is generally quite difficult to determine X-ray temperatures from the extant data, due to the limited spectral coverage of most X-ray detectors.

- The measurements of the microwave background decrement itself are very hard. Herbig *et al.* obtained their data with the best system currently available, so they probably have the most credible result.
- The method really works best for perfectly spherical clusters (where θ_{xy} can just be replaced by L^2). Irregularly shaped clusters can give spurious results.
- Since the SZ effect is itself a frequency shift, it can be simulated by a moving cluster. In this case, $\Delta T \propto v/c$. For a cluster moving with respect to the CMB reference frame at $\approx$ 1000 km s^{-1}, ΔT will have the same value as for a nonmoving plasma with $T_e \approx 1$ kev.

All these complicating effects mean that distances derived from the SZ effect are highly suspect and model dependent. The most serious of these complications involves fine-scale structure in the X-ray emitting plasma. Recent data from new X-ray satellites – *ROSAT* can detect this fine-scale structure and *ASCA* can measure temperatures directly (provided the signal strength is high) should offer improved SZ distances to some clusters in the near future.

Gravitational lensing

Gravitational lensing occurs whenever the light from a distant point source passes very near to a massive object. The space around that object is distorted and the light path can take on several different trajectories to the observer. Hence the observer sees not only multiple objects but sometimes amplified objects, depending upon the nature of the mass concentration. The details of this were first laid out by Refsdal (1964). For light from a distant point source encountering an isothermal sphere, the critical radius for lensing and amplification is given by

$$r_{crit} \propto 4\pi(\sigma_m/c)^2 D_{LS}/D_{OS} \tag{2.20}$$

where σ_m is the velocity dispersion of the lensing mass. To a reasonable degree of approximation (Chapter 3) clusters of galaxies have a potential similar to an isothermal sphere, hence in principle they can be gravitational lenses. D_{LS} is the angular diameter distance between the lens and the source, and D_{OS} is the angular diameter distance between the observer and the source. In more descriptive terms, r_{crit} is related to differences in the potential that the multiple light paths take. These potential differences give rise to delays in arrival times of the light from these multiple components that reach an observer. D_{LS} can be approximately determined if the redshifts of both the lensing source and the lensed object are unknown (the redshift of the lens is generally unknown), and in principle, σ_m can be measured as well.

To date, one system has been discovered that lends itself to this kind of analysis. The distant quasar Q 0957 + 561 (Figure 2.23; see plate section) is lensed by a foreground cluster of galaxies of known redshift. The two brightest images of the quasar are known as A and B. For a particular model of the lensing geometry and the mass distribution of the cluster it can be shown (Falco, Gorenstein, and Shapiro 1991; Surpi, Harari, and Frieman 1996) that

$$H_0 \propto \sigma_m^2 2\Delta\tau_{AB}^{-1} \tag{2.21}$$

where $\Delta\tau_{AB}^{-1}$ is the measured time delay. The principal advantage of using lensed QSOs is that, in general, QSOs are variable in their luminosity output at all wavelengths. Continuous monitoring of this system at both radio and optical wavelengths can then determine $\Delta\tau_{AB}^{-1}$. Radio observations are less susceptible to sampling gaps caused by poor weather.

Although this method of determining H_0 is quite promising because it is based on real physics; at present there are three main limitations:

- Only one source (Q 0957 + 561) has so far been discovered that is suitable, and it is dubious to derive H_0 using the statistics of one event.
- The time delay is actually quite difficult to measure because it is so long. For Q 0957 + 561, $\Delta\tau_{AB}^{-1}$ is of order 1.5 years. In the optical band, the components are sufficiently faint that they cannot be measured in moonlight; this produces significant gaps and irregular sampling in the timing data. Reconstruction of the intrinsic time delay from irregularly sampled data is difficult. Because of this, the radio data provides the best means for estimating the time delay. In fact, for Q 0957 + 561 there has been somewhat of a controversy regarding the value for $\Delta\tau_{AB}^{-1}$, as values of either 415 or 535 days can fit the timing data. Very recently, Kundic *et al.* (1997) have presented convincing data in favor of 415 days.
- The intrinsic density distribution of the lens itself must be known. For clusters of galaxies (Chapter 3) this is almost impossible to know. Detailed maps of the X-ray distribution can help in this regard. In addition to Q 0957 + 561, faint background galaxies are also lensed. As these galaxies are not point sources, their multiple images are not point sources; instead they are thin arcs or arclets (Figure 2.24). In principle the size of these arcs is related to the mass distribution in the lens. Fischer *et al.* (1997) have used observations of lensed galaxies to determine the mass distribution of the lensing cluster towards Q 0957 + 561. Interestingly, their results do not agree very well with the mass distribution inferred from the X-ray map of the cluster.

Because of the difficulty in determining the mass distribution of the lens, derivations of H_0 from this method are not yet credible. For the Q 0957 + 561 system, the strongest statement which can be made from the observed time delay and reasonable modeling of the mass distribution is that $H_0 \leq 90 \pm 30$ km s^{-1} (Kochanek 1991).

Superluminal motion in radio sources

It has been known for 20 years that some radio sources have small-scale components which seem to be separating from one another at a velocity that exceeds the speed of light. This is now understood to be an illusion caused by the relativistic acceleration of a plasma down a beam pipe pointed at the observer. The knots which are seen in the radio jets are shocks in this relativistic flow. These knots often exhibit centimeter waveband variability which is due to the passage of these

Figure 2.24 Spectacular image of arclets and rings representing gravitational lensing associated with the cluster Abell 2218. CCD image taken with the Hubble Space Telescope. This image was created with support to Space Telescope Science Institute, operated by the Association of Universities for Research in Astronomy, Inc., from NASA contract NAS5–26555, and is reproduced with permission from AURA/STScI. Digital renditions of images produced by AURA/STScI are obtainable royalty-free.

shocks through the optically thick surface of the flow. During this passage the flux increases, followed by adiabatic energy loss and a decline in the flux. The increase in the percentage polarization that accompanies such activity is associated with the shock compression of an initially tangled magnetic field, establishing a "preferred direction," and causing a significant percentage polarization for observers viewing radiation emitted in the plane of the compression **in the flow frame**. A stationary observer on the "$1/\gamma$-cone" of the flow sees maximum possible superluminal motion and high percentage polarization, because aberration causes radiation emitted in the plane of a shock traveling along the flow to be swung into the line of sight.

Novel work by Phillip Hughes and his collaborators at the University of Michigan has used this relationship between increased polarization and maximum superluminal motion to construct a geometrical shock model which allows determination of the distance to the source. Its flow dynamics are described using the analytic jump conditions for shocks in a relativistic gas. Predicted polarized flux light curves are obtained by performing radiation transfer calculations through the plasma at many epochs. Although the models contain a large number of free parameters, they can be well constrained because of the wealth of information contained in high time resolution, multifrequency flux and polarization data. In particular, the shape of the total flux profile is strongly influenced by time-delay effects, and thus by viewing angle, whereas the degree of polarization is sensitively dependent on relativistic aberration, hence on flow speed and viewing angle. Models of two well-defined outbursts in the source BL Lac that occurred in the early 1980s have an optimal fit to both the light curve profile and percentage polarization for an angle of view of almost 40°. With known angle of view and flow speed, the **apparent speed** of structures can be calculated, so the angular separation rate of source components can be predicted as a function of cosmological distance. Comparison with the actual rate, determined by very long baseline interferometry (VLBI), allows the distance to be determined, and when combined with the known redshift, H_0 may be estimated. An initial application of this technique yielded a value somewhat in excess of 100 km s^{-1} Mpc^{-1}. Although more refined models admit a somewhat smaller viewing angle, and smaller value of H_0, an important point is that, at these large angles of view (which have received strong support from independent modeling of VLBI data), *no* source speed is compatible with values of H_0 close to 50 km s^{-1} Mpc^{-1}, because the apparent component speed is too small, and superluminal motion in BL Lac would simply not be observed (Hughes, Aller, and Aller 1991).

Virial masses

This is a straightforward procedure, but due to the presence of dark matter, it is unlikely to be applicable to any real astrophysical source. The dynamical mass of a rotating cloud of gas is

$$M_{dyn} \propto V_c^2 R \tag{2.22}$$

where R is some characteristic physical scale; R is related to distance via $D\theta$ where θ is the angular size of the gas cloud. If we imagine this cloud has no dark matter in it and no stars, such that the gas comprises 100% of the dynamical mass, then the distance follows directly, as we can use the observed flux F_o of emission from that gas (assume that it is neutral hydrogen). In this case

$$M_{gas} = M_{dyn} = 4\pi D^2 F_o = V_c^2 D\theta \tag{2.23}$$

and only one value of D satisfies the observational constraints provided by F_o and V_c. This technique has been applied to one gas-rich system to date by Staveley-Smith *et al.* (1990), giving an upper limit on H_0 of 70 ± 7. If other gas-rich systems can be detected and if they have reasonable dynamics, this method may provide a statistically interesting measure of H_0.

In sum, consideration of the possible forms of direct astrophysical distance measures has yielded some promising candidates. The principle limitations are (1) the availability of real astrophysical sources that are ideal and (2) resulting values of H_0 remain model dependent. However, the mere fact these astrophysical distance measures are returning values of H_0 that do lie in the range 50–100 is quite encouraging, hence continued pursuit in this direction is a viable and appealing alternative to having to derive H_0 via the cumbersome distance scale ladder.

References

AARONSON, M., HUCHRA, J., MOULD, J., SULLIVAN, W., SCHOMMER, R., and BOTHUN, G. 1980 *Astrophysical Journal* **329**, 12

AARONSON, M., BOTHUN, G., MOULD, J., HUCHRA, J., SCHOMMER, R., and CORNELL, M. 1986 *Astrophysical Journal* **302**, 536

ABRAHAM, R. and VAN DEN BERGH, S. 1995 *Astrophysical Journal* **438**, 214

ARNETT, D. 1969 *Astrophysics and Space Science* **5**, 180

BAUM, W. 1955 *Publications of the Astronomical Society of the Pacific* **67**, 328

BERNSTEIN, G. *et al.* 1994 *Astronomical Journal* **107**, 1962

BIVIANO, A., GIURICIN, G., MARDIROSSIAN, F., and MEZZETTI, M. 1990 *Astrophysical Journal Supplements* **74**, 325

BOLTE, M. and HOGAN, C. 1995 *Nature* **376**, 399

BOTHUN, G. and MOULD, J. 1987 *Astrophysical Journal* **313**, 629

BOTHUN, G. and THOMPSON, I. 1988 *Astronomical Journal* **96**, 877

BOTHUN, G., HARRIS, H., and HESSER, J. 1992 *Publications of the Astronomical Society of the Pacific* **109**, 1220

BOTTINELLI, L., GOUGUENHEIM, L., PATUREL, G., and TEERIKORPI, P. 1991 *Astronomy and Astrophysics* **252**, 550

BRANCH, D. and TAMMANN, G. 1992 *Annual Reviews of Astronomy and Astrophysics* **30**, 359

BROWN, T., FERGUSON, H., and DAVIDSEN, A. 1995 *Astrophysical Journal Letters* **454**, L15

CAPACCIOLI, M., DELLA VALLE, M., ROSINO, L., and D'ONOFRIO, M. 1989 *Astronomical Journal* **97**, 1622

CAPACCIOLI, M., DELLA VALLE, M., D'ONOFRIO, M., ROSINO, L., and TURATTO, M. 1990 *Astrophysical Journal* **350**, 110

CARNEY, B. 1992 *Societa Astronomica Italiana, Memorie* **63**, 409

CHABOYER, B., DEMARQUE, P., KERNAN, P., and KRAUSS, L. 1996 *Science* **271**, 957

CHABOYER, B., DEMARQUE, P., and SARAJEDINI, A., 1996 *Astrophysical Journal* **459**, 558

CHAMCHAM, K. and HENDRY M. 1996 preprint

CHRISTIAN, C. and HEASLEY J. 1991 *Astronomical Journal* **101**, 848

CIARDULLO, R. and JACOBY, G. 1992 *Astrophysical Journal* **338**, 268

CIARDULLO, R., JACOBY, G., and HARRIS, W. 1991 *Astrophysical Journal* **383**, 487

COWAN, J., THIELMANN, F., and TRURAN, J. 1991 *Physics Reports* **298**, 267

DELL'ANTONIO, I., BOTHUN, G., and GELLER, M. 1996 *Astronomical Journal* **112**, 1759

DELLA VALLE, M. 1988 in *The Extragalactic Distance Scale*, Proceedings of the ASP 100th Anniversary Symposium, Astronomical Society of the Pacific, 1988, pp. 72–73

DORMAN, B. 1992 *Astrophysical Journal Supplements* **80**, 701

DRESSLER, A. *et al.* 1987 *Astrophysical Journal* **313**, 42

DUNLOP, J., PEACOCK, J., SPINRAD, H., DAY, A., JIMENEZ, R., STERN, D., and WINDHORST, R. 1996 *Nature* **381**, 581

EASTMAN, R. and KIRSHNER, R. 1989 *Astrophysical Journal* **374**, 771

EASTMAN, R., SCHMIDT, B., and KIRSHNER, R. 1996 *Astrophysical Journal* **466**, 911

ELSTON, R. and SILVA, D. 1992 *Astronomical Journal* **104**, 1360

FALCO, E., GORENSTEIN, M., and SHAPIRO, I. 1991 *Astrophysical Journal* **372**, 364

FEAST, M. 1988 *Astronomical Society of the Pacific Conference Proceedings* **4**, 9

FEAST, M. 1994 *Monthly Notices of the Royal Astronomical Society* **266**, 255

FEAST, M. and CATCHPOLE, R. 1997 *Monthly Notices of the Royal Astronomical Society* **286**, 1

FEDERSPIEL, M., SANDAGE, A., and TAMMANN, G. 1994 *Astrophysical Journal* **430**, 29
FERRARESE, L. *et al.* 1996 *Astrophysical Journal Letters* **468**, L95
FISCHER, P., BERNSTEIN, G., RHEE, G., and TYSON, A. 1997 *Astronomical Journal* **113**, 521
FLANNERY, B. and JOHNSON, C. 1982 *Astrophysical Journal* **263**, 166
FRANX, M. and DE ZEEUW T. 1992 *Astrophysical Journal* **392**, 47
FREEDMAN, W. 1988 *Astronomical Journal* **96**, 1248
FREEDMAN, W. and MADORE, B. 1990 *Astrophysical Journal* **365**, 186
FREEDMAN, W., MADORE, B., and WILSON, C. 1991 *Astrophysical Journal* **372**, 455
FREEDMAN, W. *et al.* 1994a *Astrophysical Journal Letters* **435**, L31
FREEDMAN, W. *et al.* 1994b *Astrophysical Journal* **427**, 628
FUKUGITA, M., OKAMURA, S., and YASUDA, N. 1993 *Astrophysical Journal Letters* **412**, L13
GIRAUD, E. 1987 *Astronomy and Astrophysics* **180**, 57
GONZALEZ, A. and FABER, S. 1997 *Astrophysical Journal* **485**, 80
GOULD, A. 1994a *Astrophysical Journal* **462**, 542
GOULD, A. 1994b *Astrophysical Journal* **425**, 51
GOULD, A. 1995 *Astrophysical Journal* **452**, 189
GRAHAM, J. *et al.* 1997 *Astrophysical Journal* **477**, 535
HAMUY, M. *et al.* 1995 *Astronomical Journal* **109**, 1
HARRIS, W., ALLWRIGHT, J., PRITCHET, C., and VAN DEN BERGH, S. 1991 *Astrophysical Journal Supplements* **76**, 115
HAWLEY, S., JEFFREYS, W., BARNES, T., and LAI, W. 1986 *Astrophysical Journal* **302**, 626
HERBIG, T., LAWRENCE, C., READHEAD, A., and GULKIS, S. 1995 *Astrophysical Journal Letters* **449**, L5
HUGHES, P., ALLER, H., and ALLER, M. 1991 *Astrophysical Journal* **374**, 57
HUGHES, S. and WOOD, P. 1990 *Astronomical Journal* **99**, 784
HUMPHREYS, R., AARONSON, M., LEBOFSKY, M., MCALARY, C., STROM, S., and CAPPS, R. 1986 *Astronomical Journal* **91**, 808
JACOBY, G., CIARDULLO, R., and FORD, H. 1990 *Astrophysical Journal* **356**, 332
JACOBY, G. *et al.* 1992 *Publications of the Astronomical Society of the Pacific* **104**, 599
JIMENEZ, R. and PADOAN, P. 1996 *Astrophysical Journal Letters* **463**, L17
KELSON, D. *et al.* 1996 *Astrophysical Journal* **463**, 26
KENNICUTT, R., FREEDMAN, W., and MOULD, J. 1995 *Astronomical Journal* **110**, 1476
KHOKHLOV, A., MULLER, E., and HOFLICH, P. 1993 *Astronomy and Astrophysics* **270**, 223
KOCHANEK, C. 1991 *Astrophysical Journal* **382**, 58
KUNDIC, T. *et al.* 1997 *Astrophysical Journal* **482**, 75
LANDY, S. and SZALAY, A. 1992 Astrophysical Journal **391**, 494
LANEY, C. and STOBIE, R. 1988 *Astronomical Society of the Pacific Conference Proceedings* **4**, 9
LUTZ, T. and KELKER, D. 1975 *Publications of the Astronomical Society of the Pacific* **87**, 617
MADORE, B. and FREEDMAN, W. 1991 *Publications of the Astronomical Society of the Pacific* **103**, 933
MADORE, B., MCALARY, C., MCLAREN, R., WELCH, D., NEUGEBAUER, G., and MATTHEWS, K. 1985 *Astrophysical Journal* **294**, 560
MOULD, J. and KRISTIAN, G. 1986 *Astrophysical Journal* **305**, 591
MOULD, J., HAN, M., and BOTHUN, G. 1989 *Astrophysical Journal* **347**, 112
MOULD, J., GRAHAM, J., MATTHEWS, K., NEUGEBAUER, G., and ELIAS, J. 1990 *Astrophysical Journal* **349**, 503
NIELSEN, E. *et al.* 1997 *Astrophysical Journal* **583**, 745

NOMOTO, K., SUGIMOTO, D., and NEO, S. 1976 *Astrophysics and Space Science* **39**, L37
PANAGIA, N., GILMOZZI, R., MACCHETTO, F., ADORF, H.-M., and KIRSHNER, R. 1991 *Astrophysical Journal Letters* **380**, L23
PERLMUTTER, S. *et al.* 1995 *Astrophysical Journal Letters* **440**, L41
PERRYMAN, M. *et al.* 1995 *Astronomy and Astrophysics* **304**, 69
PERRYMAN, M. *et al.* 1997 *Astronomy and Astrophysics* **323**, 49
PERSIC, M. and SALUCCI, P. 1991 *Astrophysical Journal* **368**, 60
PHILLIPS, M. 1993 *Astrophysical Journal Letters* **413**, L105
PIERCE, M. 1994 *Astrophysical Journal* **430**, 53
PIERCE, M. and JACOBY, G. 1995 *Astronomical Journal* **110**, 2885
PIERCE, M. and TULLY, B. 1988 *Astrophysical Journal* **330**, 579
PRICHET, C. 1988 in *The Extragalactic Distance Scale*, Proceedings of the ASP 100th Anniversary Symposium, Astronomical Society of the Pacific, 1988, pp. 59–61
PRITCHET, C. and VAN DEN BERGH, S. 1987 *Astrophysical Journal* **316**, 517
RAWSON, K. *et al.* 1997 *Astrophysical Journal* **490**, 517
REFSDAL, S. 1964 *Monthly Notices of the Royal Astronomical Society* **128**, 295
RICHER, H., CRABTREE, D., and PRITCHET, C. 1990 *Astrophysical Journal* **355**, 448
RICHMOND, M. *et al.* 1995 *Astronomical Journal* **109**, 2121
RIESS, A., PRESS, W., and KIRSHNER, R. 1996 *Astrophysical Journal* **473**, 88
RIX, H. and ZARITSKY, D. 1995 *Astrophysical Journal* **447**, 82
SAHA, A., HOESSEL, J., and KRIST, J. 1992 *Astronomical Journal* **103**, 84
SAHA, A. *et al.* 1996 *Astrophysical Journal* **466**, 55
SANDAGE, A. 1970 *Astrophysical Journal* **162**, 841
SANDAGE, A. 1972 *Astrophysical Journal* **173**, 475
SANDAGE, A. 1994a *Astrophysical Journal* **430**, 1
SANDAGE, A. 1994b *Astrophysical Journal* **430**, 13
SANDAGE, A. 1996a *Astronomical Journal* **111**, 1
SANDAGE, A. 1996b *Astronomical Journal* **111**, 18
SANDAGE, A. 1996c preprint
SANDAGE, A. and TAMMANN, G. 1993 *Astrophysical Journal* **415**, 1
SANDAGE, A. and TAMMANN, G. 1995 *Astrophysical Journal* **446**, 1
SANDAGE, S. and TAMMANN, G. 1997 preprint
SANDAGE, A., TAMMANN, G., and FEDERSPIEL, M. 1995 *Astrophysical Journal* **452**, 1
SANDAGE, A. *et al.* 1992 *Astrophysical Journal Letters* **401**, L7
SANDAGE, A. *et al.* 1994 *Astrophysical Journal Letters* **423**, L13
SANDAGE, A. *et al.* 1996 *Astrophysical Journal Letters* **460**, L15
SCHAEFER, B. 1995 *Astrophysical Journal Letters* **447**, L13
SCHMIDT, E. 1984 *Astrophysical Journal* **285**, 501
SCHMIDT, B., KIRSHNER, R., and EASTMAN, R. 1992 *Astrophysical Journal* **395**, 366
SCHMIDT, B. *et al.* 1994 *Astrophysical Journal* **432**, 42
SCHOMMER, R., OLSZEWSKI, E., and AARONSON, M. 1984 *Astrophysical Journal Letters* **285**, L53
SILBERMANN, N. *et al.* 1996 *Astrophysical Journal* **470**, 1
SIMON, N. and CLEMENT, C. 1993 *Astrophysical Journal* **410**, 526
SMITH, L., SANDAGE, A., LYNDEN-BELL, D., and NORTON, R. 1963 *Astronomical Journal* **68**, 293
SPRAYBERRY, D., BERNSTEIN, G., IMPEY, C., and BOTHUN, G. 1995 *Astrophysical Journal* **438**, 72
STAVELEY-SMITH, L. *et al.* 1990 *Astrophysical Journal* **364**, 23

STEIDEL, C., GIAVALISCO, M., PETTINI, M., DICKINSON, M., and ADELBERGER, K. 1996 *Astrophysical Journal Letters* **462**, 17

STETSON, P., VANDENBERG, D., and BOLTE, M. 1996 *Publications of the Astronomical Society of the Pacific* **108**, 560

STORM, J., NORDSTROM, B., CARNEY, B., and ANDERSON, J. 1994 *Astronomy and Astrophysics* **291**, 121

SURPI, G., HARARI, D., and FRIEMAN, J. 1996 *Astrophysical Journal* **464**, 54

SWEIGART, A. 1997 *Astrophysical Journal Letters* **461**, L23

TAMMANN, G. and SANDAGE, A. 1995 *Astrophysical Journal* **452**, 16

TONRY, J. 1991 *Astrophysical Journal Letters* **373**, 1

TONRY, J., AJHAR, E., and LUPPINO, G. 1990 *Astronomical Journal* **100**, 1416

TONRY, J., AJHAR, E., and LUPPINO, G. 1991 *Astronomical Journal* **101**, 1942

TONRY, J., BLAKESLEE, J., AJHAR, E., and DRESLER, A. 1997 *Astrophysical Journal* **475**, 399

VAN DEN BERGH, S. 1991 *Publications of the Astronomical Society of the Pacific* **103**, 609

VAN DEN BERGH, S. 1995a *Astrophysical Journal* **446**, 39

VAN DEN BERGH, S. 1995b *Science* **270**, 1942

VAN DEN BERGH D. and Poll, H. 1989 *Astronomical Journal* **98**, 1451

VISVANATHAN, N. 1989 *Astrophysical Journal* **346**, 629

VON HIPPEL, T., BOTHUN, G., and SCHOMMER, R. 1997 preprint

VON HIPPEL, T., GILMORE, G., and JONES, D. 1996 *Monthly Notices of the Royal Astronomical Society* **273**, 39

WALKER, A. 1991 *Astronomical Journal* **103**, 1166

WALKER, A. 1992 *Astrophysical Journal Letters* **390**, L81

WELCH, D., MADORE, B., MCALARY, C., and MCLAREN, R. 1986 *Astrophysical Journal* **305**, 583

WHEELER, J. and HARKNESS, R. 1990 *Report on the Progress of Physics* **53**, 1467

WILSON, C., FREEDMAN, W., and MADORE, B. 1990 *Astronomical Journal* **99**, 149

WINGET, D., HANSEN, C., LIBEBERT, J., VAN HORN, H., FONTAINE, G., NATHER, R., KEPLER, O., and LAMB, D. 1987 *Astrophysical Journal Letters* **315**, L77

WORTHEY, G. 1993 *Astrophysical Journal* **409**, 530

YASUDA, N., FUKUGITA, M., and OKAMURA, S. 1997 *Astrophysical Journal Supplements* **108**, 417

ZEPF, S., ASHMAN, K., and GEISLER, D. 1995 *Astrophysical Journal* **443**, 570

CHAPTER THREE

Structure in the Universe: Galaxies, Clusters, Superclusters, Walls and Voids

3.1 Overview and caveats

Regardless of the specific origin of structure in the Universe, its existence allows us to probe the distribution of matter on various size scales from which origin scenarios can be reconstructed. At a minimum, observations clearly show that (1) the mass in the Universe is confined to discrete units called galaxies, (2) these galaxies are not distributed uniformly but are highly clustered on a variety of size scales, and (3) there are large regions that are apparently devoid of mass. Due to this complexity, it is rather difficult to determine what represents a fair volume of sampling for this structure. Nevertheless, an accurate measurement of the mass distribution over a fair volume will provide us with observational constraints on ρ, the other cosmological parameter of interest. Furthermore, the distribution of mass on both small and large scales provides a stringent constraint on any physical theory which is used to predict the formation and growth of such structure. This is discussed in detail in Chapter 5.

The measurement of large-scale structure (LSS) from observations is subject to two important caveats:

1. Any departure from homogeneity can retard the expansion locally.

2. There are serious concerns over how one best measures LSS.

On a large scale the Universe exhibits uniform isotropic expansion. Any departure from homogeneity produces a region of overdensity (as well as a region of underdensity) which can serve to retard the expansion locally. This effect is well known in the nearby Universe as our expansion motion away from the Virgo cluster is being retarded by its presence. In the early Universe these mass inhomogeneities, if they existed, would have a similar local effect on the copious photon field.

Suppose there exists a density contrast $\delta\rho/\rho = \delta_x$ over some comoving scale x. The physical size of this region depends upon epoch in the Universe, hence it is driven by the scale factor a. Thus the "radius" is ax. The total amount of mass within that radius is $\delta_x M$. The gravitational potential which arises from this density enhancement is fully specified in a Newtonian manner:

$$\phi = GM/R = G\delta_x M/ax \tag{3.1}$$

Since $\delta_x M$ is not a quantity that is easily inferred from observations, it is better to rewrite equation (3.1) in terms of the matter density ρ. Since we have a density contrast, the relevant density is $\delta_x \rho_b$ where ρ_b is the average background mass density. We now have

$$\phi = (G/ax)(4\pi/3)(ax)^3 \delta_x \rho_b \tag{3.2}$$

At very early times, the Universe is essentially critical, with the critical density defined in equation (1.35). From this we can substitute for ρ_b to yield

$$\phi = \tfrac{1}{2}\Omega H^2 (ax)^2 \delta_x \tag{3.3}$$

The scale factor a grows as $H^{-2/3}$. In addition, density fluctuations grow linearly with the scale factor and the overall rate of growth is a function of Ω (Chapter 5). So we have δ_x directly proportional to a and H^2 goes as a^{-3}. Substituting into equation (3.3) shows that there is *no* time dependence on the gravitational potential which is associated with the mass fluctuation. The fluctuation grows against an expanding background in such a manner as to keep ϕ constant.

Radiation which is contained within this fluctuation but escapes this particular potential is gravitationally redshifted by an amount proportional to ϕ, the depth of the potential well that the photons have to climb out of. This loss of energy causes those photons to have slightly lower temperature than the photons that did not escape from this potential. In this way, the CMB radiation has temperature fluctuations which are directly proportional to the density fluctuations. This is known as the Sachs–Wolfe effect (Sachs and Wolfe 1967). If density fluctuations in the early Universe were present (to produce the structure we observe today), the Sachs–Wolfe effect is an inescapable consequence. What remains to be determined is the overall amplitude of the temperature fluctuations.

For simplicity we assume the temperature fluctuations have a linear dependence on the potential:

$$\delta T/T \sim (axH)^2 \delta_x \tag{3.4}$$

for the present epoch, and we use an averaging scale $a_0 x$ equal to the Hubble length (cH_0^{-1}) so that

$$\delta T/T \sim \delta_x(0) \tag{3.5}$$

The Hubble length defines a distance over which light can travel given the expansion age of the Universe. This region is called a horizon as it represents a causally connected portion of the Universe. Material that exists outside of an observer's horizon cannot communicate with that observer unless communication occurs faster than the speed of light. However, since the Universe is expanding, all observer horizons increase with time, hence that material will eventually be inside an observer's horizon. Note that, in an open Universe, H_0 tends towards zero and the horizon (cH_0^{-1}) tends towards infinity. Furthermore, the angular size of our horizon is a strong function of redshift (and Ω). At the time of recombination, when the CMB signal can be observed, the angular size of the horizon is $\approx 2°$.

The *Cosmic Microwave Background Explorer* (*COBE*) had an angular resolution of $\approx 7.5°$, hence it made measurements over many scales of average size $a_0 x$ (e.g., over many horizons). The measured *COBE* anisotropy, on this scale, is $\delta T/T \approx 10^{-5}$. So within any one horizon, $\delta T/T$ is ~ 0, indicating that conditions of homogeneity and isotropy are maintained, but there is a small fluctuation signal when averaged over many horizons. The challenge is to understand how the rich variety of structure we observe today can be reconciled with these small temperature fluctuations. In principle, structure on all scales has arisen from a perturbation spectrum which had an average density contrast of 10^{-5} at the time of recombination. Thus, under the gravitational instability model for the formation of LSS (Chapter 5), the primordial density fluctuation spectrum is directly reflected by the fluctuations in the CMB.

There are serious concerns over how one best measures LSS in the local Universe. The simplest way is to use galaxies as tracers of the mass. By using redshift surveys to determine the three-dimensional galaxy distribution, one gets a direct measure of the mass distribution under the assumption that galaxies are fair tracers of this distribution. On small scales, there is a well-established relation between local galaxy density and galaxy type in the sense that dense systems like elliptical galaxies are preferentially found in regions of high galaxy density (e.g., clusters) of galaxies whereas less dense systems (disk galaxies) are found in lower density regions. Hence some care must be taken in redshift surveys to sample a range of galaxy types to ensure that all environments are probed.

In Chapter 6 we discuss the properties of low surface brightness (LSB) galaxies. In general, these galaxies are missing from optical redshift surveys as it is difficult to acquire a good optical spectrum of such a diffuse system. If LSB galaxies are distributed in a fundamentally different manner than galaxies of higher surface brightness, then bias could be present in optical redshift surveys. This point has been investigated in some detail (Bothun *et al.* 1986; Bothun *et al.* 1993; Impey *et al.* 1996), and on scales, greater than or equal to $\sim 2h^{-1}$ Mpc there is no bias. This is true also for redshift surveys that use either emission line galaxies (e.g., Salzer *et al.* 1988) or *IRAS* galaxies (e.g., Fisher *et al.* 1995). The conclusion is that all different kinds of galaxies adequately trace structure on large scales, and in fact, it is this very insensitivity of large-scale structure to galaxy properties that allows it to be determined from redshift surveys. The only bias is that elliptical galaxies are preferentially confined to virialized cores and therefore are not good tracers of structure on scales larger than individual galaxy clusters. However, this also means that elliptical galaxies are rare and are a minority population of redshift surveys.

In principle there is a more direct way to measure the mass distribution. Any mass overdensity will produce a perturbation of the expansion velocity of a galaxy which is near it. The extreme case, a virialized cluster, represents a situation where the perturbation is so severe that the expansion velocity is completely overcome, so the cluster in fact does not expand with the rest of the Universe. Since this perturbation is gravitationally induced, its effect on neighboring galaxies decreases as R^{-2}. Any difference between expansion velocity and observed velocity is called the **peculiar velocity**. If an all-sky peculiar velocity map of a fair volume of the Universe can be made, then under the caveat that these peculiar motions are gravitationally induced, the velocity residuals in this map will point directly at mass concentrations. If there were no bias between light and mass, this peculiar velocity map would point at observable mass concentrations (e.g., clusters of galaxies). This is because both the intensity of light and the gravitational force fall off as R^{-2} – you should fall where the light is.

3.2 Hierarchical structures uncovered

In the late 1970s dedicated redshift surveys of galaxies were begun. These surveys have culminated in the discovery of large voids (regions where there are few or no galaxies), great linear chains of galaxies, great attractors which perturb the expansion motion of galaxies, and an overall complex distribution of galaxy clustering on a variety of size scales. Prior to redshift surveys, significant structure in the Universe was known just from galaxy positional data. Both Abell and Zwicky defined criteria for grouping positional points together into structures and generating catalogs of galaxy clusters. These catalogs remain a basic source for the study of LSS.

In characterizing LSS on the basis of two-dimensional (2D) positional data, there can be much ambiguity due to the dependence of observed small-scale structure on larger-scale structures. Figure 3.1 (see plate section) shows the results of an N-body simulation involving the distribution of dark matter (Chapter 4), which is assumed to dominate the mass. Notice that the distribution is quite filamentary and is essentially one-dimensional (1D). The observed Universe also exhibits filamentary structure in the 2D positional data, but the filaments are really 2D sheets of galaxies that are not adequately present in the N-body models. Nevertheless, these filaments or sheets can give rise to apparent structure if you are looking along the line of sight of a particular filament. It is therefore important to develop good diagnostics to separate out gravitationally bound structures with the more weakly bound structures that are embedded in an overall large-scale structure component such as a filament or wall.

Unfortunately, this is rather difficult to do. Numerous studies have shown that the velocity distribution of galaxies within a structure is a very poor predictor of whether or not that structure is bound. This is because small, dense structures can generate infall from the larger, lower density region in which they are embedded. Thus the velocity distribution of the galaxies in any structure represents the convolution of those which originally formed there and those which have only recently arrived. But this issue may not even be relevant in a filamentary Universe as the structure there is mostly defined by the distribution of large, low density regions (e.g., voids).

In general terms, X-ray emission is a reasonable signature of a bound structure. The X-rays come from gas in the intracluster medium (ICM), which is heated by the gravitational potential of the structure. The origin of this gas is unclear but likely sources include gas which has been liberated from galactic potentials (presumably through tidal interactions) and leftover gas that never was part of a galactic potential. Until recently, most X-ray satellites would measure only a flux, which could then be transformed into an X-ray luminosity using the distance to the structure. X-ray luminosity, however, depends on both the X-ray temperature and the total gas mass. Although there is a loose correlation between X-ray luminosity and the observed velocity distribution, it provides at best ambiguous evidence that the structure is gravitationally bound. X-ray temperature is a more direct way to measure the depth of the potential, but until the recent *ROSAT* X-ray satellite, X-ray temperatures were difficult to measure. Within a bound system in hydrodynamic equilibrium, there will be a natural and predictable correlation between X-ray luminosity and X-ray temperature. Mergers of two subsystems with different potential well depths (temperatures) or line-of-sight projection of subsystems along a filament will tend to smear out this correlation, making difficult X-ray identification of bound systems on a scale less than 0.5 Mpc.

3.2.1 Groups of galaxies

These are small-scale structures with radii typically $\leq 0.5-1.0$ Mpc and which contain a few (3–10) prominent galaxies. They have overdensities which average around $\delta\rho/\rho \approx 80$ (Ramella *et al.* 1994). Most of these groups are only weakly gravitationally bound. Our own Local Group is a good example; by number it consists of dwarf galaxies (undetectable in more distant groups) and by mass it consists of the Milky Way and M31. M33 is a minor mass constituent. The large spiral IC 342 might be a dynamically important portion of the Local Group as its distance remains uncertain. There is also recent evidence that a galaxy known as Maffei I might also be dynamically important. Maffei I is difficult to study as it is located at a galactic latitude of only +0.6°, hence it is heavily reddened. This is unfortunate as Maffei I is undoubtedly the closest elliptical galaxy to the Milky Way. Its distance may be as near as 3 kpc and, together with IC 342, it may form a small system similar to the Local Group (Krismer, Tully, and Gioia 1995). Indeed, two dwarf galaxies associated with this system have also been discovered recently (McCall and Buta 1995).

In addition to loose, weakly bound groups there are tightly bound compact groups. These compact groups are curious in that they have dynamical timescales (crossing times) of a few $\times 10^8$ years. These apparently short dynamical timescales have led many to conclude that these groups can't be real, but instead are accidental projections of a few galaxies. However, given the $\approx$ 100–200 compact groups which have been cataloged, this explanation seems untenable; many of these groups must be real. This reality has recently been confirmed by the presence of X-ray emission in some compact groups, as detected by the *ROSAT* satellite. Work by Diaferio, Geller, and Ramella (1995), Ramella *et al.* (1995) and Pildis (1995) has shown that the presence of X-ray gas in compact groups correlates most strongly with that group's membership in a larger grouping of galaxies. X-ray emitting compact groups also tend to have a low spiral fraction.

In general, groups of galaxies are fairly spiral-rich. Elliptical galaxies constitute a small fraction of the total group population, except in the most compact and dense groups. Since the dynamical timescales in compact groups are relatively short, it is quite possible that the elliptical galaxies which are in compact groups formed as the result of the merger of two or more spiral galaxies (e.g., Zepf and Whitmore 1993). Indeed, in some compact groups there is evidence for diffuse intracluster light (Figure 3.2) which may represent a premerger stage. However, many compact groups do not show evidence for diffuse intracluster light, which either implies little dynamical evolution (apparently in contradiction to the short inferred dynamical timescale) or their recent formation into a bound unit. For the present, the best way to study the formation and evolution of groups (bound or not) is via *N*-body/hydrodynamical simulations of the collapses of large-scale structures. Most of these simulations indicate that group formation is a consequence of what happens on larger scales, but the simulations generally lack the resolution to study the group properties in detail (Bertschinger and Juszkiewicz 1988; Weinberg, Hernquist, and Katz 1997). Observationally it is rare to find a group that is not connected to a larger-scale structure.

3.2.2 Clusters of galaxies

These are prominent structures which are easily identified on large-scale photographic plates. The best-studied nearby rich cluster is the Coma cluster of galaxies (Figure 3.3). Clusters of galaxies are subjectively classified according to a "richness" criterion. This criterion is basically the number of "bright" galaxies that appear within a fixed angular radius on the Palomar or ESO Sky Survey plates. A typical cluster detected in this manner has 100–1000 "bright" galaxies. Based on this somewhat loose criterion, George Abell and collaborators have cataloged approximately 4500 clusters of galaxies (Abell, Corwin, and Olowin 1989). Although some of these cataloged clusters are not real, being accidental projections of groups along the line of sight, most are gravitationally bound units. These clusters of galaxies generally have three components, but only two of them are accounted for in the typical fit of the cluster density profile:

$$\Sigma_r = \frac{\Sigma(0)}{(1 + r/r_0)^\beta} \tag{3.6}$$

Figure 3.2 CCD image of Hickson Compact Group 92 taken by the author using the NTT telescope at the ESO observatory. Much of the group appears to contain diffuse intergalactic light.

where $\Sigma(0)$ is the central surface density of galaxies, r_0 is the core radius and β is the slope of the power law that characterizes the radial falloff in surface density Σ_r. Figure 3.4 shows a selection of surface density profiles for clusters. Clusters have three structural components:

1. *A core* (r_0): This is a region where the projected surface density of galaxies is relatively flat. Not all clusters exhibit this feature. For those that do, the central densities can be quite high. These densities can be estimated by measuring $\Sigma(0)$, r_0 and the central velocity dispersion. For an isothermal sphere, the relation between r_0 and ρ_0 is

$$r_0 = \left(\frac{9\sigma_v^2}{4\pi G\rho_0} \right)^{1/2}$$

Figure 3.3 CCD image of the center of the Coma cluster: the richest nearby cluster, its center is dominated by many elliptical and S0 galaxies.

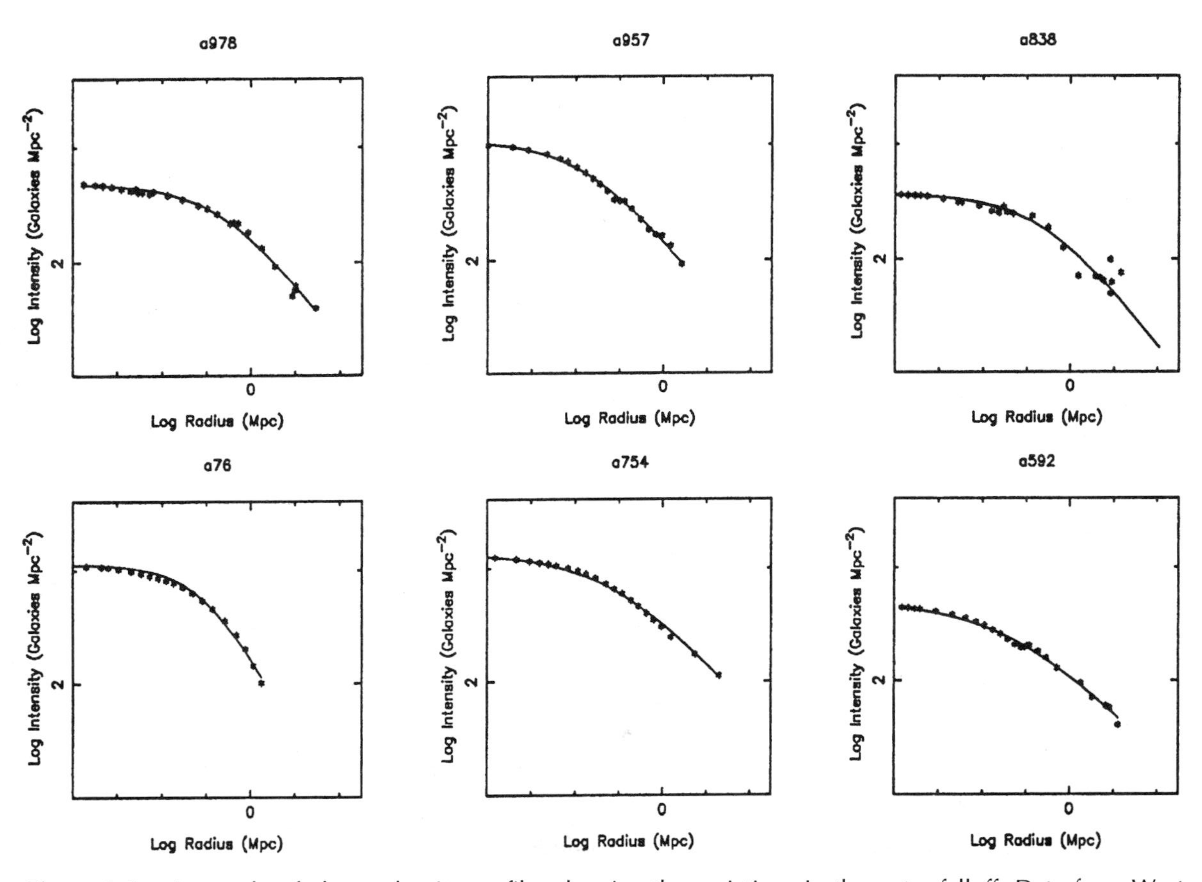

Figure 3.4 A sample of cluster density profiles showing the variations in the outer falloff. Data from West and Bothun (1990).

In the case of Coma the core radius is ≈ 300 kpc and the central density is ≈ $0.008 M_\odot$ pc^{-3}. This central density is 10^5 to 10^6 times higher than the average density of the Universe. In this high density region, spiral galaxies are almost never found.

2. *The β region*: This is where the surface density profile begins to fall off. In practice, one can define the "cluster" as the radius which encloses a density contrast of 100. This radius depends upon the slope β in equation (3.6). Clusters of galaxies show large variations in this slope. Some have very extended regions that contain many galaxies, whereas others have sufficiently steep profiles that most of the galaxies are contained in the r_0 region. If the cluster contains any spiral galaxies, they will most likely be found outside of this region.
3. *The infalling region*: Although not normally viewed as a structural component to clusters, we discuss it here because delayed infall from this region will augment both the r_0 and β regions over a Hubble time. Most of the infalling galaxies will be spirals. This infall causes slow evolution of the density profile as cluster relaxation times are long (see below). Without accurate distances, however, it is impossible to determine which galaxies are in the β region and which are still infalling.

Evidence that clusters of galaxies are gravitationally bound

Evidence that clusters of galaxies are gravitationally bound comes in a variety of forms:

1. The galaxy density is quite high and leads to values of overdensity in the range 100–1000. This range of overdensity is sufficient for the velocities of member galaxies to become virialized. The typical three-dimensional (3D) velocity dispersion is 1500 km s^{-1} whereas the expansion velocity over the scale of the cluster is 100–200 km s^{-1}.
2. Most clusters are sources of intense X-ray emission with typical temperatures of 2–10 keV. Such high temperatures are best obtained as a result of gas in hydrostatic equilibrium with respect to a deep gravitational potential. This X-ray emission is also well bounded and usually has a density profile that falls off in a manner similar to, or slightly shallower than, the galaxy surface density profile (although we note that cluster X-ray profiles are somewhat hard to measure due to low S/N in the X-ray data).
3. Deep CCD surveys of the sky have shown the existence of clusters of galaxies at $z \approx 1$ that look very much like the present-day core of the Coma cluster (Castander *et al.* 1994; Postman *et al.* 1996). This indicates that (a) some rich cluster cores did form early on and (b) in these rich clusters there has been little evolution of the core population over the last few billion years. This is not to say that augmentation of cluster cores by delayed infall is not occurring, but rather that out to $z = 1$ you can find examples of cluster cores which are as dense as seen in Coma.

Cluster timescales

Dynamical timescale This can be thought of as the time it takes for the cluster to communicate with itself through its own potential. There are many different ways to estimate this timescale but they all scale as $\rho_{clust}^{-1/2}$, which is essentially the gravitational free-fall time. The most convenient way to define the dynamical timescale is in terms of the crossing time. This crossing time is

$$\tau_{cr} \propto (R_{cl}/V_3) \tag{3.7}$$

where R_{cl} is the characteristic cluster radius and V_3 is the three-dimensional velocity dispersion. For any system of mass particles influenced by small larger-scale potential, application of the virial theorem can be made after one dynamical timescale has elapsed. This allows the system's mass to be estimated from R_{cl} and V_3 as

$$M_{cl} \propto R_{cl}V_3^2 \tag{3.8}$$

For a spherical system

$$M_{cl} \propto R_{cl}^3\rho_{clust} \tag{3.9}$$

Substituting this into equation (3.8) yields

$$\rho_{clust} \propto (V^2/R^2) \tag{3.10}$$

meaning that $R/V \propto \rho_{clust}^{-1/2}$. This result can also be arrived at by considering the time it would take for a sound wave to cross the cluster or the timescale of gravitational free-fall collapse. Equation (3.11) is a convenient parameterization of the dynamical timescale in terms of observables:

$$\tau_{dyn} = 6 \times 10^{11}\left(\frac{R}{\text{Mpc}}\right)\left(\frac{V_r}{1000\ \text{km s}^{-1}}\right) \text{years} \tag{3.11}$$

where V_r is the radial component of the three-dimensional velocity dispersion. A typical cluster has $R = 1$ and $V_r = 1000$ km s^{-1}, which leads to $\tau_{dyn} \sim 10^9$ years or roughly 10% of the expansion age.

Relaxation timescale The relaxation timescale is the characteristic timescale over which there occurs equipartition of energy among all the particles in the potential. This produces mass segregation, with the lighter galaxies having larger v than the heavier galaxies, as all galaxies have the same value of $1/2mv^2$. As the mass range for "bright" cluster galaxies is $\approx$ 100, the corresponding range in v will be 10. Thus mass segregation should be relatively easy to verify as a cluster of galaxies would exhibit its brightest galaxies in the cluster core and its fainter galaxies on the outside. Relaxation can also dynamically heat the very lightest galaxies to escape velocity. So far, there has been no convincing evidence that mass segregation in any cluster has occurred. Although some clusters (see below) are dominated by a central massive galaxy which continues to grow as a result of the assimilation (cannibalism) of other cluster members, this is not a relaxation effect. The lack of observed mass segregation helps to constrain the overall ages for clusters of galaxies.

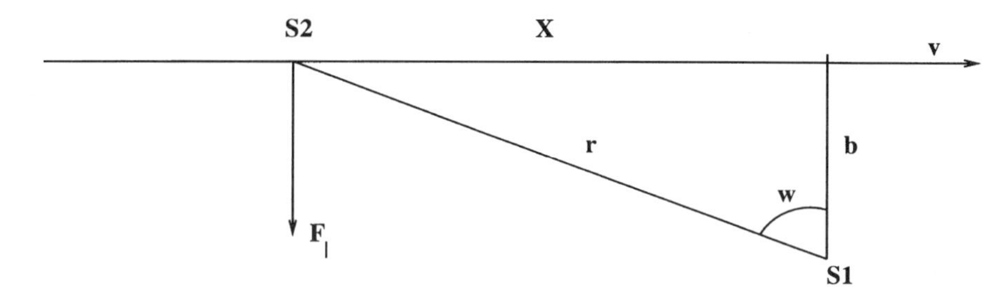

Figure 3.5 Encounter geometry for two stars which gravitationally scatter off one another.

Relaxation is motivated by two kinds of interactions: short-range two-body encounters between the individual cluster members and long-range forces due to the smooth cluster potential. In a gravitating system of N bodies there are two reasonable physical limits that can be described: (1) for small N, a body clearly moves through a field in which the mass is concentrated in the other $N-1$ bodies; (2) for large N, the potential is smooth as it is a mean potential generated by all the other particles. For clusters of galaxies, it is unclear whether they are in the limit of large or small N.

For large N, we can derive the relaxation timescale by considering a system of equal mass particles moving through a smooth potential. We now show that the relaxation timescale increases with increasing N. Figure 3.5 specifies the encounter geometry of two equal mass objects (assumed to be point masses) that gravitationally scatter off one another. For the sake of this derivation, consider the objects as stars and the potential as a collection of equal mass stars (e.g., a galaxy). The physical question we wish to address is what happens to a star as it orbits around a galaxy at some radius and gravitationally scatters off other nearby stars, hence experiencing velocity perturbations.

For this approximation we consider all encounters to be small, i.e., $\delta v/v \ll 1$. In this case we can consider that the perturbing star, S1, remains stationary and the perturbed star, S2, remains on a straight line trajectory. From the geometry of Figure 3.5 we can then specify the force law as

$$F_{\perp} = \frac{Gm^2}{(b^2 + x^2)} \cos w \tag{3.12}$$

and since $\cos w = b/r$, then

$$F_{\perp} = \frac{Gm^2 b}{(b^2 + x^2)^{3/2}} \tag{3.13}$$

We now set the distance X to be vt, so

$$F_{\perp} \sim \frac{Gm^2}{b^2}\left[1 + \left(\frac{vt}{b}\right)^2\right]^{-3/2} \tag{3.14}$$

Remembering that Newton says $F_\perp = m\dot{v}_\perp$, setting $s = vt/b$ $(ds/dt = v/b)$ and integrating over all time then yields

$$|\delta V_\perp| \sim \frac{Gm}{bv}\int_{-\infty}^{+\infty}(1+s^2)^{-3/2}\,ds \tag{3.15}$$

which can be evaluated to yield

$$|\delta V_\perp| = \frac{2Gm}{bv} \tag{3.16}$$

where we see that the perturbation in velocity depends on the initial mass and velocity of the perturbed star and its minimum distance from the perturbing star, which one might have guessed a priori.

Now consider a galaxy of radius R. The surface density of stars in that galaxy is $N/\pi R^2$. Upon one orbit of the galaxy, the star has a number of encounters δn with impact parameters in the range b to $b + db$, specified by

$$\delta n = \frac{N}{\pi R^2}2\pi b\,db = \frac{2N}{R^2}b\,db \tag{3.17}$$

The total number of encounters may be obtained by integrating over db, so if the range of db were from 0 to R, one would recover $\delta n = N$. Since the perturbations are randomly oriented, then $\langle \delta v_\perp \rangle = 0$. We are interested in the amplitude of $\delta v_\perp$, so

$$(\delta v_\perp)^2 = \left(\frac{2Gm}{bv}\right)^2\left(\frac{2N}{R^2}b\,db\right) \tag{3.18}$$

This expression is only valid in the limit where $\delta v_\perp / v \leq 1$, which is satisfied when $b = 2Gm/\delta v v_\perp \geq b_{min}$ where $b_{min} \equiv Gm/v^2$. Now integrating equation (3.7) over all possible impact parameters, b_{min} to R, yields

$$\Delta V_\perp^2 = \int_{b_{min}}^{R} 8N\left(\frac{Gm}{vR}\right)^2\frac{1}{b}\,db \tag{3.19}$$

Defining $\Lambda \equiv R/b_{min}$ (the ratio of maximum to minimum impact parameters) then yields

$$\Delta V_\perp^2 = 8N\left(\frac{Gm}{vR}\right)^2 \ln \Lambda \tag{3.20}$$

For a galaxy of equal mass stars, we can define the orbital velocity of a star at some radius R in the galaxy to be $v^2 = GNm/R$. Eliminating R from equation (3.20) then yields

$$\Delta V_\perp^2 = 8N\left(\frac{Gm}{v}\right)^2 \ln \Lambda \frac{v^4}{G^2N^2M^2} \tag{3.21}$$

We are thus led to the basic result

$$\frac{\Delta V_\perp^2}{v^2} = \frac{8 \ln \Lambda}{N} \tag{3.22}$$

We operationally define a relaxation time as the amount of time it takes for δv to be equal to v. That is, after enough encounters, the perturbations will be of the same amplitude as the initial velocity. According to equation (3.22), this will occur after a number of encounters given by $N/8 \ln \Lambda$, which we define as n_{relax}. The relaxation timescale is then just given by $n_{relax}\, t_{cross}$ where t_{cross} is the time it takes for the star to cross (orbit) the galaxy and is equivalent to a dynamical timescale, e.g., $(G\rho)^{-1/2}$. Using the definition of $\Lambda = R/b_{min}$ and $b_{min} \equiv Gm/V^2$ then yields

$$\Lambda = \frac{Rv^2}{GM} \tag{3.23}$$

which is just N. So the relaxation timescale is

$$t_{relax} = \frac{1}{8}\left(\frac{N}{\ln N}\right) t_{cross} \tag{3.24}$$

where t_{cross} is essentially R/V. Hence, although counterintuitive, long-range relaxation times slowly increase with increasing N.

If t_{relax} for some stellar system is larger than the age of the Universe, the system is said to be collisionless, which means that stars move under the influence of the mean potential generated by all the other particles. For instance, a galaxy has $N \approx 10^{11}$ and $R/V \approx 3 \times 10^8$. This yields $t_{relax} \approx 10^{18}$ years, which is eight orders of magnitude older than the Universe. A cluster of galaxies has $N \approx 10^3$ and $R/V \approx 10^9$. This yields $t_{relax} \approx 10^{11}$ years, hence mass segregation should not be observed in clusters of galaxies.

A full derivation of the two-body relaxation time can be found in Binney and Tremaine (1987). Following that, we can express the two-body relaxation timescale by the parameterized equation

$$t_{relax(2B)} = 2 \times 10^{10}\left(\frac{V_r^3}{1000 \text{ km s}^{-1}}\right)\left(\frac{M_g}{10^{12} M_\odot}\right)^{-2}\left(\frac{N_d}{1000 \text{ gal Mpc}^{-3}}\right)^{-1} \tag{3.25}$$

For a very massive galaxy, $M_g \sim 10^{12} M_\odot$ in a dense galaxy cluster ($N_d \sim 1000$ galaxies per cubic megaparsec) the two-body relaxation timescale is 20 billion years. These conditions describe a cluster like Coma. In dense clusters such as this there is evidence that the most massive and the brightest galaxies have settled to the bottom of the cluster potential (e.g., Beers and Geller 1983; Oegerle and Hill 1994). Alternatively, they could have merely formed in place at that location. Whether it is long-range forces (3.24) or two-body relaxation (3.25) that dominates probably depends upon the mass function of the galaxies in the cluster. A cluster with a few massive galaxies in it may have the shorter two-body relaxation timescale, in which case those few massive galaxies will migrate to the center of the cluster potential and remain there. The essential point, however, is that relaxation times for clusters generally exceed the expansion age of the Universe.

Collision timescale The collision timescale can be easily derived by treating a cluster of galaxies as a thermodynamic state in a closed volume. In this volume there is a number density of galaxies N_d, a characteristic cross section πR^2_{gal}, and a characteristic velocity dispersion V_3. The product $[(N_d)(\pi R^2_{gal})(V_3)]^{-1}$ defines a timescale which is the mean time between direct collisions of the particles. Dense systems with large galaxies and high velocity dispersions have shorter collision times. A convenient parameterization is

$$\tau_{coll} = 10^9 \left(\frac{V_r}{1000 \text{ km s}^{-1}} \right) \left(\frac{N_d}{1000 \text{ gal Mpc}^{-3}} \right) \left(\frac{R^2_{gal}}{100 \text{ kpc}^2} \right)^{-1} \tag{3.26}$$

For a galaxy with $R_{gal} \sim 10$ kpc, the collision timescale in the dense core of the Coma cluster is, interestingly, comparable to the dynamical timescale! This suggests that galaxy evolution can be severely altered in such a dense environment.

Cooling timescale When cold gas from a galaxy in a cluster is liberated from the galaxy potential by some process, it heats up as it comes into virial equilibrium with the cluster potential. When the gas is ideal, its temperature is set up by the virial condition

$$\tfrac{3}{2}kT = \tfrac{1}{2}M_{cl}V_3^2 \tag{3.27}$$

The observed values of M_{cl} and V_3 predict temperatures in the keV range and we expect most clusters to be filled with an X-ray emitting plasma. The cooling of that plasma is dominated by collisional processes with the hot electrons and the ions. This cooling is primarily through metallic lines, so the heavy element abundance of the plasma is important. The cooling timescale depends upon the electron density n_e, the initial temperature, and the cooling coefficient appropriate for a particular gas composition. There is an order of magnitude difference in cooling coefficient for 0.1 solar metallicity gas and primordial (zero metallicity) gas (Silk and Wyse 1992). X-ray spectra of clusters, however, indicate that the ICM has a mean metallicity which is approximately solar. Then a convenient expression for the cooling timescale (Raymond, Cox, and Smith 1976) is

$$\tau_{cool} = 9 \times 10^7 (T_8)^{1/2} (n_e)^{-1} \text{ years} \tag{3.28}$$

where T_8 is the measured temperature in units of 10^8 K and n_e is measured in particles per cubic centimeter. Most clusters have $n_e \leq 10^{-3}$, hence when heated to a few T_8, they will remain at that temperature over a Hubble time.

Dynamical friction timescale Dynamical friction is an odd astrophysical process and can only occur in specialized environments. It is caused when a smaller mass object passes nearby a larger mass object which itself is surrounded by a gravitationally bound halo of uniformly light particles. The typical situation is a large galaxy surrounded by a halo of $M_\odot$ stars. If the lower mass galaxy passes through this halo, that passage will create a gravitational wake, which causes the low mass stars to line up behind the passing galaxy. This in turn exerts a small gravitational force on the smaller body, causing it to lose some energy. Dynamical friction represents a kind of frictional drag which causes the smaller body's motion to damp out. If the smaller body is on an orbit that makes repeated passages through that halo, its orbit will decay over time; it will spiral in and be accreted by the larger object, causing the larger object to grow in mass and size. During this process, tidal forces can act to strip more stars out of the smaller body, building up the size of the halo.

Dynamical friction acts as a damping mechanism to remove orbital energy from cluster galaxies. Binney and Tremaine (1987) demonstrate that a galaxy of mass M_g which moves with velocity v experiences a deceleration due to dynamical friction given by

$$\frac{dv}{dt} = -\frac{4\pi G^2 M_g \rho_b f \ln \Lambda}{v^2} \tag{3.29}$$

From this, we can make a straightforward derivation of the dynamical friction timescale to see whether it is relevant for clusters of galaxies. The characteristic timescale over which energy is lost via dynamical friction is defined as

$$\tau_{df}^{-1} \equiv \left(\frac{1}{v} \frac{dv}{dt} \right) = \frac{4\pi G^2 M_g \rho_b f \ln \Lambda}{v_{rel}^3} \tag{3.30}$$

where f is a numerical value related to the distribution of the orbits in the potential, ρ_b is the background density of light particles (e.g., $M_\odot$ stars), v_{rel} is the relative velocity of the companion galaxy, Λ (not to be confused with the cosmological constant) is related to the impact parameter b_{min} as

$$\Lambda = \frac{b_{min} v_{rel}^2}{G(M_g + m_b)} \tag{3.31}$$

where m_b is the mass of the background particles.

For simplicity we assume circular orbits, in which case $f = 0.4$ (Binney and Tremaine 1987), and a flat rotation curve which satisfies the equation

$$\rho_b = \frac{1}{4\pi R^2} \frac{dM(r)}{dR} \tag{3.32}$$

where $M(r)$ follows from the virial theorem as

$$M(r) = \frac{v_{rel}^2 R}{G} \tag{3.33}$$

Substituting this into equation (3.30) yields

$$\tau_{df}^{-1} = 0.4 \frac{G M_g \ln \Lambda}{R^2 v_{rel}} \tag{3.34}$$

Recalling that $\tau_{cross} = R/V$, parameterizing M_g by $(M/L)L_g$ and making use of equation (3.33) leads to

$$\tau_{df} = 2.5 \frac{\tau_{cross} v_{rel}^2 R}{G \ln \Lambda L_g} \left(\frac{M}{L} \right)^{-1} \tag{3.35}$$

which expresses the dynamical friction timescale as some multiple of the dynamical timescale. In astrophysical units, equation (3.34) can be written as

$$\tau_{df} = \frac{5 \times 10^{10}\ \mathrm{yr}}{\ln \Lambda} \left(\frac{R}{300\ \mathrm{kpc}} \right)^2 \left(\frac{v_{rel}}{1000\ \mathrm{km\ s^{-1}}} \right) \left(\frac{M/L}{10} \right)^{-1} \left(\frac{L}{10^{11} L_\odot} \right)^{-1} \tag{3.36}$$

Cluster environments in which smaller galaxies have $v_{rel} \leq 300\ \mathrm{km\ s^{-1}}$ and distances ≤ 200 kpc have values of τ_{df} that are less than H_0^{-1}. Hence there are situations where dynamical friction could substantially alter the orbits of cluster galaxies and change the overall population of the cluster. But how prevalent is this?

In a study of central cluster environments, Bothun and Schombert (1990) suggest that 25% of all rich clusters are experiencing an augmentation to the luminosity or mass of the central dominant galaxy as dynamical friction causes the orbits of these central cluster galaxies to decay. Further support for this process comes from the observation that many galaxies in the central regions of clusters appear to have truncated surface brightness profiles. This is most likely due to the tidal liberation of stellar material during the close passage to the central dominant cluster galaxy. In order to prevent the entire depletion of this population over a Hubble time, it is necessary that the galaxies currently in the centers of these clusters are on relatively elongated orbits. For most galaxy clusters, dynamical friction is avoided if the orbits of the individual members are highly radial. Studies of other galaxy clusters by Gebhardt and Beers (1991) and Merrified and Kent (1991) have reached similar conclusions.

3.2.3 Clusters of clusters: superclusters

As clusters of galaxies were cataloged it was noticed that some clusters were concentrated in certain regions of the sky. This can clearly be seen in the Abell cluster catalog just by plotting the spatial distribution of clusters. Figure 3.6 shows a 2D positional plot for galaxies near the Coma cluster. Notice the apparent filamentary structure in which clusters are embedded. Figure 3.7 shows the all-sky positional data for all cataloged galaxies. The densest distribution of light points forms a linear feature, roughly perpendicular to the plane of the Milky Way (the black "zone of avoidance" feature running down the middle). This feature is known as the supergalactic plane in which the Virgo cluster (V) is embedded. In the positional data, it is clearly a very large-scale feature. Superclusters of galaxies also appear in the Zwicky catalog but were not identified by Zwicky as such. In fact, Zwicky vigorously argued against the ideas of second-order clustering or the systematic clustering of clusters to produce larger-scale clusters. In the Zwicky catalog, superclusters show up as large, diffuse clusters which contain multiple condensations. Only when redshifts were obtained did the true scale of these structures become apparent.

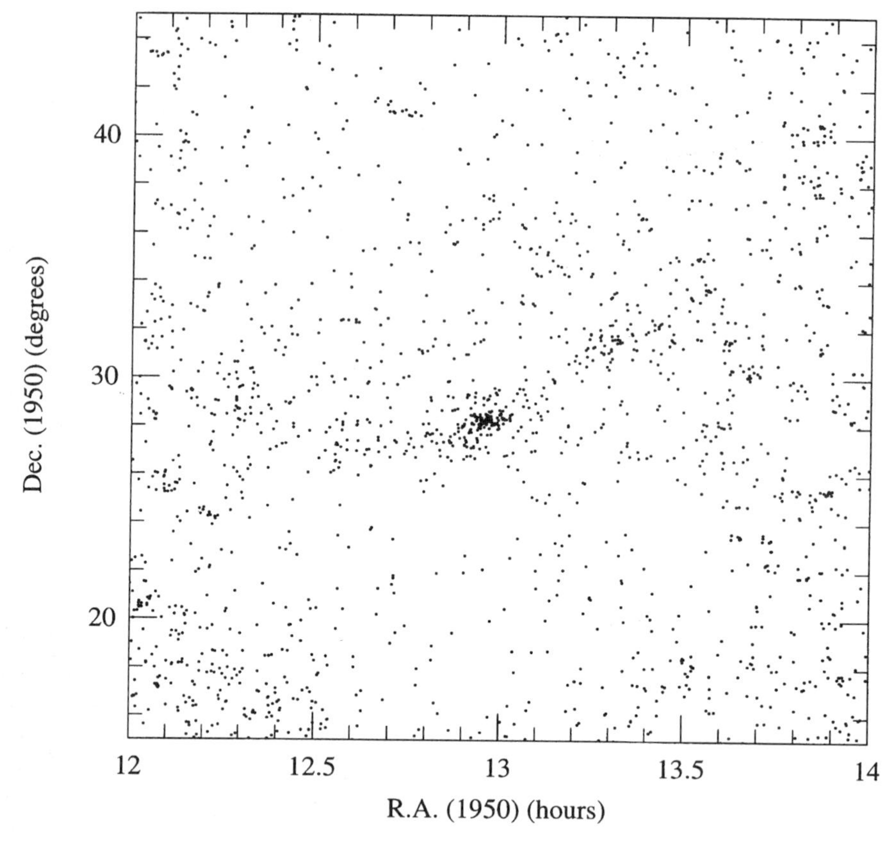

Figure 3.6 Spatial plot of the position of galaxies in the Zwicky catalog located within 30° of the center of the Coma cluster. Approximately 1900 galaxies are shown here and 200 alone define the Coma cluster. There is even a hint of the Great Wall structure.

Superclusters are large structures whose radii are difficult to define in physically meaningful terms. A typical supercluster might contain 3–10 Abell clusters. Most identifiable superclusters are elongated in shape and some appear to form a large-scale filamentary pattern. The dynamical timescale for superclusters exceeds the expansion timescale of the Universe by several factors. A supercluster is best described as a collection of virialized condensations (e.g., clusters) that are mixed into an overall low density structure. The clusters themselves fill very little of the volume of the structure. Outside the virialized condensations there is no X-ray emission and the velocity dispersion of the galaxies is low.

But what about still larger-scale structures, superclusters of superclusters? A pseudo three-dimensional plot of identified superclusters is shown in Figure 3.8. Since our census of superclusters is rather incomplete, Figure 3.8 is only suggestive. With a little imagination one can discern large-scale clustering of superclusters. One of the strangest observations that has been made, however, is the apparent very

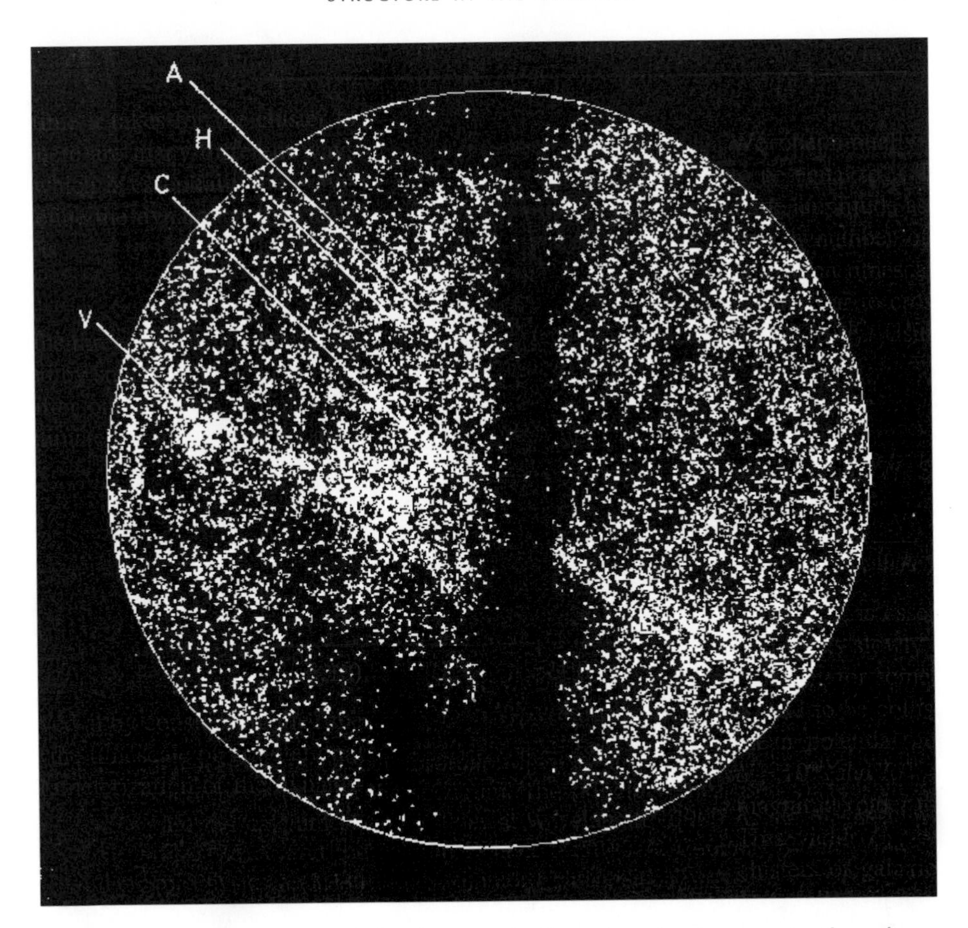

Figure 3.7 Combined spatial distribution of all cataloged galaxies onto the sphere of the sky defined by galactic coordinates. The dark band running down the center is the plane of the Milky Way, through which distant galaxies cannot be seen. The flattened distribution almost perpendicular to the plane is the Local Supercluster. Courtesy of Alan Dressler, Carnegie Institute of Washington.

large-scale alignment of clusters and superclusters. Most superclusters are highly flattened structures, so a major axis can be defined by some position angle on the plane of the sky. In a study of these position angles, West (1989, 1991, 1994) has produced some evidence that there is alignment on scales of 50–100 Mpc. And in some cases, this large-scale alignment can be matched to the smaller-scale alignment of the position angle of the radio emission axis of individual bright radio galaxies in the clusters of galaxies which are embedded in the flattened supercluster. This is circumstantial evidence for the memory retention of structure collapse over six orders of magnitude in spatial scale. It is difficult to know what to make of these alignment coincidences, but the implications are quite provocative with respect to structure formation, as gravitational collapse and dynamical relaxation should erase the memory of structure on larger scales.

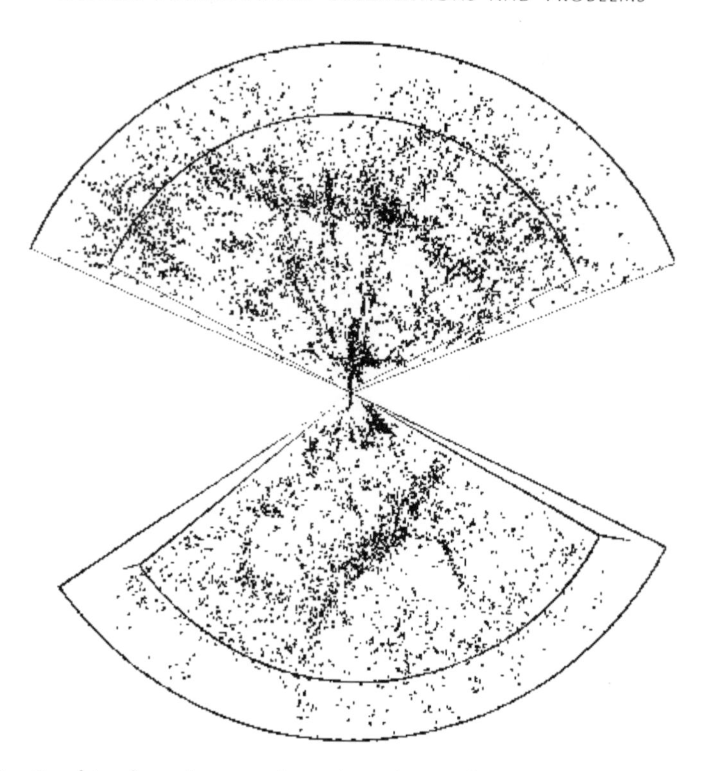

Figure 3.8 Combined northern and southern hemisphere redshift surveys initiated by M. Geller and J. Huchra. These surveys best define the coherency of structure on rather large scales. Courtesy of Dr Margaret Geller, Harvard-Smithsonian Center for Astrophysics.

3.4 There's more out there than just light

The previous discussion has concentrated on structure as it manifests itself via light-emitting objects (e.g., galaxies). In Chapter 4 we will explicitly consider the evidence for large amounts of nonluminous material in the Universe. However, it makes some sense to introduce that topic here before moving on to a discussion of redshift surveys. In particular, if there is a difference between the distribution of mass and the distribution of light (see below), what we measure as structure in redshift surveys may not identically correspond to the underlying large-scale mass distribution. And even the light we measure may not be representative of the distribution of all **baryonic** material.

In a compelling paper, Persic and Salucci (1992) compare the baryonic mass from light-emitting objects with that expected from primordial nucleosynthesis (Chapters 4 and 6). In brief, the observed abundance of light elements in the Universe (e.g., Walker *et al.* 1991) suggests that the baryonic density Ω_b is

$$0.01 \leq \Omega_b h^2 \leq 0.015$$

where h is defined as $h = H_0/100$. The visible contribution from baryons to Ω_b is given by the sum

$$\Omega_b = \Omega_{galaxies} + \Omega_{groups} + \Omega_{clusters}$$

where the latter two terms are meant to incorporate intracluster gas that is present in groups and clusters (and which usually emits X-rays). The inventory of Persic and Salucci yields a total of $\Omega_b = (2.2 \pm 0.6) \times 10^{-3} h^{-3/2}$; for a reasonable range of h this means that most of the baryons are dark (Chapter 6 covers this in detail).

In contrast, the work of White *et al.* (1993) uses X-ray observations of rich clusters of galaxies (e.g., the Coma cluster) to infer that $\Omega_b/\Omega \approx 0.1$–$0.3$ and most of the baryonic material is in the ICM. This observation is in apparent conflict with the nucleosynthesis constraint if $\Omega = 1.0$ (Carr 1994), explored in greater detail in Chapter 4. But in evaluating the results from redshift surveys, bear in mind that if $\Omega = 1.0$, luminous baryons contribute $\leq 1\%$ to the total mass density of the Universe. Then one might very well expect the light and mass distributions to be fundamentally different. For the case of $\Omega \leq 0.1$, it is likely that most of the mass is baryonic and the dark baryons should be located in the same regions as luminous galaxies. In fact, most of them would be in the halos of those galaxies.

3.5 Redshift surveys: defining physical structures

The previous manifestations of structure in the Universe were all based on the distribution of galaxies as projected onto the plane of the sky – the state of the field in 1980. Further progress required moving beyond projected galaxy distributions and into the realm of three-dimensional data sets. This meant the acquisition of galaxy redshifts, a somewhat daunting observational challenge. If the galaxies have lots of neutral hydrogen, then the Arecibo radio telescope will be able to detect that galaxy and get a redshift in approximately 10 min of observing time, for galaxies with velocities $\leq 10\,000$ km s^{-1}. Unfortunately, the Arecibo telescope can only observe galaxies that are located between declinations of 0 and 40°. Nonetheless, throughout the 1980s, Arecibo staff astronomers, specifically Martha Haynes and Riccardo Giovanelli, measured redshifts for thousands of galaxies that are accessible in the Arecibo declination strip. Those observations were the first to define a large-scale feature called the Pisces–Perseus supercluster (Giovanelli and Haynes 1985; Giovanelli, Haynes, and Chincarini, 1986), which is a large, "sheet-like" structure with mean redshift 5500 km s^{-1}.

Optical redshifts of galaxies can be much harder to obtain. If the galaxy has active star formation, it will be rich in emission lines from H II regions. Emission line redshifts require little integration time as only the emission line needs to be detected instead of the continuum light from the galaxy. In this case, using a telescope of diameter ≈ 1.5 m, a redshift can be obtained in about 5 min of integration time. Approximately 20% of all galaxies have sufficient emission line strength to allow for this rapid determination of redshift. For most galaxies, redshifts must be determined from the absorption lines which come from the integrated contribution

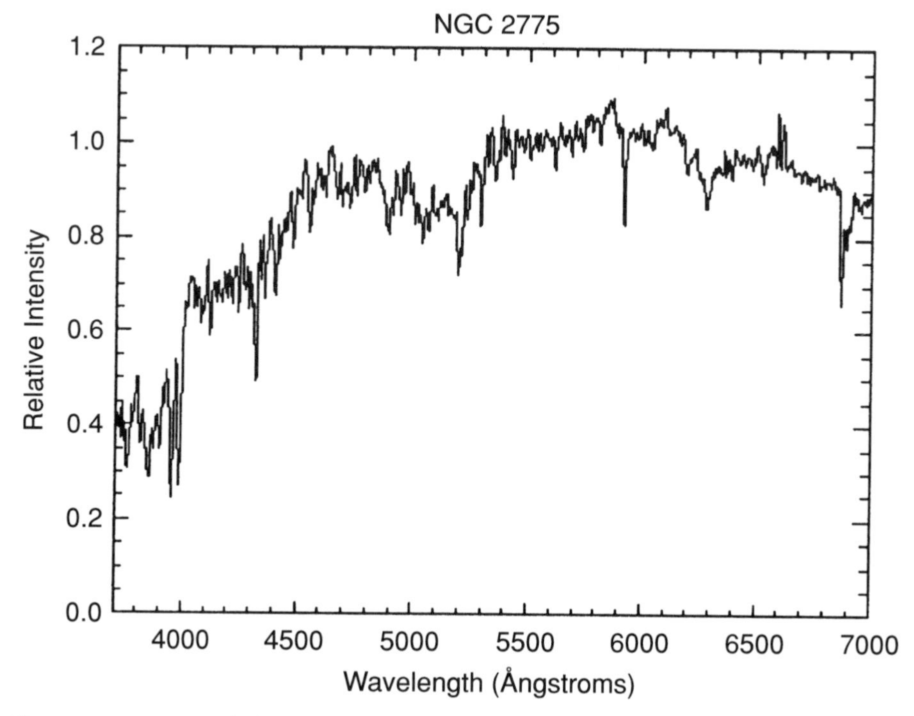

Figure 3.9 A typical absorption line spectrum of a galaxy showing the many absorption lines in the spectrum. The principal lines used for redshift determination are the calcium H and K lines at wavelength ~3900 Å, the magnesium I complex at 5175 Å, and the sodium D lines at 5800 Å. Courtesy of Jeff Willick, Department of Physics and Astronomy, Stanford University.

of all the stellar atmospheres in the galaxy. The principle absorption lines that can be detected are the H and K lines of calcium at 3933 and 3963 Å, the magnesium I line at 5175 Å and the sodium D lines at 5800 Å. A typical absorption line spectrum of a galaxy is shown in Figure 3.9. On a 1.5 m telescope an exposure time of 15–60 min is required to get sufficient signal-to-noise (S/N) to determine an absorption line redshift. To get redshifts of thousands of galaxies in this manner required a dedicated telescope, patience and an institutional commitment to performing this kind of service for the community.

These qualities were found in the form of Marc Davis, Steve Shectman and John Huchra at the Smithsonian Astrophysical Observatory. In the late 1970s Davis and Shectman designed a redshift machine for the 60 in telescope at Mt Hopkins. This spectrograph was optimized to detect the magnesium I and sodium D lines for determining a redshift; it was known as the Z-machine. In 1979 this telescope plus detector was commissioned and the first redshift surveys were made. Huchra and Davis performed the first redshift survey which contained ≈ 2500 galaxies brighter than $b = 14.5$. After this survey was completed, Margaret Geller joined the project and brought a fresh theoretical insight that allowed this rich data set to blossom into the first clear view and characterization of the 3D galaxy distribution. Much of this was based on a redshift survey of 20 000 galaxies brighter than $b = 15.5$. This basic data set, known as the CFA redshift survey, has been second only to the observations of the CMB in terms of its historic observational value to the field of cosmology. It is also safe to say that, when this survey started, no astronomer could have predicted the complexity of the galaxy distribution that would be revealed by an obscure 1.5 m telescope located above the desert in Arizona.

The disadvantage of the CFA redshift survey instrumentation is that only one galaxy at a time could be observed. A revolution in galaxy redshift survey efficiency occurred in the mid 1980s with the development of multifiber spectrography. In essence, galaxy coordinates are machined into a "plug-board" in which optical fibers are placed. Each optical fiber is fed to the spectrograph and the CCD detector. This allows for multiple objects to be observed in a single exposure. Multiple object spectrography can generally measure between 20 and 200 objects simultaneously, which represents an enormous gain over the one galaxy at a time approach. Such an instrument is most valuable when measuring redshifts in a field where there are lots of galaxies. This could be in a galaxy cluster or in some deep field where there are lots of faint galaxies. Initial work by Couch, Shanks, and Pence (1985), Ellis *et al.* (1985) and Broadhurst, Ellis, and Shanks (1988) directly showed the tremendous advantage afforded by multiple object spectroscopy.

In tooling up to do the southern hemisphere equivalent of the CFA redshift survey, Steve Shectman took full advantage of multiple object spectroscopy and was able to measure as many redshifts as are in the CFA survey but in a much shorter period of time. However, there are two potential problems with fiber-based spectroscopy:

- Since the fibers cannot be put arbitrarily close together, there is a selection effect against detecting close pairs of galaxies.
- The S/N of the spectrum is extremely dependent upon the central surface brightness of the galaxy. A typical fiber has an angular extent of 1–5 arcseconds, depending upon the camera optics. Galaxies of low surface brightness are therefore greatly underrepresented in fiber surveys.

However, these two minor problems only impact studies of the small-scale clustering of galaxies. On a large scale, their impact is negligible.

There are now about 100 000 redshifts measured between the CFA survey and a complementary survey initiated in the southern hemisphere a few years ago by Steve Shectman. In a few years, the Sloan Digital Sky Survey hopes to have measured 10^6 galaxy redshifts, although its hard to believe this increase will significantly change the view of LSS that is presently defined by $\approx$ 100 000 redshifts.

3.5.1 Hierarchical clustering patterns

The complexity in the galaxy distribution which as been revealed by these data can be qualitatively described using a more familiar analog. The distribution of population density in the United States is an excellent example of hierarchical clustering. There are large metropolitan complexes (superclusters) in which there are multiple condensations (individual large cities). Between the metropolitan areas are small towns or groups of towns, and in the case of the West Coast there are large voids. Figure 3.10 (see plate section) provides a convenient observational characterization of this structure. As lights are a good tracer of people, the density of light(s) is a good indicator of the population density. Then there are several questions one can ask about this distribution of population density, similar to the questions asked about the galaxy distribution.

- Is there a preferred scale to the clustering?
- Is there a characteristic spacing between regions of large population density?
- How big a region constitutes a fair sampling of the population distribution?
- What is the scale size over which the population density is homogeneous?

This last question is most relevant and its equivalent to the following: What is the radius of a circle such that no matter where one positioned that circle in the United States, it would contain the same number of people? The answer is that such a circle does not exist because the population is so highly clustered. Thus, within the sample offered by the size of the United States, there is no region which is homogeneous. The parallel with the Universe should be clear. We know that on a large scale the Universe is homogeneous because that is what we observe in the CMB. The most startling nature of redshift surveys is that, as larger volumes are probed, larger structures are seen. We do not yet have a strong indication that we have sampled a large enough volume to be considered a fair volume and the homogeneity scale of the galaxy distribution seems to increase with sample depth. At some scale size, an inhomogeneous galaxy distribution (and by inference only, an inhomogeneous mass distribution) is inconsistent with the nature of the CMB. There are some indications that such an inconsistency has now been observed; if so, this is a major challenge to understanding how such large-scale structure could have arisen (Chapters 4 and 5).

Redshift surveys are aimed at measuring structure in the galaxy distribution. There are two basic approaches one could adopt in revealing this structure: pencil beam surveys or slice surveys. We can return to the population analogy. In the pencil beam case we pick some small region of space and measure the redshifts of all galaxies in that region, down to some limit in apparent flux. This would be comparable to thoroughly measuring the population density of a small area in the United States. If that small area were in New York, the survey would return quite a different result than if the area were in Nevada. The analogy to the slice case would be to pick a narrow latitude range and survey everything along that latitude from the West Coast to the East Coast. A plot of population density versus longitude would then give a representation of the distribution of high and low density regions and their connectedness. Unfortunately, this topology will depend on the chosen latitude, as it is possible to choose a latitude strip which intersects no large cities.

For a highly clustered distribution it is not clear which sampling strategy is optimum for characterizing the galaxy distribution. In both cases one has to worry about not intersecting the structure. In the case where the size of a fair volume is relatively small, the slice approach to sampling is the best, provided the width of the slice exceeds the local correlation scale of individual galaxies. Then a slice strategy will probe the various clustering scales that are present. Where the size of the fair volume is large, a few pencil beams are probably the best way to identify the scale of the largest structures. However, these pencil beams will provide very little information on small and intermediate scales. Moreover, since only sparse sampling is done along the pencil beam, small-scale clustering effects can give rise to an apparently large clustering signal. That is, at some redshift, the size of the pencil beam will project to the same linear dimension as the galaxy–galaxy correlation length scale. Then two galaxies at that redshift will be detected instead of one. This can give the illusion of a large overdensity of galaxies at that redshift when in fact the enhancement is due to the small-scale clustering nature of galaxies. The slice approach does not suffer from this bias, so it is probably the better way to characterize the local topology. The following sections consider the results of both approaches as each has provided very intriguing data about the three-dimensional galaxy distribution.

3.5.2 The coming of the Stickman

The basic strategy of the slice surveys was devised by Margaret Geller at the Smithsonian Astrophysical Observatory. In addition to the survey design, Geller also developed novel and sophisticated techniques for analyzing the data in order to properly characterize the features which were seen. The first slice survey was completed in 1986 (e.g., de Lapparent, Geller, and Huchra 1986); the initial results were astounding then and they remain astounding today. Figure 3.11 (see plate section) is by now quite famous and is commonly known as the Stickman figure. It shows the essence of the complex galaxy structure defined by the slice survey. The slice dimension (equivalent to latitude in the population analogy) has been collapsed, so the axes represent angular distance on the sky and redshift distance from the Earth. In the absence of peculiar velocities, redshift distance is equivalent to physical distance. The opening angle of the survey represents the angular extent of the slice. There are several features to note:

- Any virialized structure will appear as a vertical structure pointed towards the origin (the observer). Galaxies in these structures have nearly the same physical distance from the observer, but the velocity dispersion of the structure stretches their location in redshift space. The main body of the Stickman is defined by such a structure – here the Coma cluster of galaxies.
- The apparent arms of the Stickman represent "walls." These are regions of relatively low density (so low that many pencil beams surveys can probe a wall but not intersect any galaxies) which are stretched out in angular space over

several hundred million light-years. These are the largest structures yet discovered; the very largest is known as the Great Wall (Geller and Huchra 1989). They seem to represent a concentration of galaxies on an imaginary surface. This has given rise to the viewpoint that large-scale features are the result of the expansion of voids piling up and collecting material along their boundaries (Weinberg, Ostriker, and Dekel 1989). It is somewhat difficult to account for these kinds of structures under the standard gravitational instability paradigm (see later).

- Regions in which there are few or no galaxies generally appear fairly round; they are called voids. A typical void has a "redshift" diameter of 3500–5000 km s^{-1}, i.e., a typical void is much larger than the volume defined by the Local Group–Virgo cluster concentration of galaxies.

These slice surveys have revealed strikingly coherent structures in the large-scale galaxy distribution. The sharp and dense edges of the voids strongly suggest that the process of galaxy and cluster formation is somehow triggered by the intersecting surfaces of voids. Current slice surveys suggest that the filling factor of voids in the Universe is about 75% and there is some probability distribution that can describe void sizes. But this probability distribution is based on the current data set, and if we have not yet surveyed a fair volume, we have not yet identified the largest void.

3.5.3 The void in Boötes

Identification of very large-scale voids can be an ambiguous process. In 1981 Kirshner *et al.* claimed to have detected a void of unprecedented proportion (Kirshner *et al.* 1981). This claim was based on the results of three pencil beam redshift surveys that were each separated by 35°. In all three fields there was a gap in the redshift distribution at velocities $\approx$ 15 000 km s^{-1}. A follow-up redshift survey by Kirshner *et al.* (1987) of 240 bright galaxies in the region between these three fields revealed that none had a redshift of 12 000–19 000 km s^{-1}. A uniform distribution of galaxies would predict that 31 out of 240 should have redshifts in this range. This region of space is in the constellation Boötes and the redshift gap is known as the Boötes void. The overall size scale of this, the first large void to be discovered, was $\approx$ 7500 km s^{-1}, significantly larger than the typical void size seen in the slice surveys. The question then became, Is the Boötes void the largest void there is?

The announcement of a large hole in space issued a challenge to many observers to work extra hard to fill it in. This challenge was met by a variety of workers who probed the void with objective prism surveys to search for emission line galaxies. This region of space was also well surveyed by the *IRAS* satellite. *IRAS* measures the integrated dust emission, which is heated by the ambient interstellar radiation field of galaxies as it peaks in the 20–100 μm range of the electromagnetic spectrum. The total amount of energy contained in this emission can be quite large (and in some galaxies it comprises more than 90% of the total energy released), so *IRAS*

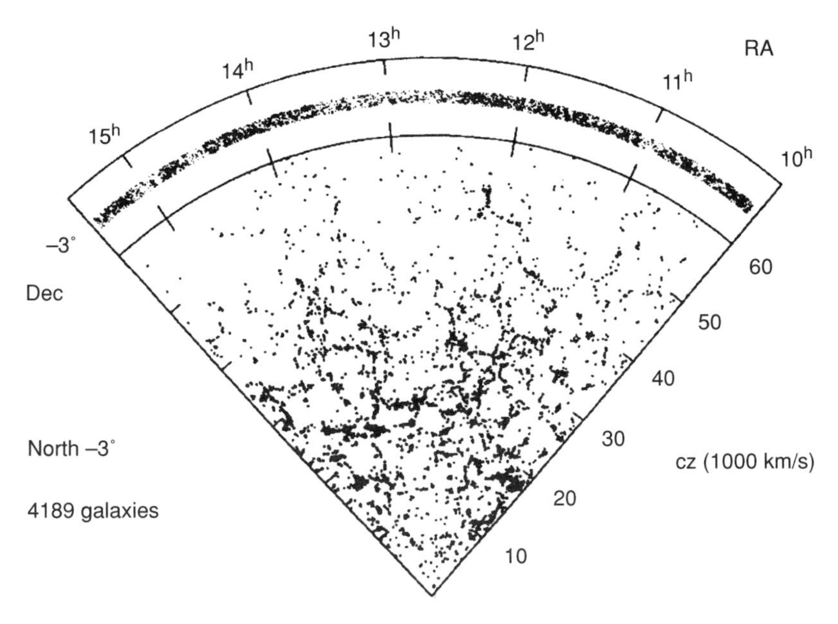

Figure 3.12 Slice diagrams of a section of the Las Campanas Redshift Survey initiated by Steve Shectman and collaborators. The void-filled Universe is quite apparent in this slice and numerous thin walls structures are also evident. Reproduced, with permission, from Shectman *et al.* (1996).

can detect galaxies out to relatively large distances. Greg Aldering and collaborators have extensively analyzed the *IRAS* data that is available in the Boötes void area. After many years of investigation, Aldering *et al.* (1997) have now discovered 53 galaxies within the boundaries of this void, as defined by Kirshner *et al.* (1987). With these data, Aldering *et al.* have redefined the density profile in the Boötes void and have now shown that the overall characteristics of the void are quite typical of those seen in the slice surveys. Hence, if larger-scale voids than those seen in the CFA slice exist, they have not been detected yet. This is consistent with the most recent results of the Las Campanas Redshift Survey (LCRS) of the southern survey. The LCRS goes deeper than the CFA surveys up to now, and the largest void seen in that survey has a diameter of ≈ 6000 km s^{-1} (Da Costa *et al.* 1996). Figure 3.12 shows some of the results of the LCRS and the by now familiar pattern of walls and voids.

3.5.4 The deeply probed Universe

Suggestions of very large-scale structure (and by inference large-scale voids) has been seen in the recent pencil beam surveys performed by David Koo and his collaborators. To date, a total of five very deep pencil beam fields have been done (Willmer *et al.* 1995) at both the north and south galactic poles. An initial analysis of about 400 galaxies (Broadhurst *et al.* 1990) yielded very surprising results.

Although the slice surveys have shown a pattern of high and low galaxy densities on scales of tens of millions of light-years (or in redshift distance 1–5000 km s^{-1}), the deep pencil beam surveys found separations between peaks in the galaxy distribution of 10 000–20 000 km s^{-1}. Moreover, these new surveys seem to identify a scale of $\approx$ 12 800 km s^{-1} that is present in the large-scale galaxy distribution. This scale is comparable to the depth of the entire survey that discovered the Great Wall of galaxies discussed above. To make matters worse, the initial analysis show that the spacing between density peaks was periodic. This kind of large-scale "cosmic picket fence" is certainly at odds with the preferred model of a homogeneous and isotropic universe. Critics (e.g., Ramella, Geller, and Huchra 1992; Bahcall 1991) suggested that the results were created by chance intersections of the pencil beam with various rich groups or clusters of galaxies, or a rare manifestation of a cellular distribution of galaxies (e.g., van de Weygaert 1991).

To resolve this issue, Koo's international team gathered data for another 1000 galaxies in both directions. Their original results were not only confirmed, but they discovered that the peaks in the galaxy distribution extended away from known rich clusters, suggesting that the observed peaks were probably due to intersections with other Great Walls that permeate all space. With more data, the initial suggestion of periodic spacing between the peaks also vanished, to the collective relief of the cosmological community. That relief can be summed up in a prophetic phrase uttered by Marc Davis in 1990: "If the galaxy distribution is truly periodic, it is safe to say we understand less than zero about the early Universe."

Knowing less than zero is an undesirable feature of any cosmological model, yet for a while, this is what the data suggested. The reconciliation between the slice results and the deep pencil beam probes calls for a topology in the galaxy distribution that is "like a sponge." There are cells within larger cells, with the largest of the cells corresponding to a spacing of 10 000–20 000 km s^{-1} seen in the pencil beam data. Within this overall cellular topology, most of space is devoid of galaxies. The intersection between cells of hierarchical sizes seems to have produced some very large-scale clustering. The evolution of this clustering pattern within the hierarchical cell sizes is a matter of much current research.

3.5.5 Correlation scales

The identification of preferred scales in the galaxy distribution provides important constraints on structure formation theories. Although the identification of physical scales is best determined by measuring the power spectrum of the galaxy distribution (Chapter 5), much information can also be learned by using a counts-in-cells technique. The sky is broken up into cells of fixed volume and the number of galaxies in each 3D cell is counted. The sizes of the cells are varied and the counts are done again until some kind of convergence or pattern is seen. The analysis of the galaxy distribution, as done with this counts-in-cells technique and as applied to the CFA redshift survey, has yielded the following scales of interest. These spatial scales serve as important constraints on structure formation theories (Chapter 5).

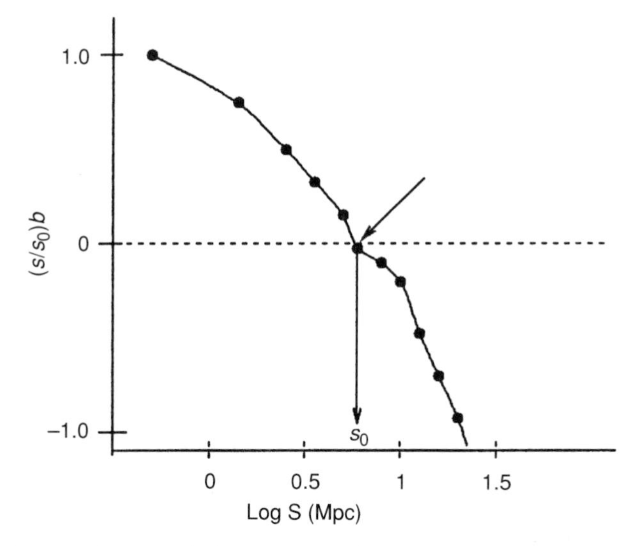

Figure 3.13 Two-point correlation function for groups of galaxies: the spatial correlation function is well approximated by the form $(s/s_0)b$, where the correlation length s_0 and the slope b of the power law are the fitting parameters. The dashed line indicates normalization of the power law fit, which defines the correlation length. The spatial scale over which the data cross this line defines s_0. The y-axis plots the logarithm of $(s/s_0)b$. The fit to a power law for this data gives slope of −1.3 and a correlation length of 8.1 Mpc. Adapted from Ramella, Geller and Huchra (1989).

- Between 75 and 80% of the survey volume is devoid of bright galaxies, which leads to two possibilities. Either these voids are preferentially inhabited by faint galaxies, or the galaxies themselves, of all types, are concentrated in the intersecting surfaces of voids.
- The luminosity function (LF) of galaxies is fitted by the parameters $M_B^* = -19.2 \pm 0.1$ and $\alpha = -1.1 \pm 0.1$ (Chapter 6). To first order, galaxy redshifts are indicative of distance, which allows conversion of apparent flux into an intrinsic luminosity. Although clustering (see below) introduces a second-order correction to the relation between observed redshift and distance, it is a minor perturbation in determining the LF. An accurate determination of the LF is another important quantitative result that can constrain theoretical models of structure formation. However, as we shall see in Chapter 6, the faint end slope of the LF can not be well determined due to surface brightness bias in existing redshift surveys and furthermore is significantly steeper than the value of $\alpha = -1.1 \pm 0.1$. In fact, a new analysis of the LF for the CFA redshift survey as a function of galaxy type has been presented by Marzke *et al.* (1995). They find that, for some galaxy types, the faint end slope is considerably steeper than $\alpha = -1.1$.
- The two-point correlation length of galaxies is $7.5h^{-1}$ Mpc. The data is shown in Figure 3.13. This correlation length is defined by the probability (in excess

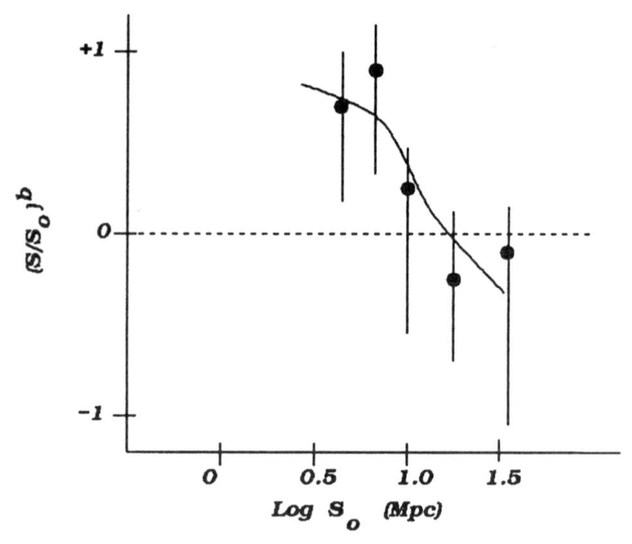

Figure 3.14 Cluster–cluster correlation function for an X-ray flux limited sample of galaxies from Bahcall and Cen (1992). The x-axis is the spatial scale s, in log (h^{-1}/Mpc) and the y-axis plots the logarithm of $(s/s_0)b$. Although the data set is noisy and the sample size is small, the plot is consistent with a correlation length of $21h^{-1}$ Mpc. This is equivalent to the correlation length found in numerical simulation of low density universes (Bahcall and Cen 1992; Mo, Jing, and White 1996).

of random) of finding a galaxy at a distance r from a given other galaxy. The radius at which this probability equals unity is identified as the correlation length. A determination of the correlation length is a good descriptor of small-scale clustering.

- The cluster–cluster correlation length is $\approx (20 \pm 5)h^{-1}$ Mpc (Postman, Huchra, and Geller 1992). This is shown in Figure 3.14. The relatively large uncertainty in this length arises from selection effects in making cluster catalogs; different samples of clusters give somewhat different values for the correlation length.

3.6 The peculiar velocity field

Although redshift surveys have provided a wealth of data that has helped to discover and describe LSS, this data is really one step removed from the more fundamental cosmological issue – the distribution of mass. Redshift surveys probe the positions of galaxies in redshift space. This is not necessarily the same as physical space, as mass concentrations serve to distort the appearance of structure when plotted in redshift space (Kaiser 1986). A trivial example is provided by the fingers-of-god signature of virialized structures in redshift space. Recall that earlier in this chapter we briefly considered evidence for significant amounts of nonluminous matter in clusters of galaxies. Although the nature of the dark matter and its possible distribution is the subject of extensive discussion in Chapter 4, we consider it here because it definitely affects the relation between redshift space and physical space. In fact, since we don't actually know what the mass distribution is, its total effect on the appearance of structure in redshift space is very difficult to ascertain.

When we considered various correlation length scales, we implicitly assumed that the light distribution is a fair and unbiased tracer of the mass distribution. Under this assumption, voids (i.e., lightless areas) are also massless areas, so the highly clustered nature of the light distribution reflects a highly clustered mass distribution. On the other hand, if the distribution of mass were smoother than the distribution of light (it's hard to imagine any circumstances where the mass distribution would actually be clumpier) then bias would be introduced, assuming that light traces mass. The simplest form of biasing is a linear relation between the galaxy density and the dark matter density:

$$\delta_{galaxies}(r) = b\delta_{darkmatter}(r)$$

where b is the linear biasing factor. Expressed in this sense, b represents a large-scale mean biasing value. Larger or smaller values of b on smaller scales can certainly exist (and probably do).

Chapter 5 will discuss the physical basis for biasing; here we treat it phenomenologically. Qualitatively, the issue of biasing stems form our lack of knowledge about the nature of dark matter and the process of structure formation. So it is clearly bold but dangerous to assume that most of this mass is actually trapped into some gravitational potential which also hosts a luminous galaxy. If this indeed were the case, then structure formation would be very efficient as all of the mass would be contained in discrete potentials and there would be little possibility for a smoother distribution of mass between them. It seems unlikely that structure formation was as efficient as this, so there is some expectation that the mass distribution should be smoother than the light distribution. In the extreme case where the mass distribution is smooth on the same scale as the galaxy distribution is clumpy, there is a clear bias between the mass and light distributions.

Unfortunately, redshift surveys alone provide inadequate constraints on the relative distributions of dark and light matter. This is because any **peculiar velocity** introduced by the presence of a mass concentration will cause a distortion in redshift space along the line of sight. The effect is similar to the previously discussed case of a virialized structure where the internal velocity dispersion is larger than the expansion velocity over the scale of the structure. Then the apparent separation of pairs of galaxies will appear much larger in redshift space than in physical space. And where there is a mean infall of some group of galaxies towards a larger mass concentration (such as the infall of the Local Group to Virgo), then some compression in redshift space relative to physical space will occur as the Hubble expansion velocity has been retarded by this gravitationally generated infall. Thus distortions in redshift space can provide clues to locations which have significant peculiar velocities, but any further progress relies on measuring the relative distances of galaxies in order to derive the amplitude of the peculiar velocity.

We define a peculiar velocity as the difference (positive or negative) between the predicted expansion velocity of a galaxy and its observed velocity. Thus, we can only measure peculiar velocities in the radial direction. The determination of peculiar velocities involves only relative distances, hence it is independent of the actual value of H_0. Peculiar velocities are determined via the following steps:

1. Choose a reference frame at rest with respect to the CMB (assumed to be an inertial frame). This reference frame could be the observed velocity transformed to the CMB frame, or a relative velocity with respect to some structure that is at rest with respect to this frame. The Coma cluster is generally chosen for this at-rest structure.
2. Using some distance measuring technique, determine the relative distance of a galaxy from the Coma cluster.
3. Use the relative distance to predict the observed velocity. For example, the CMB frame velocity of Coma is about 7200 km s^{-1}. If we measure some galaxy to have a relative distance modulus to Coma of $(m - M) = +0.75$ (e.g., a factor of 2 in distance) then its predicted observed velocity would be $7200 \times 2 = 14\,400$ km s^{-1}.
4. Compare the predicted velocity with the observed velocity. A nonzero difference (outside of the measurement errors) represents a peculiar velocity. In the Coma example, if the observed velocity (again in the CMB frame) is 16 000 km s^{-1}, then

   $$v_{pec} = v_{obs} - v_{pred} = 16\,000 - 14\,400 = 1600 \text{ km s}^{-1}$$

The physical meaning of this peculiar velocity is that this galaxy has a +1600 km s^{-1} departure from its expansion motion with respect to us. Such a departure could be generated by the infall of this galaxy to a more distant virialized structure. The 600 km s^{-1} motion of the Milky Way galaxy, as measured by the observed CMB dipole anisotropy is an example of such an induced motion (see below).

An understanding of error propagation in this methodology is vital to assessing the quality and credibility of derived peculiar velocities. This is treated in great detail by Strauss and Willick (1995). The main problem is that relative galaxy distances must be determined to fairly high accuracy. Since this is difficult to achieve with any galaxy sample, the subsequent determination of the peculiar velocity field is rather imprecise because the data is inherently noisy. Errors in the peculiar velocity are directly proportional to fractional errors in the distance determination. If you are able to measure the relative distance of a galaxy to an accuracy of 10% (which is good), it translates into a 1σ error of 500 km s^{-1} at a mean redshift of 5000 km s^{-1}. At a minimum, a secure measurement requires an S/N of at least 3σ, which translates into 1500 km s^{-1}, or a 30% deviation from Hubble expansion velocity. Although the size of this error bar can be reduced by $N^{1/2}$ if one uses a sample of galaxies all in one cluster, our overall ability to measure small (i.e., $\leq 10\%$) deviations from Hubble flow is limited. This difficulty can give the illusion that the local Hubble expansion is quite smooth (Sandage and Tammann 1995). For

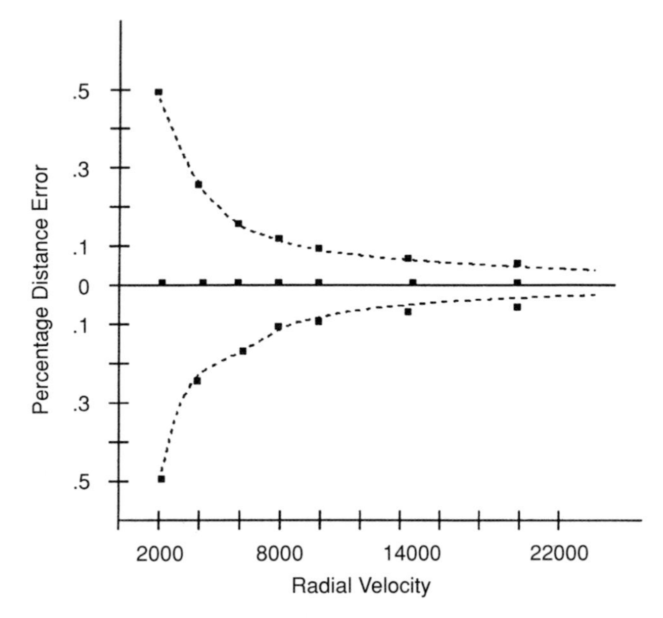

Figure 3.15 Schematic representation of Hubble flow out to 20 000 km s^{-1}, in which there is a constant 1000 km s^{-1} peculiar velocity perturbation which translates into a percentage distance error. Beyond a velocity of 10 000 km s^{-1} uncertainties in distance would produce percentage error estimates larger than those produced from this peculiar velocity perturbation.

instance, at 20 000 km s^{-1} some galaxy might have a 1000 km s^{-1} perturbation due to its proximity to a mass concentration, but it will be close to the Hubble line when viewed from Earth. This is schematically shown in Figure 3.15.

3.6.1 Inflows and outflows

Relative distance determinations for substantial samples of galaxies in regions where there are complete redshift surveys can vastly increase the discriminatory power of the data in mapping out the true mass distribution. Mapping of the peculiar velocity field constrains a combination of the large-scale matter density contrast and the cosmological density parameter Ω. Due to the presence of biasing, the large-scale matter density contrast is really measured by the large-scale luminosity density contrast multiplied by b. It is thus essential to obtain a fair volume of the Universe when measuring the peculiar velocity field. In the highly clustered Universe we have discussed, that regions of mass overdensity will cause a net inflow of material towards the center of the mass overdensity. For large regions of mass underdensity (e.g., large voids) there will be a net outflow of material form the void center. For smaller-scale voids, however, the outflow is compressed due to the effects of nearby larger-scale voids which have bigger outflows (van de Weygaert 1995).

The measurements of peculiar velocities has a rich history of producing ambiguous results with a touch of irony. The ambiguous nature of the results is a direct reflection of the methodology, which generally produces low S/N indicators of peculiar velocities. The first announced detection of systematic peculiar velocities in the local velocity field was made in 1976 by Vera Rubin and Kent Ford (Rubin *et al.* 1976). In general, this result was not believed and many papers were written to explain away the result on the basis of conspiratorial selection effects or twisted statistics. The main problem at the time with the Rubin–Ford effect was the lack of a framework in which to understand it. The possibility of large-scale motions produced by the collective gravitational acceleration of distant aggregate mass concentrations was not considered to be viable in a local Universe where smooth Hubble flow was assumed to dominate.

A year after the Rubin–Ford paper was published, observations of the CMB revealed the possible existence of a dipole anisotropy (DA). By 1979 the measurements of George Smoot and his colleagues had established the amplitude of the DA to be ≈ 600 km s^{-1}. A more modern determination of the DA vector by Smoot *et al.* (1992) based on *COBE* data has amplitude 620 km s^{-1} and director $\alpha = 10.75$ h and $\delta = -28.5°$. The DA is best understood as motion of our galaxy, amplitude ≈ 600 km s^{-1}, with respect to the frame of reference established by the CMB (which we assume to be an inertial frame). The origin of this motion is most likely due to gravitational acceleration of the Milky Way by a distant mass concentration. This discovered DA establishes that the local Hubble flow does have some noise. The Smoot *et al.* measurements also established a direction for the DA. Unfortunately, this vector is not directly pointed at the nearest known mass concentration, the Virgo cluster. Additional regions of overdensity are therefore required to fully explain the DA. The detection of the DA is quite important because it provides confirming evidence of peculiar velocities determined by other, lower S/N means. In fact, without the existence of the DA, claims of peculiar velocities in the local velocity field of amplitude a few hundred kilometers per second would probably still be met with wholesale skepticism. The firm existence of the DA now validates the general concept of large-scale peculiar velocities and confirms that the Rubin–Ford conceptual effect is real, even though their more detailed original claim has not been verified.

3.6.2 Virgocentric flow

Shortly after detection of the DA, Aaronson *et al.* (1980) presented evidence that the determination of H_0 from the distance and observed velocity of the Virgo cluster yielded systematically lower values than using more distant clusters. This difference suggested that the observed velocity of Virgo was not equal to its cosmic velocity but instead was a combination of its cosmic velocity and the infall of the Local Group towards Virgo. This phenomenon is known as Virgocentric flow. A schematic illustration of the way that a virialized structure like Virgo perturbs the local Hubble flow is shown in Figure 3.16. Virgocentric flow seriously distorts the relation between redshift and distance as multiple distances can give rise to the same observed velocity.

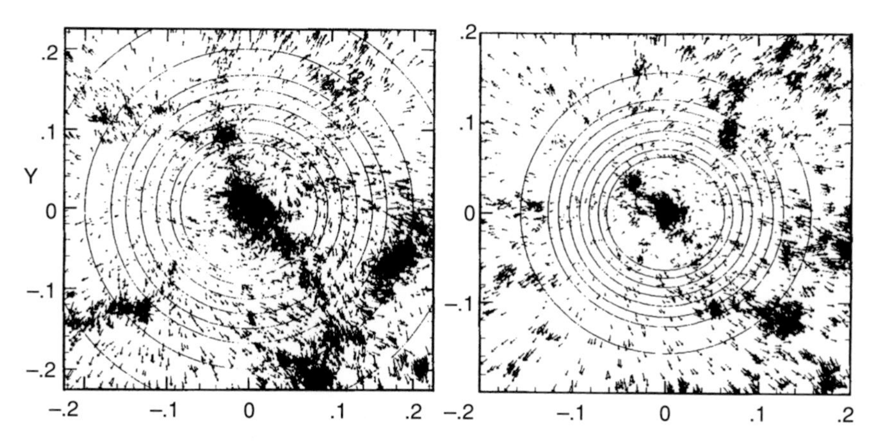

Figure 3.16 Distortion of the velocity field caused by spherically symmetric infall into a virialized structure. Here the spatial distribution of infall galaxies is plotted using heads or tails to indicate the amplitudes of their infall velocities. Note that the heads and tails always point at the center of the virialized cluster. Reproduced, with permission, from Villumsen and Davis (1986).

Let's assume for the sake of illustration (and easy math) that the distance to Virgo is 15 Mpc and that $H_0 = 100$. The cosmic velocity of Virgo is then 1500 km s^{-1}. A galaxy in Virgo at this distance which was at rest with respect to the Virgo cluster would also have a velocity of 1500 km s^{-1}. However, a galaxy located at a distance of 10 Mpc between us and Virgo could be accelerated by Virgo to a velocity of 500 km s^{-1}, so its observed velocity would be $1000 + 500 = 1500$ km s^{-1} even though its distance is 10 Mpc. Similarly, a galaxy located at 20 Mpc could have its expansion velocity with respect to us retarded by its proximity to Virgo at a value of 500 km s^{-1}, hence its observed velocity would be $2000 - 500 = 1500$ km s^{-1}. To understand this was occurring would require the measurement of relative distances with sufficient accuracy that could resolve the differences between distances of 10, 15 and 20 Mpc.

If Virgo is intrinsically at rest with respect to the CMB, a determination of the Milky Way's infall velocity will provide the necessary correction to the observed velocity of Virgo to produce its cosmic velocity. From this cosmic velocity, H_0 can be determined using the Virgo cluster distance modulus determined in the last chapter. Significant efforts to determine this infall velocity were made by Aaronson *et al.* (1982) and Davis and Peebles (1983). Those efforts continue to the present day with no real convergence on the infall velocity. Values of 250 km s^{-1} (Aaronson *et al.* 1982; Jerjen and Tammann 1993) to 350 km s^{-1} (Tonry, Ajhar, and Luppino 1989) are consistent with the data, and this range of values produces an error of almost 10% in the determination of H_0, even if we know the distance to the Virgo cluster with arbitrary accuracy.

This infall of the Local Group towards Virgo should also be reflected against some more distant reference frame. In 1986 Aaronson *et al.* were able to detect this reflex motion of the LG using a reference sample of clusters located at distances of 50–100 Mpc. Moreover, this detection also carried with it one of the initial

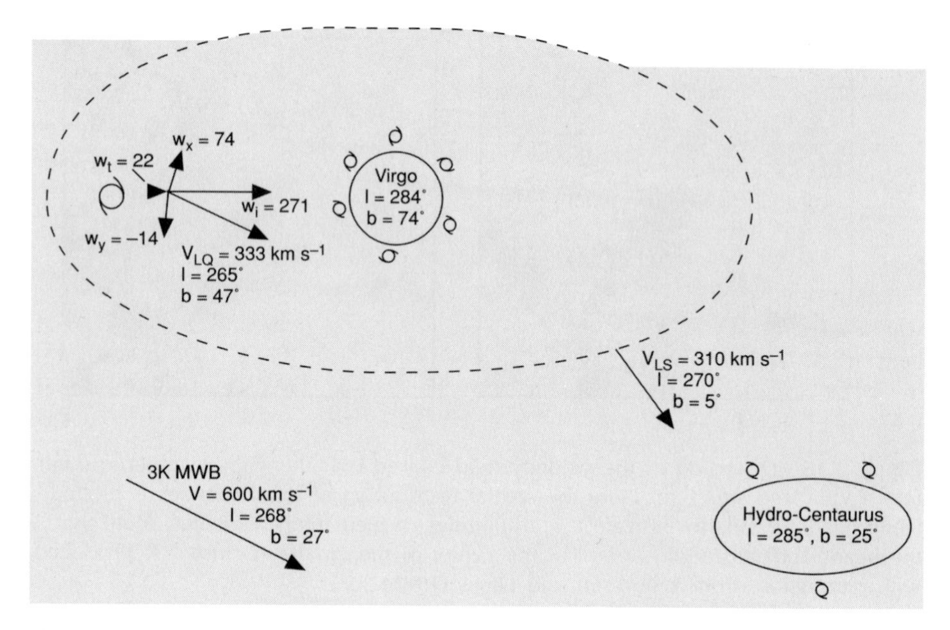

Figure 3.17 Schematic representation of the local velocity field from Aaronson *et al.* (1986). Here the Milky Way is infalling towards Virgo at approximately 300 km s^{-1} and the entire local supercluster is infalling towards Hydra–Cen at approximately 300 km s^{-1}. The vector sum of these two infall components to the motion of the Milky Way approximately accounts for the observed dipole anisotropy in the CMB.

indicators that the scale over which peculiar velocities are generated is significantly larger than the separation between Virgo and the LG. The DA does not point directly at the Virgo cluster. The vector difference between the DA and the Virgo cluster points in the general direction of the Hydra–Centaurus region. This suggests that the Local Supercluster is feeling the attraction of its nearest mass concentration, the Hydra–Cen supercluster. This situation is schematically illustrated in Figure 3.17 and shows that Virgo itself may also be moving.

Thus, a correction for Virgocentric infall alone is insufficient to recover the cosmic velocity of Virgo. An additional correction for the motion of the entire Local Supercluster towards Hydra–Cen must now be formulated and applied. In a kinematic description of the local velocity field, the motion of the LG towards Virgo can be thought of as a form of dipole anisotropy. The influence of Hydra–Cen acts as a quadrupole anisotropy in the local velocity field. (e.g., Lijle, Yahil, and Jones 1986). If mass concentrations other than Hydra–Cen are felt by the LG then higher-order anisotropies need to be considered as well.

3.6.3 Large-scale flows: continental drift in the nearby Universe?

The Aaronson *et al.* (1986) data set was restricted to the declination range of the Arecibo Observatory as they used the Tully–Fisher (TF) relation to measure the distances and V_c was obtained from 21 cm neutral hydrogen emission profiles in rotating-disk galaxies. From that data set, Aaronson *et al.* (1986) obtained a limit of $\approx$ 500 km s^{-1} on the random motion of clusters of galaxies, motion caused by the large-scale mass distribution. This argues that, for distant clusters, only small corrections are needed to transform from observed velocity to cosmic velocity. Hydra–Cen is located in the southern hemisphere (hence unobservable from Arecibo) at an observed velocity of 4500 km s^{-1}. The Aaronson *et al.* limit of 500 km s^{-1} is therefore only a 10% perturbation in the cosmic velocity of Hydra–Cen.

The Aaronson *et al.* (1986) paper was published in March. Two months before was a conference on galaxy distances and deviations from universal expansion, held on the Kona coast of the Big Island of Hawaii. The meeting was noteworthy in two respects: (1) boring presentations could be compensated by a quick dip in the ocean, 100 yards away and (2) the first data were presented that the Hydra–Cen supercluster had a much larger peculiar velocity than was allowed for by the Aaronson *et al.* (1986) data. In every respect, this data would change our perception of the local Universe forever, and ultimately it would lead to heated debates in the professional journals as to exactly how noisy was the local Hubble flow. To date, this situation has not been resolved; see the excellent and comprehensive review of Strauss and Willick (1995).

The principal fault of the Aaronson *et al.* sample was its limited sky coverage. A proper mapping of the large-scale flow pattern requires an all-sky sample. In the early 1980s a team of astronomers, led by Sandy Faber of the University of California at Santa Cruz, developed a method that was similar to the TF method but which could be applied to elliptical galaxies. Recall how the TF method works under the assumptions that the circular velocity V_c of a rotating-disk galaxy is driven by its total mass. If there is little variation in the ratio of **mass to luminosity** (M/L) then V_c is a measure of intrinsic luminosity. Elliptical galaxies are nonrotating and are supported by the internal velocity dispersion (σ_v) of their stars. More massive ellipticals have deeper gravitational potentials, hence higher values of σ_v. If M/L for ellipticals has little variation, σ_v is an indicator of intrinsic luminosity. Originally too large to make it competitive with the TF relation, there are two sources of error associated with this method:

- *Anisotropy in the orbits of the stars in ellipticals*: Just as the TF relation demands that rotating-disk galaxies are circularly symmetric, the σ_v method demands that the orbits of the stars are isotropic. This is because we only measure the radial component of σ_v. If the orbits are anisotropic, our value of σ_v will depend on the orientation of the long axis of these anisotropic orbits with respect to the observer.
- *A radial dependence of σ_v in elliptical galaxies*: This means that observational determinations of σ_v are aperture dependent. The same problem exists in the TF relation as measurements of V_c are also aperture dependent because the rotational velocity of a galaxy is a function of radial distance from the center. For spiral galaxies, however, V_c can be determined from an aperture which is larger than the galaxy itself; hence it contains the whole rotation curve. This is the main advantage of 21 cm neutral hydrogen observations. For elliptical galaxies it is

impossible to measure σ_v through a large aperture; this is because the signal comes from the integrated brightnesses of all the stars, and hence it is dominated by the central highest surface brightness regions of the elliptical.

The first source of error cannot really be overcome, although dynamical models of ellipticals are consistent with a low degree of anisotropy. The second problem was solved by the combined talents of Alan Dressler, David Burstein, Roger Davies and Donald Lynden-Bell, all of whom were members of Faber's elite team. After looking at the σ_v data for ellipticals over many years, they concluded that improvement in Faber's original method (e.g., Faber and Jackson 1976) could be obtained by selecting apertures which enclosed a constant surface brightness from one elliptical to the next. Using this method to define the aperture in which to measure σ_v gave rise to the D_n–σ relation for measuring relative distances between elliptical galaxies. The scatter in this relation is similar to that in the TF relation, so D_n–σ is now competitive with the TF relation.

The D_n–σ method has the advantage that only optical telescopes are required; hence it can be applied to any elliptical galaxy in the sky. The D_n–σ group made measurements in regions of the sky that Aaronson *et al.* could not cover (see Dressler's 1994 book *Voyage to the Great Attractor* for a detailed summary of their work). This fuller sky coverage revealed disconcertingly large deviations from expansion motion (Dressler *et al.* 1987). These results were most distressing for Aaronson *et al.* because they meant that a reliable determination of H_0 from cluster data was probably not possible. Furthermore, if Hydra–Cen is moving with respect to the CMB, it cannot be the sole source of the observed DA, so a more distant mass concentration is required if the motion is gravitational in origin. Even more interesting was the possibility that the entire region from the Milky Way to the Hydra–Cen supercluster was moving at 600 km s^{-1}. This motion is called **bulk flow** and suggests a kind of plate tectonic model for the nearby Universe, in which large regions are streaming at a constant velocity towards distant mass concentrations. But what mass concentration could produce acceleration over such a large scale?

Additional analysis of the elliptical galaxy data by Lynden-Bell *et al.* (1988) led to a model in which the idea of bulk flow was replaced with an infall pattern that was driven by a rather large mass concentration. This mass concentration has been dubbed the Great Attractor (GA) and is the subject of Dressler's book (Dressler 1994). The putative GA lies behind the Hydra–Cen supercluster at a kinematical distance of 4350 km s^{-1}. Infall of Hydra–Cen, the Local Supercluster and our galaxy toward the Great Attractor then accounts for the observed positive peculiar velocities. Thus, the LG feels both the accelerations of the Virgo cluster and the GA, and the relative normalization of these two vectors depends on Ω and the relative overdensities $\delta\rho/\rho$ of the two mass concentrations.

3.6.4 Infall around the Great Attractor: a case study

Virialized structures in the Universe produce velocity perturbations of galaxies which are located near them. The amplitude of these perturbations $\delta v/v$ depends on the background density of the Universe Ω and the density contrast $\delta\rho/\rho$ interior to the shell upon which the galaxy is located. The unperturbed velocity in this case is given by $v = H_0 r$ where r is the distance from the source causing the perturbation. The only situation that can be solved analytically (e.g., Regos and Geller 1989; Lightman and Schechter 1990) is the case of a spherical density perturbation, in which case $\delta v/v$ can be expanded in powers of $\delta\rho/\rho$ as

$$\delta v/v = f1(\Omega(\delta\rho/\rho) + f2(\Omega)(\delta\rho/\rho)^2 + \ldots \tag{3.37}$$

In the linear approximation of Peebles (1980), equation (3.37) is usually abbreviated to the form

$$\delta v/v \sim -\tfrac{1}{3}\Omega^{0.6}\,\delta\rho/\rho \tag{3.38}$$

More precise derivations by Regos and Geller (1989) and Lightman and Schechter (1990) yield values of $f1 = -1/3\Omega^{0.57}$ and $f2 = 4/63\Omega^{0.62}$. This second-order approximation and the fifth-order approximation of Regos and Geller are only valid in the regime of small $\delta\rho/\rho$. For denser regions it is necessary to employ the exact solution of Regos and Geller (1989). However, they do note that the nonlinear approximation of Yahil (1985) tracks the exact solution quite well out to overdensities of ≈ 20. Yahil's approximation is of the form

$$\delta v/v \sim -\frac{1}{3}\Omega^{0.6}\left[\frac{\delta\rho/\rho}{(1+\delta\rho/\rho)^{0.25}}\right] \tag{3.39}$$

Equation (3.39) is the standard infall equation that can be used to predict peculiar velocities as a function of radial distance from a virialized structure. To make a prediction, a value for Ω has to be assumed and the radial falloff with distance from the center of the virialized structure of $\delta\rho/\rho$ must be estimated. In general, one assumes that $\delta\rho/\rho$ falls of as $r^{-\gamma}$, where γ is taken to be 2. We can also define the turnaround radius r_t of a virialized structure as the radius at which the inward peculiar velocity of a shell exactly cancels its outward expansion, as seen by the center of the virialized structure. This condition requires $\delta v/v = -1.0$. In an $\Omega = 1.0$ Universe, the mean density enclosed within r_t is 5.5 (assuming spherical symmetry), or $\delta\rho/\rho = 4.5$. Equation (3.39) predicts $\delta v/v = -1$ at $\delta\rho/\rho = 4.67$. The density contrast within r_t as a function of Ω is given by Regos and Geller (1989). In an $\Omega = 0.1$ Universe, the density contrast is 29, hence even the nonlinear approximation breaks down before r_t is reached. In that sense, infall patterns are easier to treat in the case of $\Omega = 1.0$ since nonlinear effects do not set in quite so rapidly.

We now have a framework for analyzing peculiar velocities in terms of a spherical infall model. In principle it can also determine or constrain Ω. More stringently, the analysis really only constrains $\Omega^{0.6}/b$, where b is the bias parameter used in linear biasing theory. If on the scale of the observations, mass is a perfect tracer of light, then $b = 1$ and Ω can be recovered directly. But studies of peculiar velocities for galaxies in the vicinity of the Hydra–Cen supercluster convincingly show this procedure is not very straightforward. As such, the history of these studies is an excellent case study of the complex and ambiguous nature of the peculiar velocity field which makes the overall pattern difficult to decipher.

The Hydra–Cen supercluster consists of two main mass concentrations, the Hydra cluster and the Centaurus cluster. They are denoted by the letters C and H in Figure 3.8, where Centaurus is a part of the supergalactic plane. The Hydra cluster appears to be unperturbed and is at rest with respect to the CMB, and it has a cosmic velocity of 4055 km s^{-1}. On the other hand, the velocity histogram of Centaurus is strongly bimodal with one peak occurring at $v \approx 3000$ km s^{-1} (Cen30) and the other at $v \approx 4500$ km s^{-1} (Cen45). Using a sample of elliptical galaxies, Faber *et al.* (1989) report kinematic distances of 2220 ± 250 km s^{-1} for Cen30 and 3175 ± 335 km s^{-1} for Cen45. These distances are substantially less than their redshift distances, which indicates that Cen30 and Cen45 *both* have positive peculiar velocities. In conflict with this result are the observations of Lucey and Carter (1988), who measure kinematic distances for a similar sample of elliptical galaxies to determine peculiar velocities of 3115 ± 280 km s^{-1} for Cen30 and 2675 ± 435 km s^{-1} for Cen45. These results indicate that Cen30 is at rest and Cen45 is infalling at $v \approx 2000$ km s^{-1}. The opposite case was found from an analysis of the relative distances to spiral galaxies. Using the TF relation, Aaronson *et al.* (1989) derived mean peculiar velocities of 2835 ± 250 km s^{-1} for Cen30 and 4260 ± 450 km s^{-1}. In this case, Cen30 exhibits a velocity residual of ≈ 600 km s^{-1}, whereas Cen45 appears to be at rest. There is widespread disagreement among the various data sets.

The basis of the GA-driven inflow model of Lynden-Bell *et al.* (1988) is that both Cen30 and Cen45 have peculiar velocities with amplitudes of ≈ 1100 km s^{-1}. If the observed DA of 600 km s^{-1} for the Milky Way is also the result of GA-driven inflow, we have two values of $\delta v/v$ at two different distances. With some assumptions, these data can be input to equation (3.39) to fix the distance of the GA. A GA-driven acceleration of the LG with an amplitude of 600 km s^{-1} requires $r_t \approx 13$ Mpc or a mass ≈ 20 times that of the Virgo cluster. For $\Omega = 1.0$ this corresponds to a mass of $1.3 \times 10^{16} M_{\odot}$ within $r = 44$ Mpc. For $\Omega = 0.1$ the enclosed mass is reduced by a factor of almost 3. For the smaller-scale case of the Virgo cluster, an LG infall velocity of 250 km s^{-1} requires the Virgo cluster to have a turnaround radius of 4.5–5 Mpc.

As of 1991 there were two alternatives to the GA-driven inflow model that remained consistent with the data: (1) Cen30 is at rest (at $v \approx 3000$ km s^{-1}) and drives an acceleration of the Local Supercluster towards it; this was schematically shown in Figure 3.17; (2) Cen30 has the same peculiar velocity (e.g., 600 km s^{-1}) as the Milky Way, and hence it is participating in the same flow pattern. The simultaneous existence of three very different physical models to the same data set illustrate how ambiguous this problem is. However, a GA-driven infall model should be symmetric on both the front and back, and as of 1991, only data existed on the front. The discovery of infall on the back of the GA would therefore confirm that model.

In 1992 and 1993 Aaronson and coworkers published two papers which definitively showed there was no backside infall towards the hypothesized GA (Bothun *et al.* 1992a; Mould *et al.* 1993). Other investigators soon reached similar conclusions (e.g., Mathewson, Ford, and Buchhorn 1992). Figure 3.18 summarizes the situation

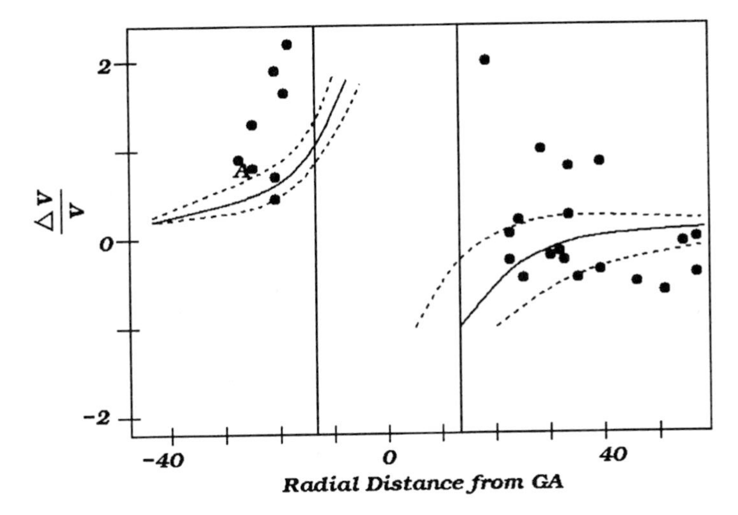

Figure 3.18 Individual galaxy peculiar velocities plotted in comparison with the spherically symmetric infall model that would be generated by a Great Attractor located at a kinematic distance of 4350 km s^{-1} from the Milky Way. The solid curved lines define the front and back caustic infall surfaces and the dashed lines represent 1σ error surfaces. The solid vertical lines represent r_t for the Great Attractor. The letter A denotes the Antlia cluster. The model has been normalized such that the Great Attractor accounts for the entire observed dipole anisotropy in the CMB. Adapted from Figure 4 in Bothun *et al.* (1992a).

as it presents a comparison of the model infall pattern (3.39) with the data of Bothun *et al.* (1992a). The GA center distance has been set to 4350 km s^{-1}, as proposed by Lynden-Bell *et al.* (1988). The solid vertical lines define r_t. Points outside of r_t are expected to fall off along the solid lines, which represent the imaginary caustic surface of the virialized structure. The model does not go to 0 at $r = 43.5$; this is because we have normalized the model to a value of 600 km s^{-1} at $r = 43.5$ as we assume the DA is generated by the GA. The dotted lines around the model show the effects of distance errors on the quantity $\delta v/v$.

If infall towards the GA center were responsible for producing the observed amplitude of LG acceleration, we would expect the observed data points to fall within the error surfaces surrounding the basic infall model. It is obvious that on the front the data are generally at much higher peculiar velocity than the model would predict. On the back, although some points fall within the model predictions, many do not and many have the wrong sign for the observed peculiar velocity. To better examine the data set, it is convenient to break it up into three samples, which we denote as F (far), N (near), and T (transverse). Sample F is located at a mean distance of 35 Mpc on the back of the GA center. At this distance the model predicts $V_p = -705 \pm 40$ km s^{-1}. The data yield $V_p = -717 \pm 462$ km s^{-1}, a 1.5σ detection of backside infall, which we do not consider significant. Sample N is located at a mean distance of 21.5 Mpc from the GA on the front side. The model predicts $V_p = +1006 \pm 30$ km s^{-1} but the data yield $V_p = +2366 \pm 334$ km s^{-1}. Hence

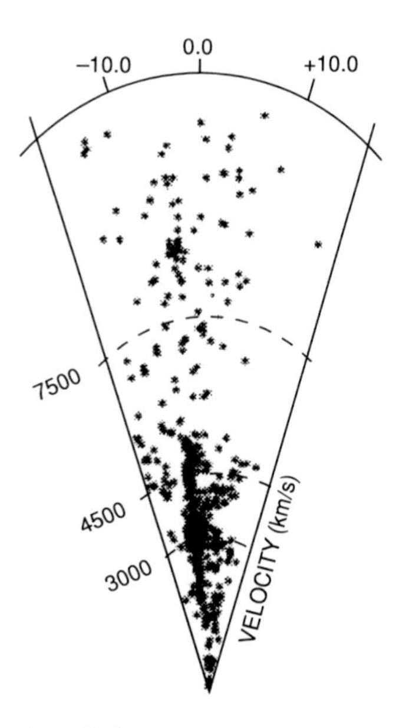

Figure 3.19 Cone diagram for all known velocities that are in the putative Great Attractor region of the sky. The large "finger" is the Centaurus cluster. Reproduced from Bothun *et al.* (1992a).

the F and N samples do not indicate any of the required symmetry associated with an infall pattern driven by a virialized structure. Finally, there are the sample T galaxies. These are galaxies whose radial infall to the GA is perpendicular to our line of sight, so V_p should be zero; but the measured V_p for this sample is 851 ± 181 km s^{-1}. The transverse sample has a significant radial component to its peculiar velocity when, in the context of symmetric spherical infall towards the GA center, none should have been observed.

On balance, the available data seems to rule out the GA model as a viable way to explain the local peculiar velocity field. By itself, the GA model was always worrisome as one could not optically identify a large concentration of light that would correspond to the GA. Although the GA is at somewhat low galactic latitude, its angular extent is large enough to allow its detection in a redshift survey without being obscured by our own galaxy. At the GA center distance of 4350 km s^{-1}, r_t subtends an angle of 16.5°. Figure 3.19 shows the redshift cone diagram in the proposed GA direction for all galaxies with redshifts that are located within the projected turnaround radius of the GA. The very obvious virialized finger belongs to the Centaurus cluster, a structure that was originally measured to have $V_p \approx 1100$ km s^{-1} and provides the foundation of the GA model. In Figure 3.19 the GA center is located at (0, 4350). There is no hint of any virialized structure at this location. Since spherical infall is best generated by virialized structures, the Centaurus cluster is an excellent candidate for contributing to the observed CMB DA. It is therefore difficult to understand the observations which suggest the Centaurus cluster moves in the same 600 km s^{-1} flow as the LG.

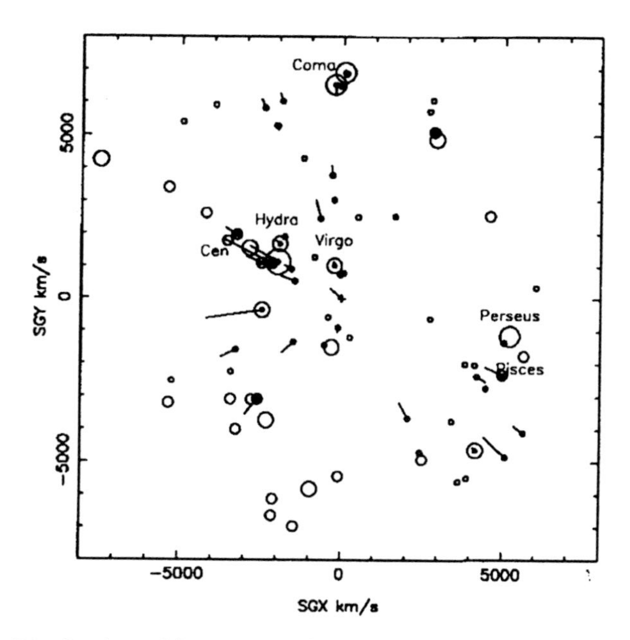

Figure 3.20 Distribution of known peculiar velocities for nearby groups and clusters as projected on the plane of the Local Supercluster. Nominal distances are in km s^{-1} and the amplitude and direction of any measured peculiar velocity is indicated by the arrows that are tagged onto some groups and clusters. Adapted from Mould *et al.* (1993).

To date, the situation with respect to the origin, scale and evolution of peculiar velocities remains confusing. The acquisition of more data by Mould *et al.* (1991, 1993) suggests a picture that is hybrid between Aaronson *et al.* (1986) and the D_n–σ group. In this scenario the Hubble flow, as delineated by nearby clusters of galaxies, is relatively quiet, except for samples in the vicinity of large superclusters such as Hydra–Cen or Perseus–Pisces. At these locations, substantial large-scale flows are seen with up to 50% of the observed velocity being due to peculiar motions. Figure 3.20 presents a representation of the flow pattern as it has been measured for nearby clusters of galaxies. This flow pattern remains consistent with either bulk flow models or infall-driven models with no clear resolution in sight (Strauss and Willick 1995).

3.7 The physical structure of voids and walls: a case study

3.7.1 Are voids real structures?

As emphasized earlier, the appearance of structure in redshift space may not be an accurate representation of their structure in physical space. This is particularly relevant to the apparently thin nature of the Great Wall as well as the apparent

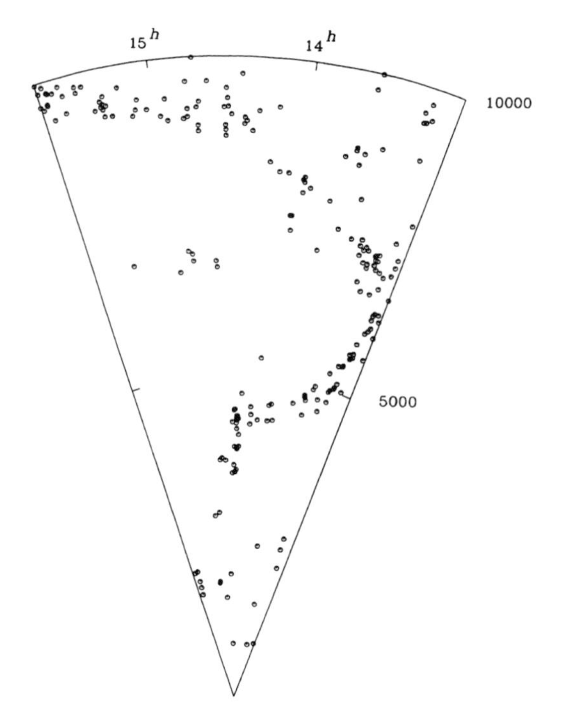

Figure 3.21 Structure of the large void seen in the first CFA slice data in redshift space. Adapted from Bothun *et al.* (1992b).

sharpness of the void boundaries. In general, these structures are of low galaxy density and are preferentially inhabited by spiral galaxies. If there are enough spiral galaxies available, the measures of relative distances using the TF relation can be used to compare the physical size of the structure with its apparent size in redshift space. For instance, the apparent voids that are seen in redshift space could be manifestations of smaller structures which have large outflow velocities away from the center. In this case, the physical size of the void would be significantly smaller than the size inferred from the redshift ratio of the near and far sides. However, if the voids are truly massless, some outflow is expected but that is generally small (Regos and Geller 1991).

The physical structure of one of the larger voids detected in the CFA slice survey was measured in 1992 by Geller and collaborators (Bothun *et al.* 1992b). Figure 3.21 shows the structure of this void in redshift space. In redshift space, its diameter is ≈ 5000 km s^{-1}. Relative distances between spirals located on the near and far redshift edges of the void were measured using the I-band TF relation. The observed scatter in the TF relation for this sample is fairly small, $\sigma = 0.18$ mag, and this allows for fairly accurate measures of relative distance. Three principal results emerged from this study:

- The mean redshift ratio of the far and near sides of the void is 2.16 ± 0.08 and the ratio determined from the relative distance measures is 2.22 ± 0.10. Thus the physical size of the void is the same as its redshift size. This means the voids are physical structures, not artifacts in redshift space; they likely formed at early times and have been expanding at the same rate as the rest of the Universe, thus preserving their character.

- Consistent with the first result, no large net outflow was detected at the void boundaries. The formal 1σ upper limit on outflow is $\approx 5\%$ of the void diameter.
- The nearside component of the void which is closest to the Coma cluster has an infall velocity of $\approx 900 \pm 250$ km s^{-1} at mean distance 26 h^{-1} Mpc from that structure. This is a substantially higher infall than for the LG, which is located about 15 Mpc away from the center of the Virgo Cluster. To first order, this indicates that Coma is a much more massive cluster than Virgo. Streaming velocities (towards virialized clusters) along the void wall of this amplitude are predicted by most models in which voids exist near clusters of galaxies (Regos and Geller 1991). These streaming velocities give rise to the appearance in redshift space that the void wall smoothly connects to the virialized structure.

3.7.2 Are walls really thin?

A large I-band TF sample of 175 spiral galaxies located in the Great Wall (GW) structure has been obtained and analyzed by Dell'Antonio and coworkers (Dell'Antonio, Bothun, and Geller 1996; Dell'Antonio, Geller, and Bothun 1996). For this larger sample, the scatter in the TF relation is $\sigma = 0.30$ mag. Coupled with the sample size, this scatter is sufficient to determine the physical thickness of the wall, which is necessary to confirm it is a 2D structure, not a 1D filament spread out in redshift space by peculiar velocities whose origin would be unknown. Using this data set, Dell'Antonio and coworkers examine four main issues: (1) the amount of infall onto the GW, (2) the large-scale peculiar velocity of the GW itself, (3) the thickness of the GW and (4) the amount of shear along the entire length of the structure. These issues are all highly relevant to the stability and longevity of the GW.

The results of this study can be summarized as follows:

- The observed velocity dispersion in the GW is ≈ 500 km s^{-1} (Ramella, Geller, and Huchra 1992). The TF data can place a 90% confidence upper limit on mean infall velocity of ≈ 500 km s^{-1}. This produces an upper limit on GW wall thickness of $11h^{-1}$ Mpc, which strongly confirms that it is a cold, 2D structure in physical space as well as redshift space.
- Over the angular extent of the wall, the shear velocity is constrained to be -70 ± 210 km s^{-1}. The lack of significant shear indicates the GW to be a quiet, rather unevolving structure. Furthermore, both the low infall and shear velocities (each consistent with zero) indicates that the GW is not subject to any significant perturbation from clusters that are embedded within it. As the angular extent of this structure corresponds to a linear size of a few hundred megaparsecs, the low degree of shear can place a limit on the power spectrum (Chapter 5) of large-scale fluctuations (e.g., Feldman and Watkins 1995).

- A larger data set is required to more confidently determine whether the GW is at rest with respect to the CMB. The current data do not suggest the GW has significant large-scale motion but the possibility that it participates in the same bulk flow pattern as the LG can only be ruled out with 70% confidence.

3.8 A really, really big flow?

The results of these two case studies help to further define the relation between structure seen in redshift space to the structure which exists in physical space. From these studies we now know that voids are real structures devoid of mass and that walls are cold, quiescent 2D structures. We also know that virialized structures produce infall patterns and that the local velocity field is perturbed by the distribution of these virialized structures. However, we still don't know which structures really dominate the observed kinematics in the local velocity field. Adding to this confusion is the remarkable observation of Lauer and Postman (1994). In their startling paper, Lauer and Postman (LP) used distance estimates to elliptical galaxies to conclude that bulk motion is occurring over a scale of 15 000 km s^{-1}. This implies a mass inhomogeneity on a scale of 150–300 Mpc, roughly 5% of the horizon size (c/H_0). This is also larger than the volume over which we presently measure H_0, which then seriously calls into question whether a precision measurement of H_0 can even be made. Unfortunately, the direction of this bulk motion, as determined by LP, is inconsistent with the direction of the CMB DA; this is very disconcerting as well as just plain odd. Hence the LP result has serious implications with respect to structure formation models (Chapter 5) and the measured anisotropy of the CMB. As emphasized by Strauss *et al.* (1995), the LP result is largely incompatible with any existing structure formation models, and if the observations are correct, an unprecedented large-scale flow has been discovered.

Given the fundamental problem potentially posed by the LP result, several groups are actively seeking to verify or exclude it. Branchini, Plionis, and Sciama (1995) use the Abell cluster catalogs as a tracer of the mass from which they reconstruct the density field to predict peculiar velocities of clusters of galaxies over a scale of 18 000 km s^{-1}. Fitting these peculiar velocities to a dipole term recovers the amplitude and direction of the LG velocity, which is consistent with the CMB DA and inconsistent with the LP result. However, the data don't completely rule out the LP amplitude and direction, due to the large distance-dependent errors which are present in the LP sample. Furthermore, the Branchini *et al.* method assumes the Abell catalog has a uniform sampling of clusters across the sky, and this is a very dangerous assumption. In particular, the southern compilation of the Abell cluster catalog relied on better photographic material than the older, northern compilation.

Other studies have also failed to verify the original LP result. For example, the studies of the Great Wall by Dell'Antonio and coworkers are inconsistent with the LP bulk flow vector but only at the 75% level. Recent observations by Giovanelli *et al.* (1996) using relative TF distances to spiral galaxies have also failed to find significant bulk flow on the LP scale. Their results (Figure 3.22) indicate that the

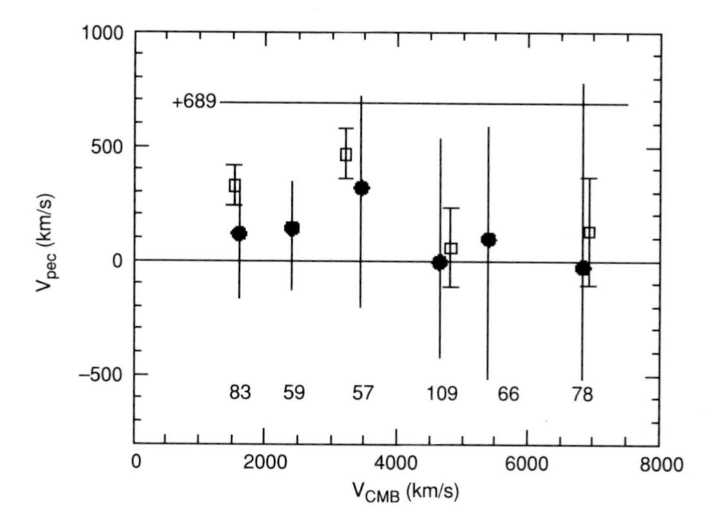

Figure 3.22 Peculiar velocity: the average peculiar velocity in this data declines to zero at an observed redshift of 6000 km s^{-1}. This behavior is inconsistent with the Lauer and Postman (1994) result indicated by the +689 constant velocity line. The number below each data bin marks the number of galaxies in the sample. Reproduced, with permission, from Giovanelli *et al.* (1996).

large-scale peculiar velocity field declines towards zero by redshift 6000 km s^{-1}. A decline to zero on this scale agrees well with previous studies (Da Costa *et al.* 1996).

The failure to verify the LP result in some sense is comforting, for if $\approx$ 600–700 km s^{-1} bulk flow is occurring on these large scales, this would signal a return to the cosmology of Marc Davis in which we know less than zero. However, this situation really is not yet settled, as there is nothing obviously wrong with the LP data set or the analysis. One has to appeal to systematic errors in the LP distance indicators, or in the properties of the galaxies used to define the distances, and these errors have not yet been identified.

3.9 A determination of H_0 using the Virgo and Coma clusters

The identification of large-scale peculiar velocities is a potentially serious obstacle to the recovery of H_0 from local samples of galaxies. In fact, Turner, Cen, and Ostriker (1992) suggest that could be fatal and say, perhaps prophetically:

> Even if the local expansion rate is known to be 80 ± 8 km/s/Mpc out to $30h^{-1}$ Mpc in the North Galactic Cap, the 95% confidence limits on the true global value of H_0 is 50–128 km/sec/Mpc in a CDM model.

To counter this with observations, Mould (1996) has developed a simple but effective model which assumes that H_0 is 50% higher when measured using a local sample of galaxies compared to using samples that are located at the distance of the Coma cluster or beyond. Figure 3.23 summarizes the data used in comparison with

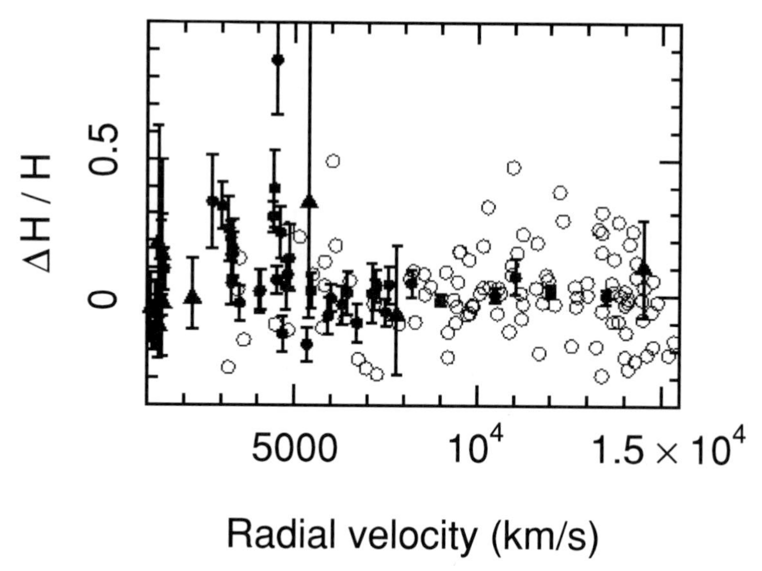

Figure 3.23 Deviations from a uniform Hubble flow compiled by Mould (1996): (●) clusters of galaxies with Tully–Fisher distances, (▲) EPM data of Schmidt *et al.* (1994), (○) brightest cluster members from Lauer and Postman (1994). There is no evidence that a significantly different value of the Hubble constant pertains for samples located inside and outside the distance of the Coma cluster at $v = 7200$ km $s^{\angle 1}$. Moreover, it is clear that beyond 5000 km s^{-1}, the noise in determining H_0 has greatly diminished. Note the considerable scatter in the Lauer and Postman sample. Courtesy of Dr Jeremy Mould, Mount Stromolo and Siding Springs Observatories, New South Wales, Australia.

the model. Although the data is noisy (due to errors in distance estimates and to peculiar velocities), the mean value reaches $\Delta H/H = 1$ by $v \approx 7000$ km s^{-1}. This strongly suggests that a global value of H_0 can be recovered by using samples of galaxies at these distances or by accurately measuring the relative distance modulus between Virgo and Coma. We now proceed to do exactly that, but first we consider why using the Virgo cluster alone is not the ideal route for determining H_0.

The first step is to determine the cosmic velocities of Virgo and Coma. The observed velocity of Virgo is determined in the heliocentric frame of reference, i.e., the velocity with respect to the Sun. Since the Sun is rotating around the galaxy, a correction for galactic rotation must be made so that the Virgo velocity is with respect to the Milky Way galaxy. And there is a small correction for the random perturbation on the Milky Way galaxy produced by the other LG members. Both of these corrections are well understood and the transformation from heliocentric to galactocentric velocity is

$$v_{gal} = v_h + 300 \cos b \sin l \text{ km s}^{-1} \tag{3.40}$$

where b and l are coordinates in galactic latitude and longitude. The cosmic velocity of Virgo is then

$$v_{cosmic} = v_{gal} + v_{da} + v_{qa} \tag{3.41}$$

where v_{da} represents the dipole term in the local velocity field (essentially the pattern speed of infall of the LG towards Virgo) and v_{qa} is the quadrupole term that represents the **random** effects of other mass concentrations on Virgo's motion. Unfortunately, current determinations of v_{da} and v_{qa} are rather uncertain and they depend on the model.

In addition, it is now generally recognized that the Virgo cluster has substantial substructure (Yasuda, Fukugita, and Okamura 1997) which makes the determination of the mean v_h somewhat problematical. Take the brightest and most massive galaxy in Virgo, M87, its $v_h = 1292 \pm 10$ km s^{-1}. But the mean velocity of all probable members of Virgo (some 700 galaxies with velocities) is $v_h = 1150 \pm 51$ km s^{-1}. All of this translates into a rather large uncertainty in v_{cosmic} for Virgo and this does not seem like a promising route to recover H_0.

The better way to determine v_{cosmic} is to use v_{cosmic} for Coma and measure the relative distance between Virgo and Coma. Since Coma is only a few degrees away from Virgo in the plane of the sky, our radial infall towards Virgo is reflected in the observed velocity of Coma. However, since the observed velocity of Coma is ≈ 7000 km s^{-1}, an uncertainly of ± 100 km s^{-1} in our Virgocentric infall velocity is rather inconsequential. Several hundred velocities have been measured for galaxies in the Coma cluster core and the resulting mean velocity is $v_h = 6925$ km s^{-1}. If we use an infall velocity of 300 ± 100 km s^{-1} then we derive $v_{cosmic} = 7225 \pm 100$ km s^{-1}.

The relative distance modulus between Coma and Virgo has been measured by a number of different techniques and the agreement is good. The most credible methods and results are the following:

- The H-band TF measurements of Aaronson *et al.* yield $\Delta(m - M) = 3.69 \pm 0.16$. There is some concern that many of the spirals used in the sample of Aaronson *et al.* are not associated with small foreground and background groups, not with the Coma cluster. Indeed there is some evidence (e.g., Bothun *et al.* 1992b) that spirals are currently infalling to Coma as they are to Virgo. However, the Aaronson *et al.* sample is not biased towards preferential sampling of the front side of Coma but instead covers the entire range of possible infalling galaxies.
- Using the color–magnitude (CM) relation for ellipticals, as measured in the UBV passbands, Sandage (1972) derives $\Delta(m - M) = 3.66 \pm 0.14$. The relation between color and intrinsic luminosity for ellipticals is likely driven by metallicity variations as a function of the stellar mass of the elliptical. The metallicity of a star determines its atmospheric opacity at different wavelengths. This is especially true in the case of the atmospheres of red giant stars, whose light dominates that of elliptical galaxies. The CM effect is better measured in the infrared than the optical, and IR observations by Bower, Lucey, and Ellis (1992) yield $\Delta(m - M)$ 3.70 ± 0.09.
- The D_n–σ relation as applied by Dressler *et al.* (1987) yields $\Delta(m - M) = 3.65 \pm 0.20$. Evidence that Virgo and Coma ellipticals define the same intrinsic D_n–σ relation has been presented by Lucey *et al.* (1991).

These methods produce a mean Virgo–Coma relative distance modulus of $\Delta(m - M) = 3.68 \pm 0.03$. This is an extraordinarily small error bar, which reflects the overall consistency of relative distance determinations. The relative distance modulus assumes there is no foreground reddening towards either Virgo or Coma. The Coma cluster is very near the galactic north pole, and Virgo is also at fairly high latitude, so $E(\mathrm{B} - \mathrm{V}) = 0.0$ seems reasonable.

It would now seem that we are in a reasonable position to derive H_0. A relative distance modulus of $\Delta(m - M) = 3.68$ corresponds to a factor of $(10^{0.40\Delta(m-M)})^{1/2} = 5.45$ in linear distance. The predicted v_{cosmic} for Virgo is then $7225/5.45 = 1326 \pm 25$ km s^{-1}. The low error bar indicates that the uncertainty in LG infall velocity is no longer very important when v_{cosmic} is derived in this manner. In Chapter 2 we saw that the most probable distance modulus to Virgo is in the range $(m - M) = 30.9$–31.2 or linear distances of 15.1–17.4 Mpc. The lowest value of H_0 that can be obtained is $1351/17.4 = 78$. The highest value of H_0 that can be obtained is $1301/15.1 = 86$. Placing Virgo at a distance of $(m - M) = 31.5$ yields $H_0 = 68$. We thus conclude a firm lower limit of 70 can be placed on H_0. A likely upper limit is 90, achieved using $(m - M) = 30.8$. This is the distance that results if the distance to LMC is $(m - M) = 18.35$. It is our prediction that more precise determinations of H_0 will converge on values between 70 and 90. For $\Omega = 0.1$–1.0 this range in H_0 limits the expansion age of the Universe to between 7.5 and 13.5 billion years. This seems to be significantly less than the age of the oldest starts in globular clusters. Such an age range leaves the door wide open for nonzero Λ cosmologies, and this may be the most significant result that has been obtained after 30 years of trying to determine H_0.

References

AARONSON, M., HUCHRA, J., MOULD, J., SULLIVAN, W., SCHOMMER, R., and BOTHUN, G. 1980 *Astrophysical Journal* **239**, 12

AARONSON, M., HUCHRA, J., MOULD, J., SCHECHTER, P., and TULLY, B. 1982 *Astrophysical Journal* **258**, 64

AARONSON, M., BOTHUN, G., MOULD, J., HUCHRA, J., SCHOMMER, R., and CORNELL, M. 1986 *Astrophysical Journal* **302**, 536

AARONSON, M. *et al.* 1989 *Astrophysical Journal* **338**, 654

ABELL, G., CORWIN, H., and OLOWIN, R. 1989 *Astrophysical Journal Supplements* **70**, 1

ALDERING, G., BOTHUN, G., MARZKE, R., and KIRSHNER, R. 1997 preprint

BAHCALL, N. 1991 *Astrophysical Journal* **376**, 43

BAHCALL, N. and CEN, R. 1992 *Astrophysical Journal Letters* **398**, L81

BEERS, T. and GELLER, M. 1983 *Astrophysical Journal* **274**, 491

BERTSCHINGER, E. and JUSZKIEWICZ, R. 1988 *Astrophysical Journal (Letters)* **334**, L59

BINNEY, J. and TREMAINE, S. 1987 *Galactic Dynamics*, Princeton NJ: Princeton University Press

BOTHUN, G. and SCHOMBERT, J. 1990 *Astrophysical Journal* **360**, 436

BOTHUN, G., BEERS, T., MOULD, J., and HUCHRA, J. 1986 *Astrophysical Journal* **308**, 510

BOTHUN, G., SCHOMMER, R., WILLIAMS, T., MOULD, J., and HUCHRA, J. 1992a *Astrophysical Journal* **388**, 253

BOTHUN, G., GELLER, M., KURTZ, M., HUCHRA, J., and SCHILD, R. 1992b *Astrophysical Journal* **395**, 349

BOTHUN, G., SCHOMBERT, J., IMPEY, C., SPRAYBERRY, D., and MCGAUGH, S. 1993 *Astronomical Journal* **106**, 530

BOWER, R., LUCEY, J., and ELLIS, R. 1992 *Monthly Notices of the Royal Astronomical Society* **254**, 589

BRANCHINI, E., PLIONIS, M., and SCIAMA, D. 1996 *Astrophysical Journal Letters* **461**, L17

BROADHURST, T., ELLIS, R., and SHANKS, T. 1988 *Monthly Notices of the Royal Astronomical Society* **235**, 827

BROADHURST, T., ELLIS, R., KOO, D., and SZALAY, A., 1990 *Nature* **343**, 726

CARR, B. 1994 *Annual Reviews of Astronomy and Astrophysics* **32**, 531

CASTANDER, F., ELLIS, R., FRENK, C., DRESSLER, A., and GUNN, J. 1994 *Astrophysical Journal Letters* **424**, L79

COUCH, W., SHANKS, T., and PENCE, W. 1985 *Monthly Notices of the Royal Astronomical Society* **213**, 215

DA COSTA, L. *et al.* 1996 *Astrophysical Journal Letters* **468**, L5

DAVIS, M. and PEEBLES, P.J.E. *Astrophysical Journal* **267**, 465

DE LAPPARENT, V., GELLER, M., and HUCHRA, J. 1986 *Astrophysical Journal Letters* **302**, L1

DELL'ANTONIO, I., BOTHUN, G., and GELLER, M. 1996 *Astrophysical Journal* **112**, 1759

DELL'ANTONIO, I., GELLER, M., and BOTHUN, G. 1996 *Astronomical Journal* **112**, 1780

DIAFERIO, A., GELLER, M., and RAMELLA, M. 1995 *Astronomical Journal* **109**, 2293

DRESSLER, A. 1994 *Voyage to the Great Attractor*, New York: Knopf

DRESSLER, A. *et al.* 1987 *Astrophysical Journal* **313**, 42

ELLIS, R., COUCH, W., MACLAREN, I., and KOO, D. 1985 *Monthly Notices of the Royal Astronomical Society* **212**, 687

FABER, S. and JACKSON, R. 1976 *Astrophysical Journal* **204**, 668

FABER, S. *et al.* 1989 *Astrophysical Journal Supplements* **69**, 763

FELDMAN, H. and WATKINS, R. 1994 *Astrophysical Journal Letters* **430**, L17

FISHER, K., LAHAV, O., HOFFMAN, Y., LYNDEN-BELL, D., and ZAROUBI, S. 1995 *Monthly Notices of the Royal Astronomical Society* **272**, 885

GEBHARDT, K. and BEERS, T. 1991 *Astrophysical Journal* **383**, 72

GELLER, M. and HUCHRA, J. 1989 *Science* **246**, 897

GIOVANELLI, R. and HAYNES, M. 1985 *Astronomical Journal* **90**, 2445

GIOVANELLI, R., HAYNES, M., and CHINCARINI, G. 1986 *Astrophysical Journal* **300**, 77

GIOVANELLI, R. *et al.* 1996 *Astrophysical Journal Letters* **464**, L99

IMPEY, C., SPRAYBERRY, D., IRWIN, M., and BOTHUN, G. 1996 *Astrophysical Journal Supplements* **105**, 209

JERJEN, H. and TAMMANN, G. 1993 *Astronomy and Astrophysics* **276**, 1

KAISER, N. 1986 *Monthly Notices of the Royal Astronomical Society* **219**, 785

KIRSHNER, R., OEMLER, A., SCHECHTER, P., and SHECTMAN, S. 1981 *Astrophysical Journal Letters* **248**, L57

KIRSHNER, R., OEMLER, A., SCHECHTER, P., and SHECTMAN, S. 1987 *Astrophysical Journal* **314**, 493

KRISMER, M., TULLY, B., and GIOIA, I. 1995 *Astronomical Journal* **110**, 1584

LAUER, T. and POSTMAN, R. 1994 *Astrophysical Journal* **425**, 418

LIGHTMAN, A. and SCHECHTER, P. 1990 *Astrophysical Journal Supplements* **74**, 831

LILJE, P., YAHIL, A., and JONES, B. 1986 *Astrophysical Journal* **307**, 91

LUCEY, J. and CARTER, D. 1988 *Monthly Notices of the Royal Astronomical Society* **235**, 1137

LUCEY, J., GUZMAN, R., CARTER, D., and TERLEVICH, R. 1991 *Monthly Notices of the Royal Astronomical Society* **253**, 584

LYNDEN-BELL, D. *et al.* 1988 *Astronomical Journal* **326**, 19

MCCALL, M. and BUTA, R. 1995 *Astrophysical Journal* **109**, 2460

MARZKE, R., GELLER, M., DA COSTA, L., and HUCHRA, J. 1995 *Astronomical Journal* **110**, 477

MATHEWSON, D., FORD, V., and BUCHHORN, M. 1992 *Astrophysical Journal Supplements* **81**, 413

MERRIFIELD, M. and KENT, S. 1991 *Astronomical Journal* **101**, 783

MO, H., JING, Y. and WHITE, S. 1996 *Monthly Notices of the Royal Astronomical Society* **282**, 1096

MOULD, J. 1996 in *The MSSSO Heron Island Workshop on Peculiar Velocities in the Universe*, http://qso.lanl.gov/~heron/Mould/mould.html

MOULD, J. *et al.* 1991 *Astrophysical Journal* **383**, 467

MOULD, J. *et al.* 1993 *Astrophysical Journal* **409**, 14

OEGERLE, W. and HILL, J. 1994 *Astronomical Journal* **107**, 857

PEEBLES, P.J.E. 1980 *The Large Scale Structure of the Universe*, Princeton NJ: Princeton University Press

PERSIC, M. and SALUCCI, P. 1992 *Monthly Notices of the Royal Astronomical Society* **258**, 14P

PILDIS, R. 1995 *Astrophysical Journal* **455**, 492

POSTMAN, M., HUCHRA, J., and GELLER, M. 1992 *Astrophysical Journal* **384**, 404

POSTMAN, M. *et al.* 1996 *Astronomical Journal* **111**, 615

RAMELLA, M., GELLER, M., and HUCHRA, J. 1989 *Astrophysical Journal* **344**, 57

RAMELLA, M., GELLER, M., and HUCHRA, J. 1992 *Astrophysical Journal* **384**, 396

RAMELLA, M., DIAFERIO, A., GELLER, M., and HUCHRA, J. 1994 *Astronomical Journal* **107**, 1623

RAMELLA, M., GELLER, M., HUCHRA, J., and THORSTENSEN, J. 1995 *Astronomical Journal* **109**, 1469

RAYMOND, J., COX, D., and SMITH, B. 1976 *Astrophysical Journal* **204**, 290

REGOS, E. and GELLER, M. 1989 *Astronomical Journal* **98**, 755

REGOS, E. and GELLER, M. 1991 *Astrophysical Journal* **377**, 14

RUBIN, V., THONNARD, N., FORD, W., ROBERTS, M., and GRAHAM, J. 1976 *Astronomical Journal* **81**, 687

SACHS, R. and WOLFE, A. 1967 *Astrophysical Journal* **147**, 73

SALZER, J., ALDERING, G., BOTHUN, G., MAZZARELLA, J., and LONSDALE, C. 1988 *Astronomical Journal* **96**, 1511

SANDAGE, A. 1972 *Astrophysical Journal* **173**, 475

SANDAGE, A. and TAMMANN, G. 1993 *Astrophysical Journal* **415**, 1

SCHMIDT, B. *et al.* 1994 *Astrophysical Journal* **432**, 42

SHECTMAN *et al.* 1996 *Astrophysical Journal* **470**, 172

SILK, J. and WYSE, R. 1992 *Physics Reports* **231**, 293

SMOOT, G. *et al.* 1992 *Astrophysical Journal Letters* **396**, L1

STRAUSS, M. and WILLICK, J. 1995 *Physics Reports* **261**, 271

STRAUSS, M., CEN, R., OSTRIKER, J., LAUER, T., and POSTMAN, M. 1995 *Astrophysical Journal* **444**, 507

TONRY, J., AJHAR, E., and LUPPINO, G. 1989 *Astrophysical Journal Letters* **346**, L57

TURNER, E., CEN, R., and OSTRIKER, J. 1992 *Astrophysical Journal* **103**, 1427

VAN DE WEYGAERT, R. 1991 *Monthly Notices of the Royal Astronomical Society* **249**, 159

VAN DE WEYGAERT, R. 1994 *Astronomy and Astronomics* **283**, 361
VILLUMSEN, J. and DAVIS, M. 1986 *Astrophysical Journal* **308**, 449
WALKER, T., STEIGMAN, G., SCHRAMM D., OLIVE, K., and KANG, J. 1991 *Astrophysical Journal* **376**, 51
WEINBERG, D., HERNQUIST, L., and KATZ, N. 1997 *Astrophysical Journal* **477**, 8
WEINBERG, D., OSTRIKER, J., and DEKEL, A. 1989 *Astrophysical Journal* **336**, 9
WEST M. 1989 *Astrophysical Journal* **347**, 610
WEST, M. 1991 *Astrophysical Journal* **379**, 19
WEST, M. 1994 *Monthly Notices of the Royal Astronomical Society* **268**, 79
WEST, M. and BOTHUN, G. 1990 *Astrophysical Journal* **350**, 36
WHITE, S., NAVARRO, J., EVRARD, A., and FRENK, C. 1993 *Nature* **366**, 429
WILLMER, C. *et al.* 1995 *Astronomical Journal* **109**, 61
YAHIL, A. 1985 in *The Virgo Cluster of Galaxies*, eds O. Richter and B. Binggeli, Garching: European Southern Observatory
YASUDA, N., FUKUGITA, M., and OKAMURA, S. 1997 *Astrophysical Journal Supplements* **108**, 417
ZEPF, S. and WHITMORE, B. 1993 *Astrophysical Journal* **418**, 72

CHAPTER FOUR

Dark Matter in the Universe

4.1 Overview

Objects which emit light, whether they are cigars, lightbulbs, stars or galaxies can be characterized by their emitted energy per unit mass. This is parameterized as the **mass-to-luminosity** ratio (M/L). For cosmological purposes, it is most convenient to express M/L in terms of solar masses and luminosities. For main sequence stars it can be shown that $L \propto M^n$ where $n \approx 3.5$–4. Thus a $10M_\odot$ star has $M/L \approx 10^{-3}$, an $M_\odot$ star has $M/L = 1$ and a $0.1M_\odot$ star has $M/L \approx 1000$. The term **dark matter** refers to the the existence of objects which have extreme M/L. Identifying the nature and extent of the dark matter component in the Universe is arguably the most significant unsolved problem in all of cosmology.

The cosmological parameter ρ is insensitive to the nature of the mass in the Universe. The only requirement is that this mass gravitates. This requirement can be met if the dark matter is composed of a small space density of very massive objects or a large space density of very low mass objects. It is the total integration of this mass density over spacetime that determines ρ, hence the curvature of the Universe. Some clues about the nature of dark matter can be obtained by determining its distribution. For instance, if it can be shown that the galaxy distribution is an unbiased tracer of the mass distribution, then we can conclude that the dark matter is exclusively associated with galaxies and is not found between galaxies. In this case the dark matter must be of a form which allows it to dissipate into small-scale gravitational potentials. At the other extreme, if there is significant bias such that most of the mass is distributed between the galaxies, then it is not prone to clumping on small scales and is therefore distributed more diffusely. In general, the evidence for dark matter is a result of analyzing the motions of test particles on some scale then applying the virial theorem to estimate the mass. If the virial mass is larger than the mass estimated from the "light," this indicates the presence of gravitating mass which has no corresponding light and is therefore "dark." The

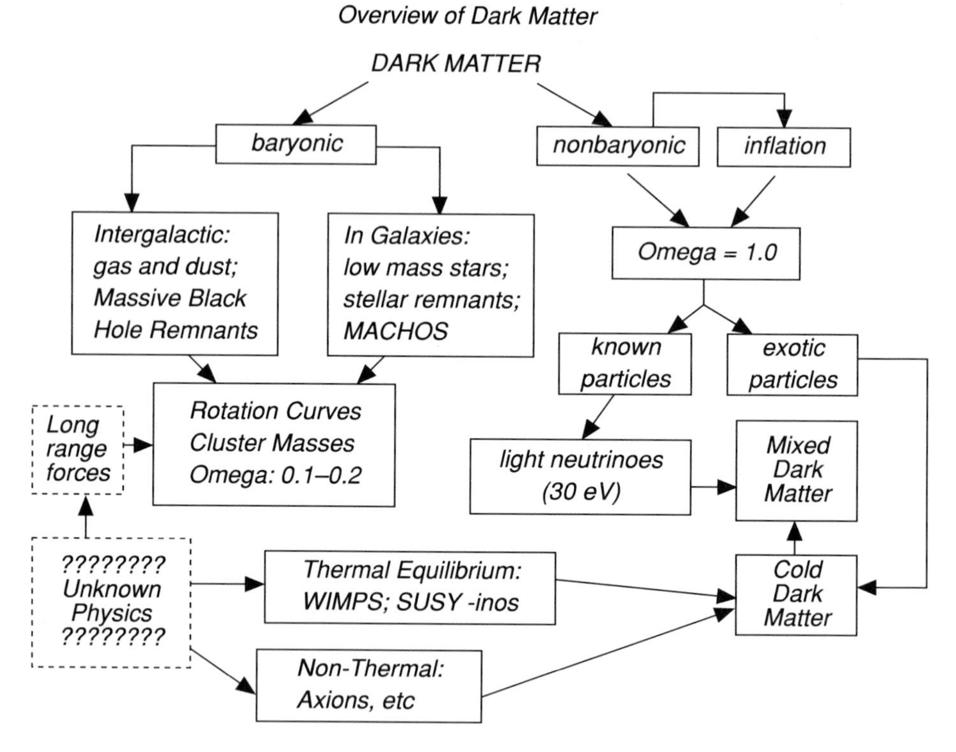

Figure 4.1 Flowchart summarizing the dark matter situation: possible candidates are driven by the actual value of Ω. The role of unknown physics in all of this is indicated by the dashed lines and boxes.

ability of dark matter to gravitate is independent of its nature. A test particle under the influence of gravity will not care whether the gravitating mass is baryonic or nonbaryonic. Figure 4.1 provides an overview of the possible kinds of dark matter which could exist and provide most of the gravitational mass in the Universe.

4.1.1 The virial theorem

Because the virial theorem is of central importance in dynamical mass estimates it is worthwhile to derive it in a rigorous manner using the moment of inertia of a system of N particles. For this system, we define the moment of inertia I as

$$I = \sum_{i=1}^{n} M_i R_i^2 = \sum_{i=1}^{n} M_i (x_i^2 + y_i^2 + z_i^2) \tag{4.1}$$

After one dynamical timescale, the time derivative of I is constant, so the second derivative is zero. Hence we can write

$$\dot{I} = \sum_{i=1}^{n} M_i(2x_i\dot{x}_i + 2y_i\dot{y}_i + 2z_i\dot{z}_i) \tag{4.2}$$

$$\ddot{I} = 0 \rightarrow \frac{1}{2}\ddot{I} = \sum_{i=1}^{n} M_i(\dot{x}_i^2 + \dot{y}_i^2 + \dot{z}_i^2) + \sum_{i=1}^{n} M_i(x_i\ddot{x}_i + y_i\ddot{y}_i + z_i\ddot{z}_i) = 0 \tag{4.3}$$

The first term on the right-hand side is $\Sigma_{i=1}^{n} M_i V_i^2$, which is twice the kinetic energy of the system or $2T$. The second term consists of a spatial coordinate times its second derivative, a displacement times a force, which is an energy. We identify the second term as $\Sigma_{i=1}^{n} \vec{R}_i^i \vec{F}$ where $\vec{F}$ is the total force. This term is the potential energy of the system or W. We thus have

$$\tfrac{1}{2}\ddot{I} = 2T + W \tag{4.4}$$

$$\tfrac{1}{2}\langle\ddot{I}\rangle = 0 = \langle 2T\rangle + \langle W\rangle \tag{4.5}$$

which is the well-known virial theorem in which the total energy of a system is zero.

For a self-gravitating N-body system

$$W = -\frac{1}{2}\sum_{i=1}^{N}\sum_{j=1}^{N}(GM_iM_j)\left(\frac{1}{r_ij}\right) \tag{4.6a}$$

If we assume that each particle has the same mass, then $M_i = M_j + M_t/N$ where M_t = total system mass. This yields

$$W = -\frac{1}{2}N(N-1)GM_i^2\left(\frac{1}{r_ij}\right) \tag{4.6b}$$

For large N this reduces to

$$W = -\frac{1}{2}\frac{GM_t^2}{R_{hms}} \tag{4.6c}$$

where R_{hms} is the harmonic mean separation between the system of N particles. The kinetic energy of this system is

$$T = \tfrac{1}{2}N(V_i^2)m_i = \tfrac{1}{2}M_tV_i^2 \tag{4.6d}$$

The time average of V_i is defined to be the RMS velocity dispersion of the system, σ_v. By the virial theorem we then have

$$2T + W = M_t\sigma_v^2 - -\frac{1}{2}\frac{GM_t^2}{R_{hms}} = 0 \tag{4.6e}$$

or

$$M_t = \frac{2\sigma_v^2 R_{hms}}{G} \tag{4.6f}$$

which indicates that, under the virial theorem, masses can be derived by measuring characteristic velocities over some characteristic scale size. In general, the virial theorem can be applied to any gravitating system after one dynamical timescale has elapsed.

4.2 Evidence for dark matter as a function of scale size

4.2.1 The Solar System

The first historical success of using perturbed motions to find matter was the discovery of Neptune from the observed perturbations in the motion of Uranus. Similar perturbations in the orbit of Neptune and Uranus were then used to predict the existence of a yet more distant planet. Those predictions were again verified with the discovery of Pluto in 1930 by Clyde Tombaugh, who also discovered evidence for superclusters. These successes spurred on efforts to discover a planet beyond Pluto, known as Planet 10, since the combined masses of Neptune and Pluto weren't quite enough to account for the observed perturbations in the orbits of the outer planets. Interestingly, the close passage of *Voyager II* near Neptune in 1989 allowed for a significant refinement of the mass of Neptune. *Voyager II* determined a mass 15% larger than the mass previously measured. This proper accounting of Neptune's mass removed the need for the existence of Planet 10 and indeed there is no longer any evidence for unaccounted mass in the Solar System (Hogg, Quinlan, and Tremaine 1991).

4.2.2 The solar neighborhood

The structure of our galaxy can be divided into three distinct kinematial regions that are defined by particular ratios of rotational velocity V_c to velocity dispersion σ_v. These regions are known as the thin disk, the thick disk and the halo. Their vertical density distribution, $\nu(z)$ assumes the form

$$\nu(z) = \nu_0 \exp(-z/z_h) \exp(-R/R_h) \tag{4.7}$$

where z_h is the vertical disk scale height and R_h is the radial disk scale length. This exponential form can be derived by assuming an infinitely thin disk (which is justified by the observations) together with an isothermal velocity distribution. In the case of a self-gravitating disk $\nu(z)$ goes as sech^2. The self-gravity in this case is provided by the sum of the stellar distribution and the dark matter distribution. The solar neighborhood is a region of radius roughly 300 light-years that contains a few thousand stars. This region contains thin disk, thick disk and halo stars, and their normalization is important to the determination of the mass density within this region.

- *The thin disk*: This is a coherent and highly rotating component. The average value of σ_v/V_c is ≤ 0.1 and the scale height is $\approx$ 100 pc. Since the total disk diameter is $\approx$ 30 kpc the thin disk has a thinness ratio of $\approx$ 0.003. Most of the molecular gas, hence most of the massive star formation in the Milky Way, is concentrated in the thin disk.

- *The thick disk*: The existence of this component is still somewhat uncertain, although it seems to be required to account for the observed star counts in the solar neighborhood as well as the observed stellar metallicity as a function of height above the galactic plane. Zinn and West (1984) also identified a population of globular clusters that has disk-like kinematics. Estimates of z_h range from 0.7 to 1.1 kpc with some as high as 1.5 kpc. These estimates vary according to the particular sample of stars that is used. A fair tracer of the thick disk population has not been unambiguously identified (von Hippel and Bothun 1993). The value of σ_z, however, is well determined at 45 ± 5 km s^{-1}, making $\sigma_v/V_c \approx 0.25$. Unfortunately, the thick disk is sometimes confused with the old thin disk, which has $z_h \approx 300$ pc. The higher value of z_h for the old thin disk stars is a consequence of their many orbits about the center of the galaxy and the cumulative effects of small gravitational scattering off other stars (Chapter 3). This acts like a diffusion process which causes the stars originally born in the very thin disk to diffuse vertically.
- *The bulge and the halo*: These are the spheroidal components of galaxies that are supported by internal velocity dispersion instead of large-scale rotation. The orbits of the stars can be anisotropic, so the spheroidal component is not necessarily always round. The bulge portion of the spheroid is a regime of high stellar density over a fairly small scale. The halo is best defined by the GCS of the galaxy. It is a large, very low density collection of old stars. These stars were the first to form in the galaxy as it was collapsing (Chapter 5). There is very little rotation in the galactic halo, so σ_v/V_c substantially exceeds 1. The overall scale size of the halo component is difficult to ascertain as there are few stars which trace it. For spheroidal systems, the falloff in surface density is

$$\Sigma_r \propto (r/r_e)^{1/4} \tag{4.8}$$

where r_e is the effective radius defined to enclosed 50% of the light. For the high density bulge, r_e is likely to be less than 2 kpc (Gilmore, Wyse, and Kuijken 1989; Sackett 1997). For the halo, r_e lies in the range 3–7 kpc (Bothun *et al.* 1991).

The solar neighborhood lies about 8.5 kpc from the center of our galaxy (Sackett 1997). At that distance, there are very few, if any, true bulge stars. Hence, the solar neighborhood stars are a mixture of thin disk, thick disk and halo stars. All three components have different kinematic and metallicity distributions and it is possible to assign nearby stars to these three components on the basis of observations. Halo stars in the solar neighborhood are extremely rare (hence the difficulty in finding the Population II main sequence from a trigonometric parallax sample) with roughly 1 out of every 500–800 stars belonging to this population. But normalization of the thick disk is quite uncertain; values of 2–15% are consistent with various samples. There is covariance between the determination of the normalization and z_h, with larger values of z_h producing lower normalizations. This covariance occurs because the fitting procedure keeps a relatively constant total of stars in the thick disk.

In a highly flattened rotating stellar system, the density distribution in the vertical (z) direction, is a measure of the surface mass density. This situation arises because Poisson's equation for a flattened system assumes the form

$$\frac{\partial^2 \phi}{\partial z^2} = 4\pi G\rho \tag{4.9}$$

As the density increases, the z-coordinate sees a larger derivative in the potential, which means it experiences a larger gravitational restoring force in that direction. In practice this gravitational restoring force can be estimated by measuring z_h and the vertical velocity dispersion σ_z for some well-defined sample of stars. This transformation from equation (4.9) to observables makes use of a variant of the collisionless Boltzmann equations. Since stars are not escaping from this system, the collisionless Boltzmann equation can be combined with the equation for continuity of mass to yield (Binney and Tremaine 1987):

$$\frac{\partial}{\partial z}\left(\frac{1}{\nu}\frac{\partial(\nu<\sigma_z>^2)}{\partial z}\right) = -4\pi G\rho \tag{4.10}$$

Combined with the vertical velocity dispersion, measurements of the density distribution of stars in the z-direction, $\nu(z)$, will constrain ρ.

The first attempt to utilize equation (4.10) to constrain the solar neighborhood mass density was by Jan Oort in 1932, who derived a mass density of $0.15 M_\odot$ pc^{-3}. Of course, Oort had no knowledge of the presence of another kinematic system, the thick disk, so he explicitly assumed that his sample of F stars and K giants was tracing out a few thin-disk scale heights. Under that assumption, Oort found that the equivalent surface density of the thin disk in an imaginary cylinder of height 700 pc was $90 M_\odot$ pc^{-2}.

A more recent determination of these parameters was made in a series of papers during the early 1980s by John Bahcall and collaborators. Again using samples of A and F stars, they derive estimates for ρ_0 of $(0.18\text{–}0.21) M_\odot$ pc^{-3} and an equivalent surface mass density out to 700 pc of $75 M_\odot$ pc^{-2}. The measured luminosity function of stars in this imaginary cylinder gives an equivalent surface luminosity density of $(15 \pm 2) L_\odot$ pc^{-2}. This is equivalent to a mean blue surface brightness of ≈ 23.0 mag arcsec^{-2}. Bahcall's sample gives a mean M/L in the solar neighborhood of 5.

Testing for the presence of dark matter in the solar neighborhood now becomes an accounting problem The possible sources of this mass in the solar neighborhood are (1) luminous stars, (2) interstellar gas, (3) stellar remnants (mostly white dwarfs) and (4) dark matter.

- *Luminous stars*: Determinations of the local luminosity function of stars can yield a mass function if the bulk of the stars are main sequence stars where the relation between luminosity and mass is well determined. This is the case in the solar neighborhood. The observed value of $\rho = 0.044 M_\odot$ pc^{-3} is well short of the required value (e.g., Gilmore, Wyse, and Kuijken 1989).

- *Gas*: Since stars form from gas with an overall efficiency of 1–20% (depending upon many factors), every star formation event should leave behind plenty of gas. The local volume mass density of gas has been measured at $\rho = 0.042 M_\odot$ pc^{-3}; not surprisingly, this is almost identical to the stellar value. The consideration of stars and gas accounts for one-half the inferred volume mass density.
- *Stellar remnants*: Since stellar remnants are no longer producing energy, they necessarily have high values of M/L and are hard to detect. This makes an accurate determination of their local space density quite difficult. Models of star formation in the solar neighborhood, combined with estimates of the initial mass function and mass loss rates in post main sequence evolution can be used to predict the remnant density. The detectability of white dwarfs also depends upon their cooling rate, so the inferred remnant density strongly depends upon the assumed age of the galactic disk in the solar neighborhood. None of these parameters is very well determined. Current values of $\rho = (0.01\text{–}0.03) M_\odot$ pc^{-3} are consistent with the observations, though most determinations cluster around the upper end. A reasonable upper limit to the density of stellar remnants is $\rho = 0.044 M_\odot$ pc^{-3}, since the galaxy is not yet old enough to have a mean remnant mass density higher than the observed mass density of low mass stars.

Adding the observed densities of stars, gas and stellar remnants gives ρ_0 in the range $(0.096\text{–}0.13) M_\odot$ pc^{-3}. This implies at least $0.05 M_\odot$ pc^{-3} of dark matter, or roughly one-third of the total dynamically inferred volume mass density. This is taken as evidence for significant amounts of dark matter in the solar neighborhood. However, we have ignored the thick disk contribution to the observed column of $75 M_\odot$ pc^{-2}.

In a long series of papers, Gerry Gilmore and collaborators have taken this assumption to task and have persuasively argued that omission of the thick disk has serious consequences on the question of dark matter in the solar neighborhood. In their most recent analysis, Kuijken and Gilmore (1991) derive a surface mass density of $(71 \pm 6) M_\odot$ pc^{-2} (i.e., the *same* as the Bahcall value) but conclude that only $(48 \pm 9) M_\odot$ pc^{-2} is associated with thin disk material near the sun. The remaining third is associated with the thick disk and halo, which has not been accounted for in our previous census. It is difficult to find fault with this analysis, hence the establishment of the thick disk has essentially resolved the Oort limit problem, thus removing any credible evidence for the presence of dark matter in the solar neighborhood.

4.2.3 Galactic scales

Mass distributions and predicted rotation curves

Figure 4.2 shows an image of a typical edge-on spiral galaxy where the thin disk and central bulge concentration are evident. If the distribution of light traced the distribution of mass, we would expect a high mass concentration that corresponded with the bulge light. Then the galaxy would be similar to the point mass approximation

Figure 4.2 CCD image of the nearly edge-on spiral NGC 253 taken by the author. Note the relatively thin disk and bright light concentration.

that governs Solar System orbits. From the virial theorem, $V_c^2 \propto M/R$, hence orbital velocity should decline as $R^{-1/2}$.

Rotation curves of galaxies were first systematically studied by Vera Rubin and her colleagues, starting in the late 1960s. Those initial observations showed that no spiral galaxies exhibited a rotation curve which scaled as $R^{-1/2}$. Instead, rotation curves were mostly flat with R. A typical rotation curve for a spiral galaxy is shown in Figure 4.3. It shows that the point mass approximation is invalid and another form for the potential is required.

The luminosity profiles of disk galaxies exhibit an exponential falloff. Does this provide a clue to the form of the potential? If the mass distribution is also exponential then we have

$$M(r) \propto \exp(R/r_h) \tag{4.11}$$

For this exponential mass distribution there is some value of R which maximizes this expression

$$V_c^2 \propto \frac{1}{R}\exp(R/r_h) \tag{4.11a}$$

This equation can be solved numerically to show that the maximum rotational velocity occurs at $R_{max} \approx 2.2r_h$. Strictly speaking, this result is only valid for a

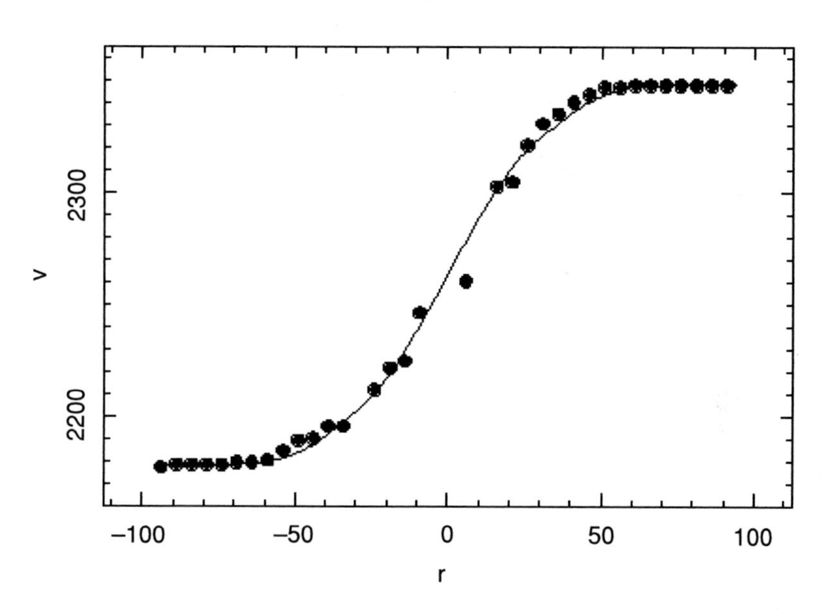

Figure 4.3 Typical optical rotation curve for a disk galaxy that shows a sharp rise then a relatively flat region out to the optical radius. Data courtesy of E. de Blok, University of Groningen.

spherical distribution of mass. But a flattened distribution yields a similar value for R_{max} and a peak rotational velocity which is 15% higher than the exponential sphere (Binney and Tremaine 1987, Fig. 2.17).

For the general case of a rotating system where the virial theorem applies, the mass enclosed by some radius r is given in equation (4.6f). For the point mass approximation, v_c goes as $r^{-1/2}$ and there is no dependence of $M(r)$ on r. Flat rotation curves indicate that v_c is not a function of r; hence $M(r)$ increases with scale. Since galaxies are obviously finite in mass, there must be a limit to this increase. Moreover, the light profile of galaxies decreases in intensity as r increases. Hence flat rotation curves demand the presence of an extended mass distribution which is not reflected in the light distribution. This extended mass distribution is generally assumed to take the form of a spherical halo. In the derivation of the dynamical friction timescale (Chapter 3) we made use of a specific halo density distribution

$$\rho(r) = \frac{1}{4\pi r^2}\frac{dM(r)}{dr} \tag{4.12}$$

Since $M(r) = v_c^2 r/G$ then $dM(r)/dr = v_c^2/G$, so that

$$\rho(r) = \frac{v_c^2}{4\pi G r^2} \tag{4.12a}$$

For this halo we also specify $M(r)$ goes as $<\rho> R^3$. Substituting this into equation (4.12a) recovers equation (4.6f).

The kind of potential that can produce the density distribution of equation (4.12) is often called an isothermal sphere. In order to achieve a balance between outward pressure and inward gravity, an isothermal sphere must satisfy the equation of hydrostatic equilibrium

$$\frac{dp}{dr} = -\rho \frac{GM(r)}{r^2} \tag{4.13}$$

For an ideal gas composed of one particle of mass m, we have $p = nkT$ and $\rho = nm$. Therefore

$$\frac{d\rho}{dr} = -\rho \frac{GM(r)}{r^2} \left(\frac{m}{kT} \right) \tag{4.14}$$

where T is the constant temperature. The usual way to solve this differential equation is to multiply both sides by r^2/ρ and differentiate each side with respect to r. Since this is a sphere, $M(r) = 4/3\pi r^3 \rho$ and $dM(r)/dr = 4\pi r^2 \rho$, so we have

$$\frac{d}{dr}\left(r^2 \frac{d \ln \rho}{dr} \right) = -\frac{Gm}{kT} 4\pi r^2 \rho \tag{4.15}$$

For this differential equation we can try a solution of the form $\rho = Cr^{-n}$. For the left-hand side we have

$$\frac{d}{dr}\left(r^2 \frac{d \ln Cr^{-n}}{dr} \right) = \frac{d}{dr}\left(r^2 \frac{1}{Cr^{-n}} - nCr^{-(n+1)} \right) = \frac{d}{dr}\left(r^2 \frac{-n}{r} \right) = -n \tag{4.16}$$

The right-hand side is

$$-\frac{Gm}{kT} 4\pi r^2 Cr^{-n} = -\frac{Gm}{kT} 4\pi Cr^{2-n} = -n \tag{4.16a}$$

which is solved as

$$C = \frac{kT}{2\pi Gm}, \qquad n = 2 \tag{4.16b}$$

Hence $\rho \propto r^{-2}$ gives rise to a potential in which v_c does not depend on r. In general, the power law exponent in the density distribution does not have to be −2 to produce a flat rotation curve, and the generalized halo density profile is

$$\rho(r) = \frac{\rho(0)}{(1 + (r/r_c)^n)} \tag{4.17}$$

where $\rho(0)$ is the central mass density and r_c is the core radius of the halo. A well-defined rotation curve, in which the luminous contribution to the mass distribution has been accounted for, can constrain $\rho(0)$, r_c and n.

Colour Plates

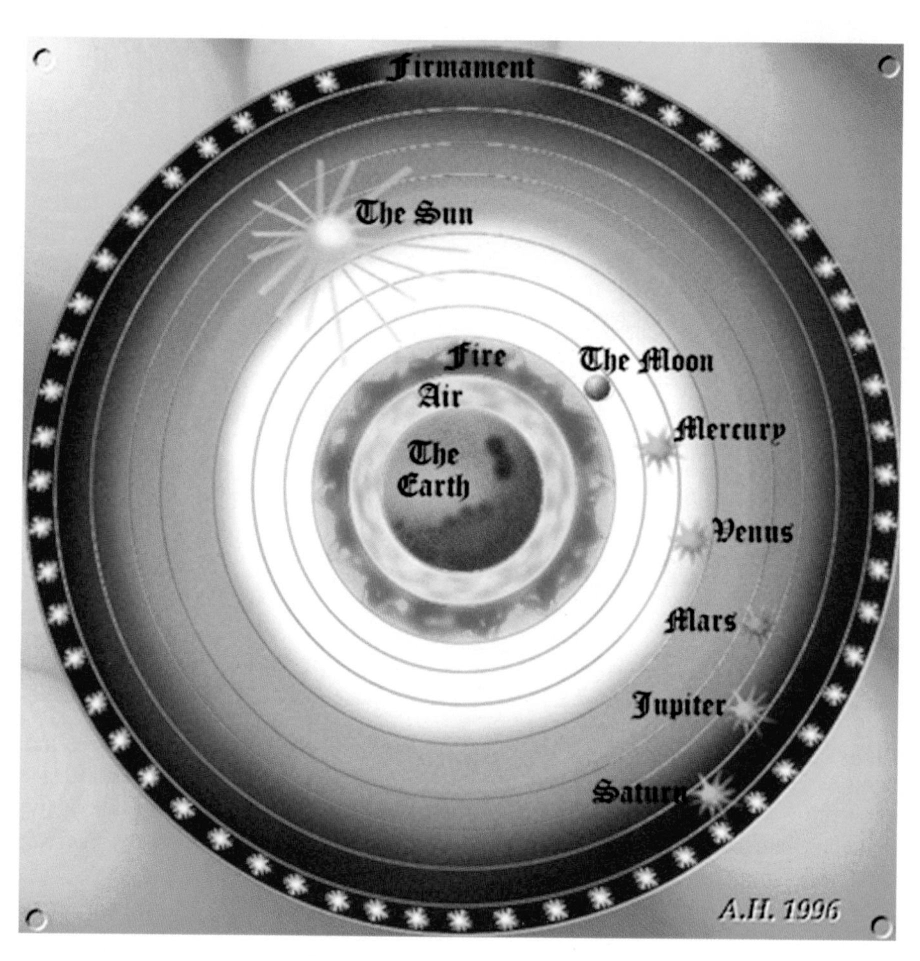

Figure 1.1 Representation of the Aristotelian system of perfect crystalline orbs in which the Earth defined the center and each of the planets, the Sun and the stars were on their own sphere in orbit about the Earth.

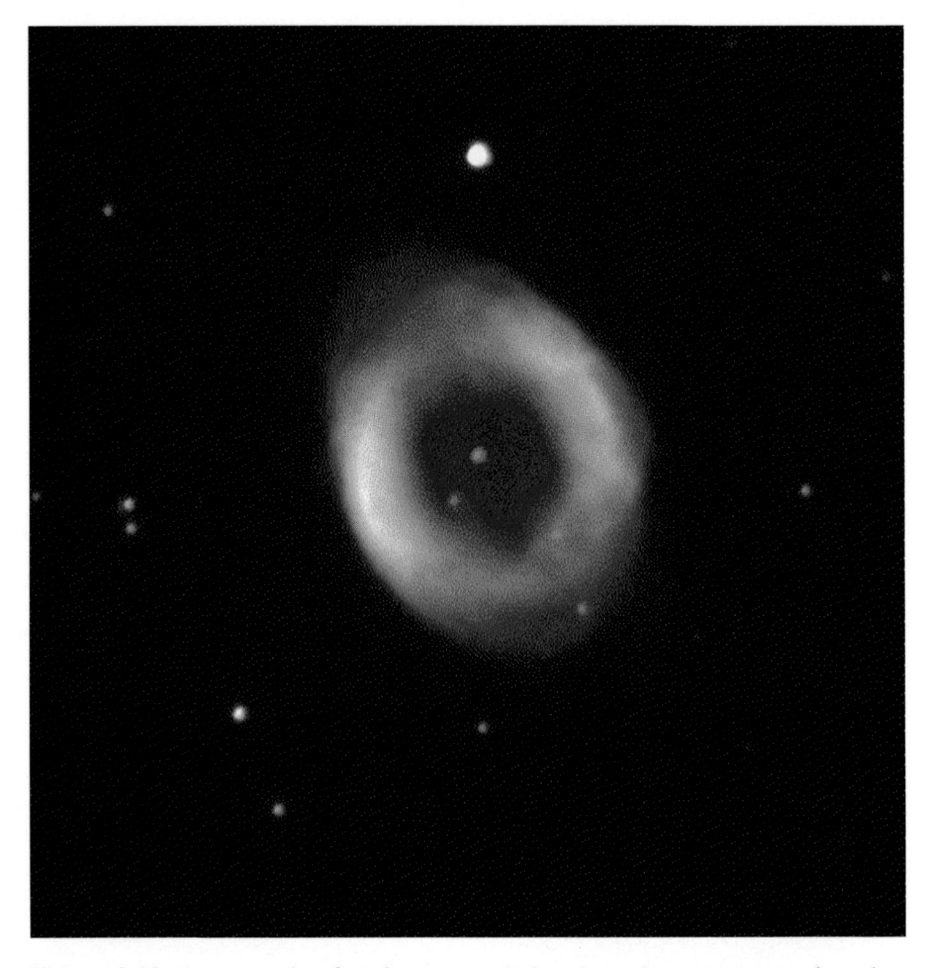

Figure 2.11 An example of a planetary nebula where the outer atmosphere has been shed and is now ionized by the central star. The mass of the central star determines the ionization rate and the subsequent emission line luminosity of the planetary nebula. This image is a true-colour CCD image obtained by Nelson Caldwell, Smithsonian Astrophysical Observatory.

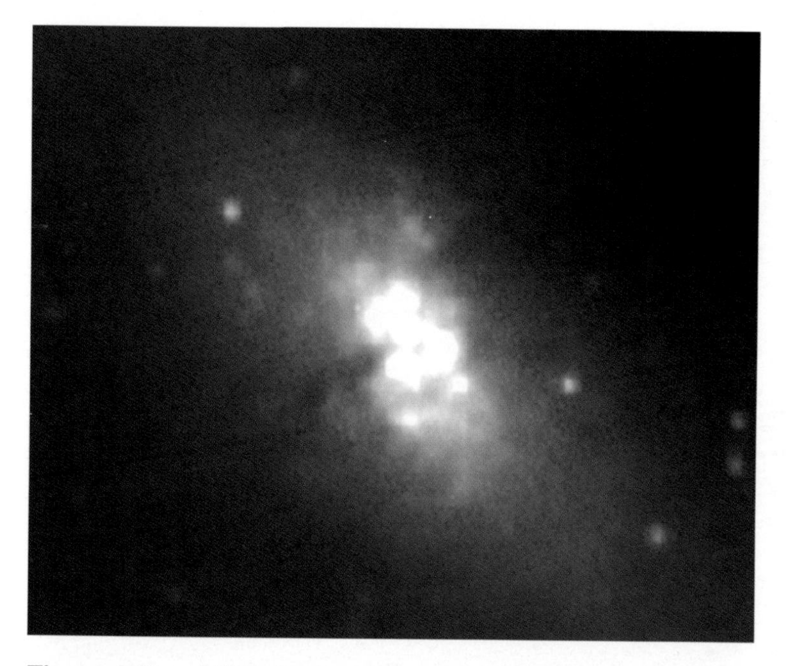

Figures 2.16 and 2.17 Images of the two nearby galaxies in which an SN la has occurred and for which there exist Cepheid-based distance determinations. Figure 2.16 (top) is a high contrast rendering of NGC 5253 in the red; the central region is dominated by blobs of star formation. Figure 2.17 (bottom) is of NGC 5128, a famous galaxy noted for its very peculiar morphology; many have suggested this galaxy is a merger remnant. Figure 2.16 is courtesy of Crystal Martin, Space Telescope Science Institute; Figure 2.17 is courtesy of Dr William Keel, Department of Physics and Astronomy, University of Alabama.

Figure 2.23 CCD image showing the double-quasar Q 0957+561. Obtained by Dr Rudolph Schild, Smithsonian Astrophysical Observatory and reproduced with permission.

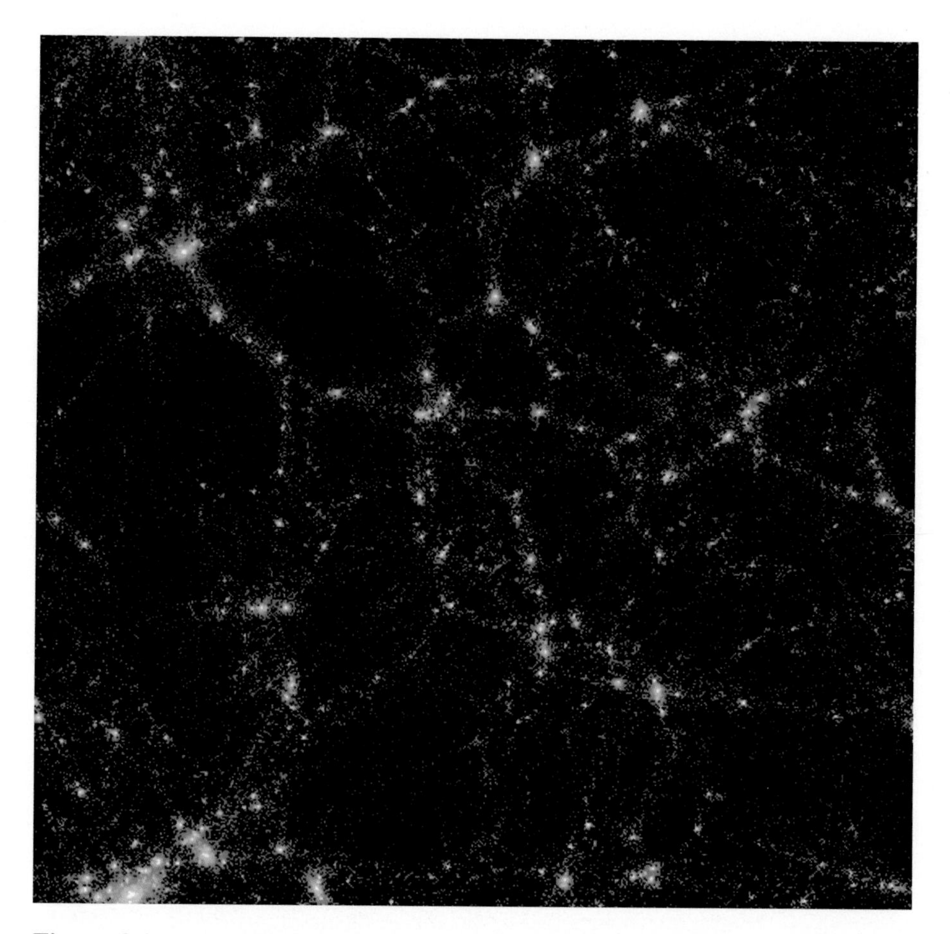

Figure 3.1 Large-scale structure in a dark matter supercomputer simulation: a void-filled universe with much filamentary structure, where clusters of galaxies appear to form at the intersections of voids. Courtesy of the HPCC group at the University of Washington, George Lake, and Tom Quinn.

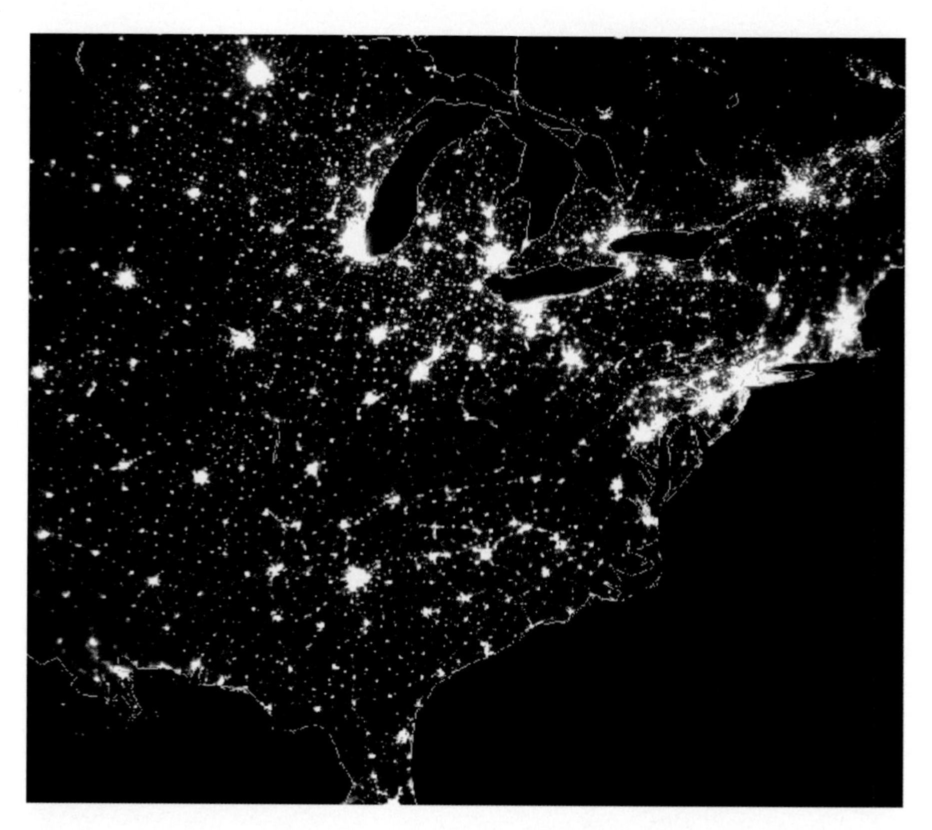

Figure 3.10 Hierarchical clustering in an image of the nighttime distribution of lights over the East Coast of the United States. Image courtesy of NOAA and reproduced with permission.

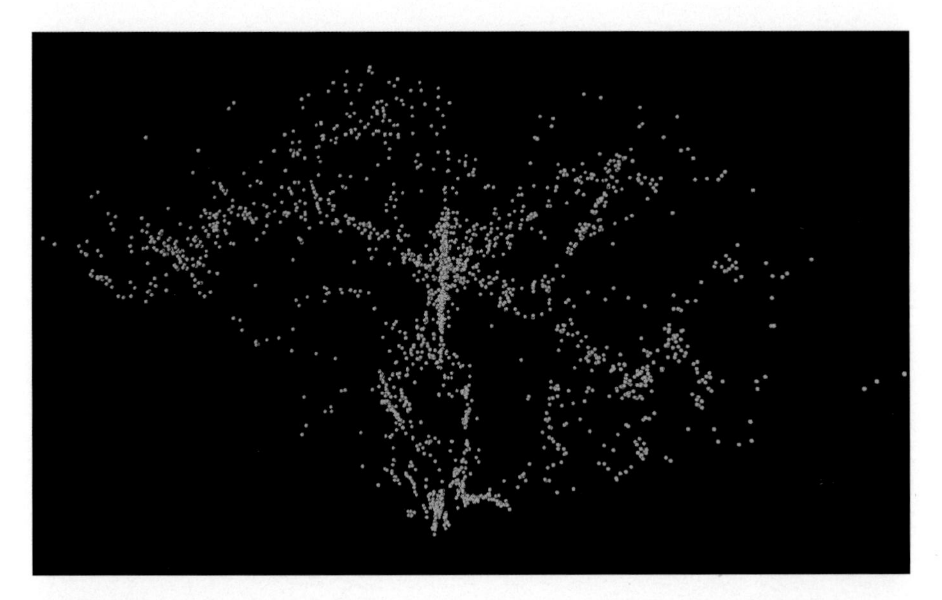

Figure 3.11 Large-scale structure is seen in a slice of the Universe first published by de Lapparent, Geller and Huchra (1986). The opening angle of the vertex represents the angular extent of the strip survey and the width in declination has been collapsed. Each galaxy has been plotted at its redshift distance from the Earth. Virilized structures, such as the Coma cluster in the center of the image, appear as linear features pointed directly at the observer. This representation of the large-scale structure clearly reveals the presence of voids. Courtesy of Dr Margaret Geller, Harvard-Smithsonian Center for Astrophysics.

Figure 5.3 The spectacular Hubble deep field showing approximately 1500 distant galaxies in a 2.5×2.5 arcminute field. This image was created with the support of the Space Telescope Science Institute, operated by the Association of Universities for Research in Astronomy, Inc., from NASA contract NAS5-26555, and is reproduced with permission from AURA/STScl. Digital renditions of images produced by AURA/STScl are obtainable royalty-free.

Observational evidence for flat rotation curves

The unambiguous identification of rotation curves in which v_c does not decline as a function of r provides very powerful evidence for the presence of a mass distribution that is like an isothermal sphere. For the case of an exponential mass distribution, the surface brightness at R_{max} is down by approximately 2.5 mag arcsec^{-2} relative to $\mu(0)$. For a typical disk galaxy this corresponds to a mean blue surface brightness level of $\approx$ 24.0 mag arcsec^{-2}, which is 1.5 mag fainter than the sky. This makes detection of stellar absorption lines and the subsequent determination of stellar rotational velocities at that radius very difficult. Without data beyond R $\approx 2.5r_h$, flat rotation curves by themselves do not provide good evidence for an extended halo mass distribution that dominates the disk dynamics. The data is still consistent with a simple exponential mass distribution. To make further progress requires the construction of rotation curves that reach well beyond $2.5r_h$ using one of the following techniques.

Optical emission lines usually arise from the ionization of hydrogen by hot stars. If there is sufficient star formation in some spiral galaxy at $r \geq 2.5r_h$, then that galaxy's rotation curve will provide a good check on the existence of an extended mass distribution. In general, there are few spirals that have such an extended region of star formation. Observations of about 100 such objects by Schommer *et al.* (1993) have revealed a mixed collection of rotation curves. Some are flat, some are still rising at the last measured point and some are falling, which indicates an end to the mass distribution. Figure 4.4 shows some examples of these kinds of rotation curves. This data is consistent with an extended halo mass distribution but not conclusive.

Conference proceedings of the late 1970s and early 1980s often contain a lively debate on the existence of dark halos around spiral galaxies. At this time, there were about 30 rotation curves of high quality but few were measured at points greater than $3r_h$, hence a lively debate fueled by the paucity of relevant data. During the mid 1980s radio interferometers such as the Westerbork Array and the Very Large Array in New Mexico began to make good two-dimensional maps of the distribution of atomic hydrogen in spiral galaxies. Up to then it had been well established that many late type spiral galaxies showed evidence for extended gas distributions relative to the stellar distribution. In some cases the gas distribution could be measured out to a radius of $\approx 10r_h$. Measurements of the extended neutral hydrogen distribution in many spiral galaxies indicated flat rotation curves over scales of $(5-10)r_h$. To date, this remains the strongest evidence for the existence of extended dark matter halos, although there are two important caveats:

- The self-gravity of the gas in these extended distributions cannot be ignored in constructing a mass model. This has important implications for the measured value of r_c.
- This result cannot strictly be generalized to other galaxies, because it is possible that the only galaxies which still have extended gas distributions are also galaxies which have extended halos.

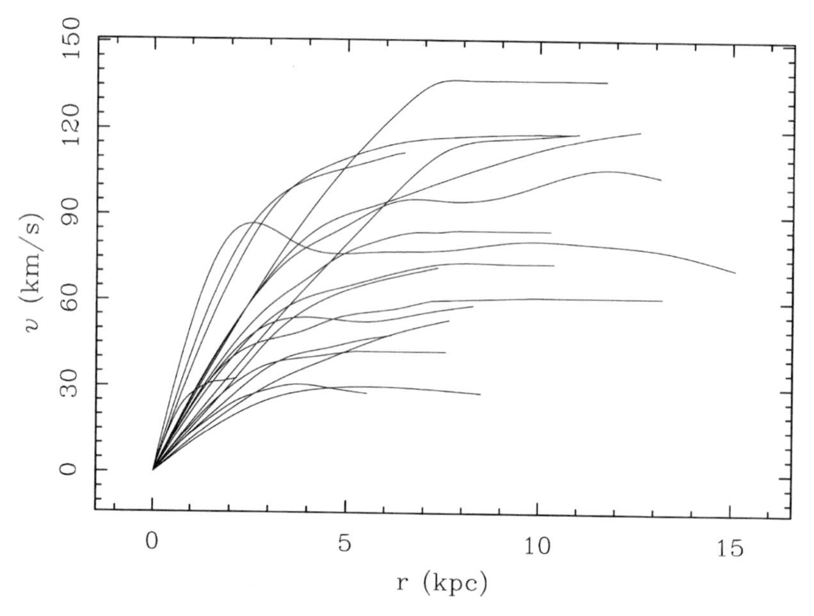

Figure 4.4 Collection of rotation curves from a variety of galaxies of varying circular velocity. All generally show a steep rise and a flat region, but there is significant variation about this basic structure. Courtesy of E. de Blok, University of Groningen.

4.3 Other dark matter indicators for galaxies

4.3.1 Hot halos around elliptical galaxies

X-ray observations of elliptical galaxies show that many have extended halos of X-ray emission whose overall size is a few times the optical radius of the galaxy. The origin of the hot gas that fills the halo is likely to come from within the optical radius. This hot gas has been driven out of the central potential, in which most of the stars are located, as a result of supernova heating of the cold gas from which those stars formed (Matsumoto *et al.* 1997). Estimates of the ratio of hot gas mass to surviving stellar mass in these ellipticals range from 0.1 to 1.0 (Forman, Jones, and Tucker 1985). As ellipticals are observed today to have very low fractional mass contents of cold gas, it seems likely that after star formation had ceased, any remaining cold gas was heated and dynamically relocated to this hot halo.

These X-ray halos have observed radial gradients in X-ray flux that indicate the mass of halo gas is in hydrostatic equilibrium with the galactic potential. Although this X-ray gas is being collisionally cooled and falling back into the galaxy, the observed inflow velocity is much less than the ambient sound speed, so the condition of hydrostatic equilibrium is still applicable. But unlike the case of the isothermal sphere, there will be a temperature gradient; hence the mass distribution $M(r)$ can only be recovered if the density *and* the temperature distributions are known. In general, the X-ray observations of the *Einstein* satellite are of insufficient quality to define the temperature profile, so $M(r)$ remains unknown, as does the actual mass of hot gas. For the extended X-ray halo around M87, the data is of high quality and indicates a mass of $\approx 10^{13} M_\odot$ out to a radius of 150–200 kpc (Feigelson *et al.* 1987; Mould, Oke, and Nemec 1987). This exceeds the stellar mass of M87 by at least an order of magnitude. In the general case of these halos, the condition of hydrostatic equilibrium and their very large extent indirectly suggest that the hot gas is prevented from escaping (and is therefore bound to the galaxy) by a dark matter halo.

4.3.2 Binary galaxies

Binary galaxies, whether they are spiral–spiral or elliptical–spiral, potentially offer the best means for measuring the total masses of galaxies. Flat rotation curves can never provide a measure of the total mass of a galaxy. But binary galaxies may be thought of as test particles that can be used to extend the rotation curve to large distances. This technique was realized as early as 1937 by Holmberg, who first applied it to a sample of binary galaxies to derive rather ambiguous and confusing results. Although the technique has much promise, there are several technical difficulties:

- Because of small-scale clustering, from a radial velocity catalog it is difficult to identify a sample of binary pairs that are physically bound into a mutual orbit. Many binaries will simply be unbound projections. Treating those cases as if they were true binaries will cause a clear bias.
- Small-scale clustering of galaxies also means that an individual galaxy may feel the gravitational tug of more than just one nearby galaxy.
- Only one projection of the relative velocity between the two galaxies can be measured. Furthermore, the orbits may be highly radial instead of circular; this seriously affects the relation between the measured mean velocity dispersion for binary pairs in some sample as well as the derived mass.
- Dynamical friction effects are potentially operative in binary galaxies; the observed relative velocities may reflect this frictional drag process instead of the total mass.

To date, the literature contains a rather large dispersion of results for binary galaxies. These results clearly depend on choice of sample and assumptions about orbits. However, the data is generally inconsistent with the point mass representation of a galaxy and suggests that galaxies have extended mass distributions.

A significant improvement on the binary galaxy approach has been made by Zaritsky *et al.* (1997), who have used a well-identified sample of small satellite galaxies in orbit about one isolated large galaxy. The isolation criterion simplifies the dynamical analysis and allows for an ensemble average of all satellites. For a sample of 115 satellites located around 69 spiral host galaxies, Zaritsky *et al.* derive a characteristic mass of $2 \times 10^{12} M_\odot$ and halo radius of ≈ 200 kpc for a luminous spiral. The data provides no evidence for a decrease in velocity dispersion out to galactocentric radii as large as 400 kpc. To date, this remains the best evidence that spirals are surrounded by very large dark halos. The interesting aspect of the Zaritsky *et al.* result is the very large halo size indicator that the actual space density of the dark matter in the galactic potential is low, much lower than the stellar density. For instance, a typical spiral galaxy has mass $2 \times 10^{11} M_\odot$ of stars confined to a disk (cylinder) of radius 15 kpc and thickness ≈ 1 kpc. The corresponding density is $\approx 0.3 M_\odot$ pc^{-3}. For a spherical dark matter halo of mass $2 \times 10^{12} M_\odot$ and radius 200 kpc, the corresponding density is $\approx 6 \times 10^{-5} M_\odot$ pc^{-3}.

4.3.3 Stellar population effects

The data discussed above has produced a large range of estimated M/L values for individual galaxies. Before discussing that, it is worthwhile to understand the range of M/L values that can arise solely from differences in stellar populations. The mass distribution of stars that arises from star formation is fitted reasonably well by a power law over the mass range M_u and M_l where the u and l subscripts refer to upper and lower mass limits. A parameterization of this power law is

$$\frac{dN}{dM} = AM^{-(x+1)} \tag{4.18}$$

where A is a normalization function. The sense of this relation is that larger values of x produce preferentially more low mass stars. The M/L value for a stellar population depends upon both x and M_l. In fact, the dependence on M_l is most critical. The sense of any power law distribution is to make a few big things and a lot of little things. The minimum mass required for sufficient core temperature to initiate nuclear reactions is $0.07 M_\odot$. If equation (4.18) holds down to $M_l = 0.01 M_\odot$, then many objects of substellar mass can be produced in a star formation event and a significant fraction of the total mass which is formed will be stored in an essentially unobservable form. These objects of substellar mass are known as **brown dwarfs**. Hence, knowledge of the faint end slope of the stellar luminosity function is directly relevant to the problem of dark matter in galaxies.

Detailed studies of faint star counts by Neil Reid and colleagues give credible evidence that a single value of x is not an appropriate description of the mass function once the mass gets below $\approx 0.3 M_\odot$. The data indicates that the mass function tends to flatten out; moreover, despite intensive observational searches, there is no evidence that objects of $0.01 M_\odot$ exist in significant numbers. In fact, only two good brown dwarf candidates have been discovered up to now (Nakajima *et al.* 1995). Interestingly, the recent discovery of planets around nearby stars by Geoff Marcy and his collaborators (Bulter *et al.* 1997) have indicated that $0.001 M_\odot$ objects are common. Hence there appears to be a real astrophysical gap between $0.001 M_\odot$ and $\approx 0.07 M_\odot$.

A description of how x is determined would fill more pages than in the whole of this book. Suffice it to say that values of 1–1.5 are consistent with the data. But there may be a circular argument involved in this determination. A value of x in the range 1–1.5 will populate the mass range $(1–2)M_\odot$. These stars have lifetimes of 10^9 to 10^{10} years and will dominate the light of a galaxy when they are either A main sequence stars or red giants. Since the lifetime of red giants and A main sequence stars is only about 10% of the total stellar age of the galaxy, then at any epoch these galaxies will have their light dominated by only a small percentage of the total stars. Moreover, as the aggregate brightness of thousands of red giants is large, the galaxy itself will be fairly luminous and easily detectable. Determinations of x are repeatedly based on samples of easily detectable galaxies, thus recovering the range 1–1.5. Since we don't understand the physics that produces x and M_l, it is worthwhile to consider two alternative star formation scenarios that would produce "dark" galaxies:

- *The low mass star dominated galaxy*: The percentage of mass that forms in stars with masses lower than $M_\odot$ is strongly dependent on x; for x in the range 1–1.5 it varies between 40 and 70%. And where x is as steep as 2.5, the percentage climbs to 95%. In this case, most of the stellar mass is in a form where the main sequence lifetime exceeds H_0^{-1} and such a galaxy will have few, if any, red giant stars. In this case, at fixed mass, when observed at the current epoch, the galaxy would have 10^3 to 10^4 times less luminosity, depending upon what M_l is and if the power law can really be extended all the way down to that mass. Such a galaxy would therefore be very red, very diffuse, have a very high M/L and be almost impossible to detect (Chapter 6). Since this galaxy never had any massive stars, it would also be quite deficient in heavy elements.
- *The remnant-dominated galaxy*: The opposite case is one where $x \leq 0$ and/or M_l is $\geq M_\odot$. In this case, star formation places most of the mass in stars with masses greater than $2M_\odot$. These are very bright initially, but don't live very long. Furthermore, the energy feedback from the massive stars and their subsequent supernova phases of evolution may well be sufficient to drive the remaining gas completely out of the galaxy. This is a case of terminal star formation in that the star formation event has been so vigorous that the remaining cold gas has been heated (perhaps to escape velocity), thus preventing further star formation. The lack of a substantial reservoir of low mass stars in this scenario means that the galaxy has a luminous phase which lasts only about 10% of a Hubble time. After that phase has ended, the galaxy is destined to fade rapidly and end up extremely diffuse, perhaps somewhat blue, with very high M/L as it would be dominated by stellar remnants (e.g., white dwarfs, neutron stars and black holes). As discussed in Chapter 6, deep galaxy counts and redshift surveys provide evidence that a population of blue galaxies at intermediate redshift may have faded to become a "dark" galaxy at $z = 0$.

4.3.4 Some measured values for *M/L*

When measuring M/L it is convenient to refer to a common band. Let us choose the blue band as the L reference. The determination of M/L depends on distance because M goes as V_c^2R and L goes as $4\pi R^2F_g$, where F_g is the observed flux (photons per square centimeter per second). For reference, the global M/L required to close the Universe for our probable range of H_0 is ≈ 2000 in the B band. The stellar population of the solar neighborhood has $M/L \approx 2$. A 10 Gyr solar metallicity population with x in the range 1–1.5 has $M/L \approx 10$. We therefore arrive at a very significant conclusion. If the Universe is closed, then the luminous portions of galaxies contribute at most $10/2000 = 0.5\%$ of the closure density.

The derived values of M/L from analysis of rotation curves, X-ray halos and binary galaxies consistently give results in the range 10–200. Most spiral galaxies average around 10–30 when r is restricted to $(3-4)r_h$. This means that if the Universe is closed, the combined light and dark mass in galaxies contributes only 10% (at most) of the closure density. On the other hand, if Ω can be determined as 0.1 (see below) then it's conceivable that all the mass in the Universe is located in the halos that define the potential wells of galaxies.

4.3.5 Clusters of galaxies

Clusters of galaxies are a few dynamical timescales old and we can apply the virial theorem to them. From this, a simple scaling argument suggests that clusters of galaxies must have more dark matter in them than individual galaxies. Virial masses scale as v_c^2R. For clusters of galaxies v_c^2 is replaced by the velocity dispersion σ_v and R refers to cluster radius, a somewhat ill-defined quantity. For galaxies, v_c is ≈ 250 km s^{-1} on a scale $R \approx 10$ kpc. For clusters σ_v is ≈ 1500 km s^{-1} on a scale $R \approx 1$ Mpc. To reproduce this much cluster mass from the sum of individual galaxies, each with $v_c = 250$ km s^{-1}, would therefore require a total population of 3600 within $R = 1$ Mpc. Real clusters, however, only have a few hundred galaxies with masses appropriate to $v_c = 250$ km s^{-1}, hence there is a notable excess of matter in clusters compared with the individual galaxies. Accounting for the observed hot X-ray gas in clusters only gains an extra factor of 2 in observed mass.

So far we have just blindly applied the virial theorem $(2T + W = 0)$ to gravitational potentials to derive masses. In this application we have not been very picky over what value of scale factor r to use. There are, however, two different physical situations which can arise: (1) a system of massless tracer particles moves through a large potential or (2) a self-gravitating system of N particles where the mass of the system is contained in those N particles. For the first case, the radius of the potential as defined by the tracer particles is appropriate to determine r. In the second case, the mean separation between the particles is a better choice to determine r. A compromise between these two extremes is to choose a quantity called the median radius r_m, which is the radius that encloses half of the system's mass. In most simple stellar systems $r_m \approx r$, so that

$$\langle v^2 \rangle \sim 0.4\frac{GM}{r} \tag{4.19}$$

which is only sightly different than the coefficient of 0.5 found in equation (4.6f).

In astrophysical units, equation (4.6f) can be expressed as

$$M_t = 4.5 \times 10^{11} \left(\frac{\sigma_v}{\text{km s}^{-1}} \right)^2 \left(\frac{R_{hms}}{\text{Mpc}} \right) M_\odot \tag{4.20}$$

Application of equation (4.20) to real clusters of galaxies has three potential sources of systematic error which can affect the derived mass.

1. Although the virial theorem is valid after only one dynamical timescale, the observational estimator σ_v of the true RMS velocity is more reliable if the system is several dynamical timescales old. Since the epoch of initial cluster collapse is not well known (Chapter 5), clusters of galaxies may only be 1–3 dynamical timescales old. The noise in the σ_v estimator depends upon $\sqrt{N}$, so this is mostly a problem for groups or small clusters.
2. Usually $\langle R_{hms} \rangle$ is calculated from the galaxy (e.g., the light) distribution in the cluster. If there is no bias between the distribution of light and mass on the scale of the cluster, this will provide a proper estimation of the length scale to be used. However, N-body simulations of cluster formation which are dominated by dark matter often show that the luminous galaxies (the baryonic material) are more centrally condensed than the dark matter distribution. Presumably this is because the baryonic material is more dissipative than the particular brand of dark matter that was chosen. If this is the case, then using $\langle R_{hms} \rangle$ from galaxy positional data clearly biases the estimation of M_t to low values. Moreover, this is a situation where the bulk of the mass lies in some smooth distribution, not in N discrete particles, and masses should be calculated from $\sigma_v^2 R$ where R is the half-mass radius of the cluster. West and Richstone (1988) used this idea to suggest that cluster virial masses, and the subsequent derivation of M/L, are too low by as much as a factor of 10. A more recent analysis by Evrard, Metzler, and Navarro (1996) shows the situation is not nearly this extreme.
3. Application of the virial theorem implicitly assumes that only one potential is being probed by σ_v. If clusters have substructure in them which has not been assimilated into the main potential, then clearly there are multiple potentials and the observed value of σ_v has no physical meaning in the context of the virial theorem. At least one excellent case study exists to illustrate this point.

Substructure and the illusion of large M/L

The Cancer cluster is a spiral-rich cluster located at a distance of about three times the distance to Virgo. Based on a few redshifts used to determine σ_v, Tift, Jewsbury, and Sargent (1973) derived an overall M/L of 1700. This is an enormously high value, and if representative, it suggests that clusters of galaxies have at least an order of magnitude more unseen matter in them than the individual galaxies. Bothun *et al.* (1983) obtained a considerably different interpretation of the Cancer cluster. Using a nearly complete redshift survey, they showed there was a correlation between position within the cluster and redshift. This is reproduced in Figure 4.5. This correlation allowed specific dynamical units to be discovered in what was

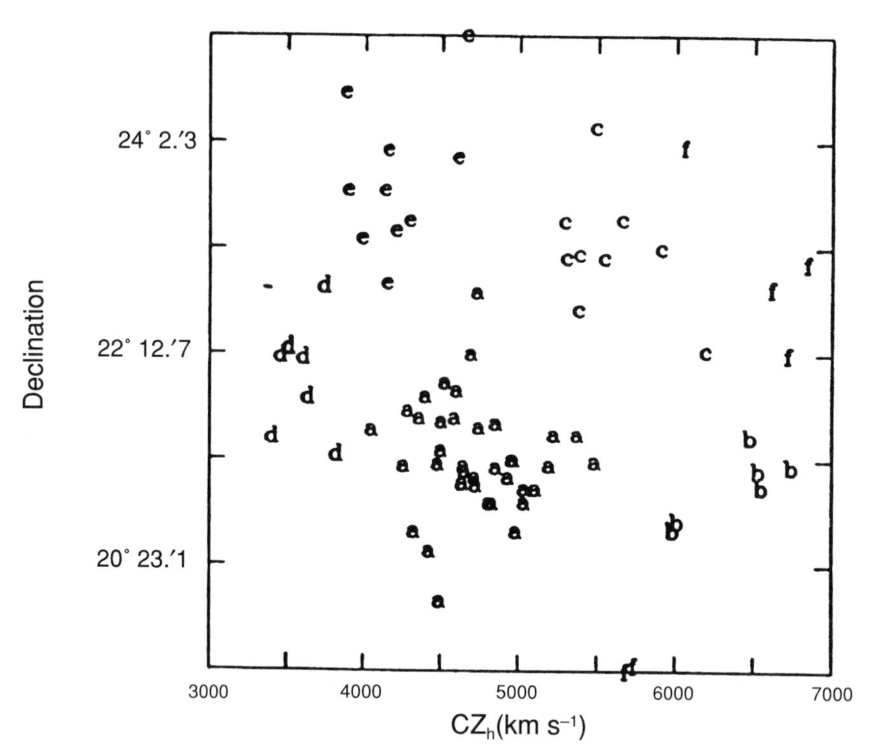

Figure 4.5 Position–velocity correlation for galaxies in the Cancer cluster. Each letter designates a different group separated in redshift space from another group. The clustering of letters together indicates that the Cancer cluster is not a real cluster but a collection of individual small groups. From Bothun *et al.* (1983).

previously considered to be a single dynamical unit. The analysis of Bothun *et al.* showed that the Cancer cluster in reality was a collection of five distinct subgroups that, when projected onto the plane of the sky, appeared to be a single cluster. Moreover, these groups are not even gravitationally bound to one another but instead are separating with the Hubble flow. Bothun *et al.* were able to lower M/L from 1700 down to 100–200, the value found for each of the individual groups.

Although Cancer is rather an extreme case, the issue of the frequency of substructure in clusters remains unresolved. Although it's likely that every cluster has some substructure if examined with sufficient resolution, the issue is how much of the cluster mass is tied up in these substructures. Substructure is very difficult to detect unambiguously. At its lowest level, substructure would manifest itself via the appearance of secondary maxima in galaxy positional data. However, galaxy positional data is usually only available for 50–200 galaxies and the resultant Poisson noise fluctuations can generate statistically insignificant secondary peaks. The combination of velocity and positional data greatly improve upon this. Position–velocity correlations, to first order, can isolate dynamical subgroups within an overall structure. But although this is a necessary condition, it is not a sufficient condition to identify substructure unambiguously.

For instance, it is now clear that most virialized clusters are surrounded by a lower density region of galaxies that are presently infalling. If their infall trajectory is projected along the line of sight to the cluster core, a position–velocity diagram will not necessarily indicate that these galaxies aren't in the core. The inclusion of an infalling population clearly raises σ_v. To resolve this ambiguity requires very accurate measures of relative distances, so the infalling group can be isolated in physical space, even though it blends with the cluster core in redshift space. In general, such distances are not available, so infall remains a worrisome complication in interpreting the measured σ_v for clusters. Indeed, as more clusters are studied in detail, the frequency of occurrence for substructure has increase from about 25–30% of all clusters in 1982 to 50–75% by now (West and Bothun 1990; West, Jones, and Forman 1995). Since the clusters themselves haven't done much dynamical evolution on the decade timescale, this increase can only be attributed to closer inspection by observers. The major physical attribute which is missing from most determinations of substructure, however, is the amount of mass involved in it relative to the main cluster.

X-ray studies of clusters of galaxies have the potential to provide more physical discrimination than position–velocity data. Potential well depth determines the virial temperature of the X-ray emitting plasma. An infalling group of lower mass than the total cluster mass will have a cooler temperature. Therefore, perhaps the best physical indicator of substructure in clusters at the level where it is dynamically important is the existence of multiple temperature components in an X-ray image of a cluster. The *Einstein* satellite could not make these measurements but *ROSAT* could and did. Indeed, the *ROSAT* observations of the Coma cluster, the quintessential example of a dense, postvirialized cluster, do show examples of multiple temperature components in the cluster core, hence revealing substructure (Briel, Henry, and Boehringer 1992). Similar substructure is also seen in the *ROSAT* data for A2151 (Huang and Sarazin 1996).

Another physical indicator of substructure is provided by gravitational lensing. Figure 2.24 shows a spectacular HST picture of the galaxy cluster A2218. Numerous arclets and rings can be seen; they are the distorted (lensed) images of resolved galaxies located behind the cluster. The orientation and degree of curvature of these features depends upon the cluster mass distribution and the amount of substructure. Analysis of the A2218 observations is most interesting because it represents the first case where substructure in the dark matter distribution might be directly inferred instead of inferring it indirectly on the basis of substructure in the light (baryonic) distribution. The analysis by Squires *et al.* (1996) shows that the inferred peak of the 2D dark matter distribution coincides with the optical and X-ray centers. Squires *et al.* also derive a total M/L for A2218 of 440 ± 80.

Recent work on gravitational lenses also underscores how critical is the very central mass distribution, in a cluster of galaxies, when determining the overall arclet morphology and extent. For instance, the presence of a central dominant cluster galaxy (a cD galaxy) can inhibit the formation of radial arcs (Miralda-Escude and Babul 1995). Detection of radial arcs (e.g., Newbury and Fahlman 1996) then limits the amount of mass that can be present in the very core of the cluster. Flores and Primack (1996) give a general treatment of lensing in which they show that the nature of the lensing is a very sensitive function of the core properties of the lensing mass itself. Fischer and Tyson (1996) show that one luminous X-ray cluster located at $z = 0.45$ shows lensing behavior that comes from two components: (1) a strong central mass concentration and (2) substructure located ≈ 1.5 Mpc from the cluster core. This strongly suggests that infall and cluster merging occur at this redshift.

In sum, because of substructure and infall, M/L determinations of clusters of galaxies made on the basis of cluster dynamics are rather unreliable. Reported values in excess of 1000 are almost certainly wrong, as substructure has not been taken adequately into account. When it is taken into account, clusters seem to define an M/L regime of 200–500, still larger than individual galaxies but not extremely so. If the Universe is closed, clusters of galaxies contribute 10–20% of the closure density, hence the dark matter must be distributed on still larger scales, occupying the space between clusters.

4.4 The large-scale distribution of dark matter

Observational determinations of the large-scale distribution of dark matter make use of the peculiar velocity formulation discussed in Chapter 3. For the local case of Virgocentric flow, the infall velocity and the position of the Local Group is driven by two competing sources of acceleration. On the one hand, the gravitational acceleration is driven by the amount of mass overdensity $\delta\rho/\rho$ in the Virgo cluster. On the other hand, this mass overdensity must compete against the average mass density of the Universe (e.g., Ω).

As a case in point, let's now consider the infall of the Local Group towards Virgo. We assume a cosmic velocity of 1500 km s^{-1} for Virgo and an LG infall velocity of 250 km s^{-1}. The observed value in light of $\delta\rho/\rho$ is ≈ 2 (Bushouse *et al.* 1985; Tully and Shaya 1984; Tonry and Davis 1981). For $\Omega = 1$ equation (3.39) yields

$$\frac{\delta v}{v} = \frac{250}{1500} = 0.17 = \frac{1}{3}\left[\frac{\delta\rho/\rho}{(1+\delta\rho/\rho)^{0.25}}\right] \tag{4.21a}$$

which is solved with $\delta\rho/\rho \approx 0.6$. This is significantly less than the observed value and readily implies that the distribution of light is more strongly clustered than the distribution of mass on this scale of ≈ 1500 km s^{-1}. For $\Omega = 0.1$ we have

$$\frac{\delta v}{v} = \frac{250}{1500} = 0.17 = \frac{1}{12}\left[\frac{\delta\rho/\rho}{(1+\delta\rho/\rho)^{0.25}}\right] \tag{4.21b}$$

which is solved with $\delta\rho/\rho \approx 3.0$. This density contrast is closer to the observed value in light ($\delta\rho/\rho \approx 2.0$) and requires little or no biasing.

This simple model shows the degeneracy between the biasing factor and the value of Ω inferred from peculiar velocity data. Although the peculiar velocity field is ideally a direct reflection of the large-scale density field, the presence of possible bias between the mass and light distributions is a complicating factor. In practice it is only the first-order component

$$\beta = \frac{\Omega_0^{0.6}}{b}$$

which can be recovered from the data. The best thing to do is clearly to select a sample in which $b \approx 1$. This is difficult to do a priori but there has been significant effort along these lines.

The most robust attempt is the POTENT effort (Bertschinger *et al.* 1990; Dekel *et al.* 1990) in which the large-scale density field, as defined in redshift surveys, is used to predict the peculiar velocity field. This is a purely local approach where peculiar velocities in one area of the sky are used to generate the local density maxima (or minima). This builds up a topological map of the dark matter distribution which can be compared to the topology based on a redshift survey when it is smoothed to the scale of the peculiar velocities.

The redshift sample that seems most suited for this purpose is a flux-limited sample of *IRAS* galaxies (Strauss *et al.* 1992). *IRAS* galaxies are generally star-forming spirals that are located in regions of modest to low galaxy density. The linchpin of this analysis is the explicit assumption that peculiar velocities are generated locally. Thus, a galaxy near a mass concentration such as the GA or the Coma cluster feels only that cluster. The LP result clearly calls this assumption into question, as the scale over which peculiar velocities are generated is unclear. When applied to the *IRAS* redshift sample, the POTENT method does return a value of $\Omega_0 = 1$, but it remains unclear if a fair volume has been sampled.

Figure 4.6 shows the comparison between the density distribution as inferred from POTENT compared to that from the *IRAS* galaxy redshift survey. The qualitative agreement is good as both maps clearly reveal the Perseus–Pisces, Coma and Hydra–Centaurus regions, the three nearest high density regions. But it is interesting to speculate that if there exist large-scale variations in dark matter densities, then galaxies located near the peaks of that distribution may have more dark matter phase mixed into their potentials than galaxies located near the valleys. This could introduce systematic error in the determination of relative distances and may be the source of systematic error in the LP result. We discuss this possibility a bit further in Chapter 5.

A lengthy but excellent review of the comparison between peculiar velocity data and gravitational instability plus CDM theory is offered by Strauss and Willick (1995). Their summary clearly shows that various attempts to measure β, through some 20 different methods, yield a range of values from 0.45 to 1.28, with no clearly preferred method dominating the measurements. Moreover, constraints on Ω from peculiar velocity data seem to depend on scale (and choice of sample). For the simplest local system, the infall of the Local Group to the Virgo cluster does support low values of Ω (e.g., 0.2–0.3), although the infall velocity is uncertain. On larger scales, all-sky samples such as the *IRAS* galaxy sample of Fisher *et al.* (1994), when smoothed with a 1200 km s^{-1} Gaussian, favor $\Omega = 1$.

From the observational point of view there are no strong constraints on Ω. Although most of the data prefer values of $\Omega \leq 0.3$, a highly dark matter dominated Universe (e.g., $\Omega = 1$) cannot be entirely ruled out. This leads to a physical divergence of cosmological models. On the one hand, the Universe might actually be relatively simple, having most or all of its mass in conventional form (e.g., baryons). On the other hand, the Universe might be dominated by some exotic form of matter that we have not detected yet. Since it's the mass distribution that determines the

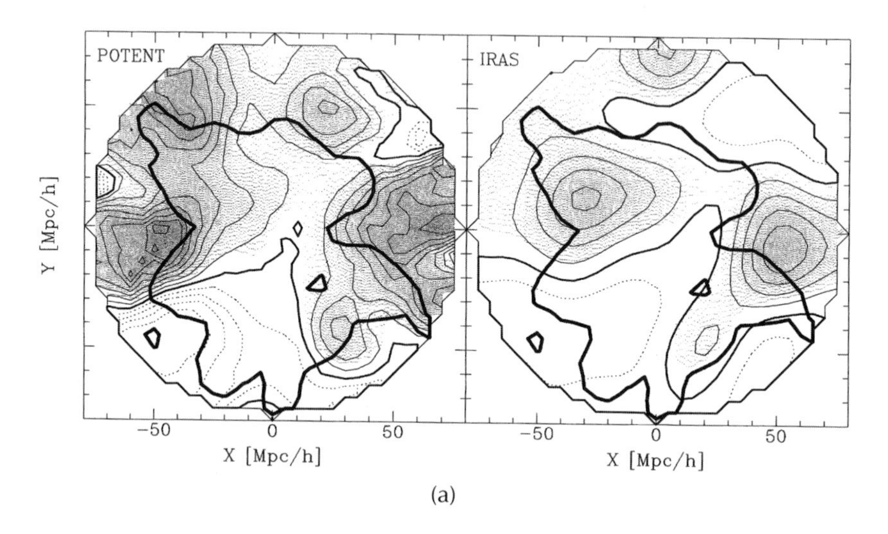

(a)

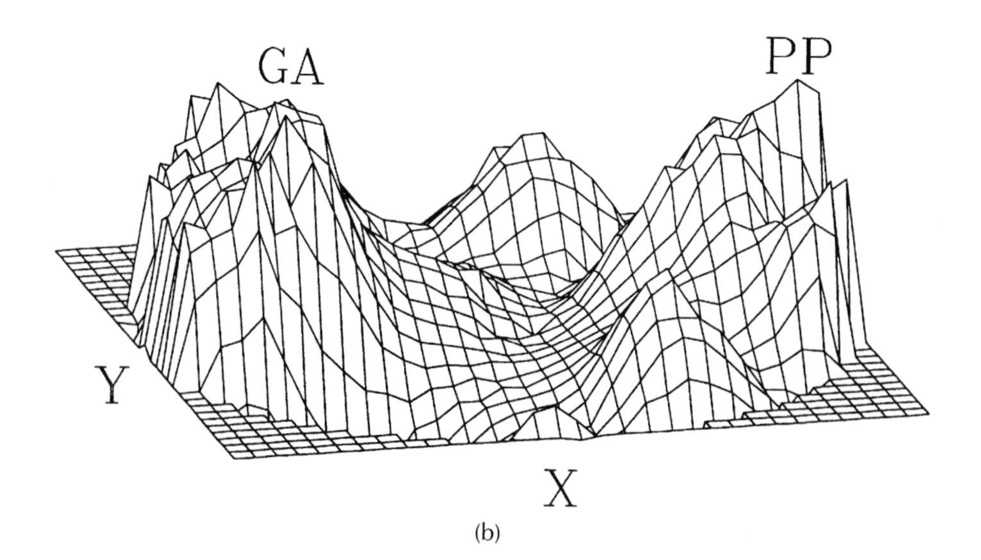

(b)

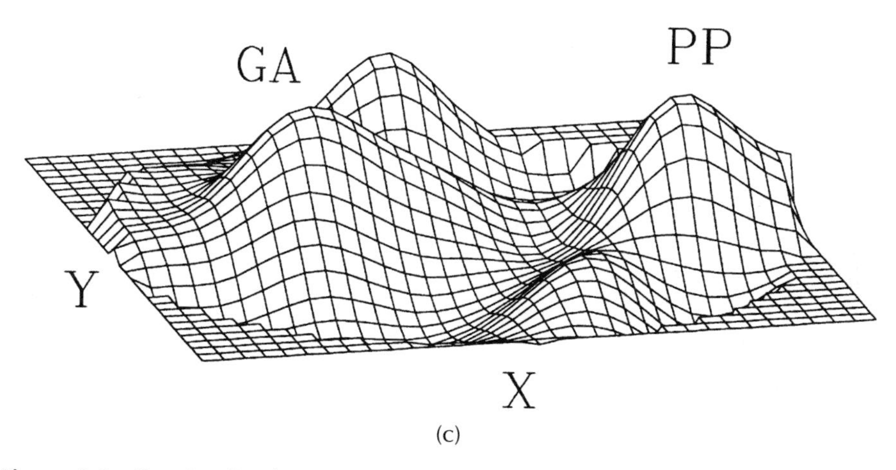

(c)

Figure 4.6 Density distributions: POTENT method versus direct redshift survey of *IRAS* galaxies. (a) Contours of density distribution; the heavy contour marks the boundary of the comparison volume of effective radius $46h^{-1}$ Mpc. (b, c) Edge-on view of the contours where the height in the surface plot is proportional to the density contrast. The data has been smoothed with a Gaussian window of radius $12h^{-1}$ Mpc. Although the POTENT density field is more strongly peaked than the density field from the *IRAS* data, the two representations are qualitatively similar in that two principal density maxima in the local Universe clearly stand out. These maxima are associated with the Hydra–Cen and Perseus–Pisces superclusters. Part (a) adapted from Dekel (1994) and Strauss and Willick (1995); parts (b) and (c), courtesy of A. Dekel, adapted from Sigad *et al.* (1997).

overall geometry of spacetime, hence our cosmological model, the solution of the dark matter problem clearly has far-reaching implications. As we will now investigate in detail, there are strong and compelling theoretical reasons to believe that $\Omega = 1$ and that dark matter dominates the large-scale behavior of the Universe.

4.5 The inflationary paradigm and $\Omega = 1$

It has been 30 years since Penzias and Wilson discovered the CMB. This observation is the very foundation of the Hot Big Bang cosmological model. In its early formulation, this model lead to a low density, baryon-dominated Universe of age ≈ 18 billion years (e.g., Sandage 1982). However, with the evidence presented above for the existence of dark matter, a fundamental alteration of this baryon-dominated cosmological model might be required. And when looked at in detail, the Hot Big Bang model does not naturally predict some aspects of the large-scale nature of the Universe. These predictive problems are enumerated below. In an effort to develop a model with more predictive power, a radical new model was proposed 15 years ago by Alan Guth and colleagues (Guth 1981). Known as the inflationary model, it postulated the existence of a phase transition in the earliest moments of the Universe, a transition which led to a brief period in which universal expansion was exponential in nature. This exponential expansion epoch is called "inflation" because, among other things, any initial curvature in the early Universe would be inflated away by the rapid expansion, thus producing a **spatially flat** Universe. Spatial flatness requires the curvature term in the Robertson–Walker metric (K in Chapter 1) to be zero. This in turn requires $\Omega = 1$ and a dark matter dominated Universe. Hence, if inflation indeed is the correct paradigm for the physics of the early Universe, it makes a fairly definite prediction regarding the mass distribution.

An excellent and detailed overview of the physics of the early Universe under this model is offered in the book *The Early Universe* by Turner and Kolb. Besides being a possible "natural" result of a phase transition, inflation also solves three problems that are not adequately addressed by the standard Big Bang model that has no inflationary epoch.

The flatness problem

Equations (1.23) and (1.28) can be combined to yield

$$\frac{k}{(HR)^2} = \frac{\rho}{\rho_c} - 1 = \Omega - 1 \tag{4.22a}$$

Since ρ is time dependent (decreasing with expansion) then Ω is also time dependent, and we can write

$$\Omega(t) = \frac{1}{1 - f(t)} \tag{4.22b}$$

where

$$f(t) = \frac{k}{R^2}\frac{3}{8\pi G\rho} \tag{4.22c}$$

At early times ρ was enormous, hence $f(t)$ was very small, Ω was very close to unity and the Universe was spatially flat. This condition is independent of the present epoch value of Ω (Ω_0). The observation that Ω_0 is in the range 0.01–1, despite the fact that the scale factor has increased enormously, has led many to suggest that $\Omega_0 = 1$ else the Universe would either have recollapsed long ago or would be curvature dominated (radius of curvature $\geq$ the Hubble length) at the present epoch. A specific value of Ω_0 in the range 0.01–1 would imply that the conditions at the present epoch are somehow imprinted on the initial conditions that determine the expansion.

The horizon problem

In an expanding Universe there are particle horizons. The size of these horizons to first order is set by the speed of light and the expansion rate, so $r_{hor} \sim cT_{exp}$. As the Universe ages (expands), the particle horizon increases and more material can come into causal contact with that particle. At early times, individual particle horizons could encompass only a fraction of the volume of the Universe. It can be shown (see the derivation in Kolb and Turner) that the horizon size at any redshift epoch is proportional to the entropy (basically the number of photons) within that horizon volume. For the matter-dominated Universe this can be expressed as

$$S_{HOR} \sim 3 \times 10^{87}(\Omega_0 h^2)^{-3/2}(1+z)^{-3/2} \tag{4.23}$$

At the present epoch ($z = 0$), we have $S_{HOR} \approx 10^{88}$ (which corresponds to a CMB photon density of about 400 cm^{-3}). At the time of recombination, $z \approx 1000$, we have $S_{HOR} \approx 10^{83}$, so the Universe at $z = 0$ consists of $\approx 10^5$ causally disconnected regions. At the time of recombination, the angular size of a particle horizon was 1–2°, yet over 360° of sky the CMB photon density is the same. Consideration of the observed abundances of light elements only exacerbates this issue. At the time of cosmological nucleosynthesis $S_{HOR} \approx 10^{63}$, yet the observed abundances of light elements today show no variation. Thus, over 10^{25} causally disconnected regions, the Universe shows homogeneity. This demands that the initial conditions of the Big Bang were homogeneous, a very improbable state.

The smoothness problem

On a large scale the Universe is extraordinarily smooth, as evidenced from the low anisotropy measured by *COBE*. Yet in this smooth Universe, there exist galaxies which are local density enhancements of order 10^4. Chapter 5 will show that density enhancements ($\delta\rho/\rho$) grow linearly with scale factor, due to gravitational effects. A natural gravitational timescale is known as the Planck time, which is 10^{-43} seconds. Galaxy formation commenced when the Universe was $\approx 10^{15}$ seconds old, hence there is potentially a 10^{58} scale in the growth of density fluctuations. If fluctuations are allowed to grow over this scale, the observed structure in the Universe could have formed out of a very initially smooth state. However, in the early Universe, purely baryonic density fluctuations are not allowed to be amplified linearly. The problem in the early Universe is due to the high radiation pressure of the photon field to which they are coupled. Then baryonic density fluctuations can grow linearly only after recombination at 10^5 seconds has occurred. In this sense, structure formation is greatly aided if there is some form of matter that can "gravitate" but which is not affected by radiation drag.

Solving flatness, horizon and smoothness

The elegance of the inflationary paradigm lies in its simultaneous solution of all three problems. The root of the paradigm is a brief period in the early Universe where it expanded exponentially instead of linearly. This exponential expansion caused the Universe to increase in scale by a factor of $\approx 10^{50}$. The trigger for this inflationary epoch is a competition in the early Universe between vacuum energy density (which acts as a source of negative pressure) and the kinetic energy density, essentially an entropy field that currently drives the uniform expansion and cooling of the Universe. It is possible to define a particular form for the scalar field that incorporates a potential energy function, usually expressed as $V(\phi)$. The scalar field ϕ is only weakly coupled to other fields that may be present. If ϕ is everywhere the same (e.g., spatially homogeneous) then it is possible to express the energy density and pressure of this fluid to first order as

$$\rho_\phi = \dot{\phi}^2/2 + V(\phi) \tag{4.24a}$$

$$p_\phi = \dot{\phi}^2/2 - V(\phi) \tag{4.24b}$$

Equations (4.24) are a partial solution to the stress–energy tensor used to describe the ϕ field. In the derivation of these equations, the early Universe is assumed to conform to the Robertson–Walker metric described in Chapter 1 and is assumed to obey Einstein's field equations. The term $\dot{\phi}^2/2$ is the kinetic energy density. If $V(\phi)$ varies slowly with ϕ and if the time derivative of ϕ is also small, then the early Universe can be characterized by $\dot{\phi}^2/2 \ll \mathrm{V}$. In this case, $\rho = -p$ and the Universe acts exactly as if a cosmological constant dominated the stress–energy tensor.

This source of negative pressure causes the Universe to undergo an exponential expansion of its scale factor. Presumably this inflationary epoch must end whenever a condition is achieved such that $V(\phi)$ is at a minimum and the potential energy of the field is then converted into kinetic energy density. In practice $V(\phi)$ will oscillate around this minimum. As the scalar field is weakly coupled to other fields, this oscillation will quickly cause the enormous amount of vacuum energy to be dumped into the kinetic energy field, which effectively reheats the Universe and fixes its entropy. After this time, the Universe expands and cools.

The actual physical conditions which determine the turn-on and turn-off phases of inflation remain obscure. Most physicists believe that inflation is like a symmetry breaking phase transition. The dominance of the vacuum energy field may then be a response to symmetry breaking at the GUTs energy scale. The end of the inflationary epoch is thought to be caused by another phase transition in which the symmetry between the weak nuclear force and the electrostatic force was broken. This occurs at an energy scale of $\approx$ 500 GeV or a timescale of $\approx 10^{-15}$ seconds. At this time, the potential has achieved a minimum, meaning that the cosmological constant has decayed to some minimum value and remained constant since then. This minimum value need not have been zero.

This period of exponential expansion has inflated out any initial curvature, hence inflation directly solves the flatness problem. In particular, inflation predicts a spatially flat Universe (to 1 part in 10^{50}). The horizon problem is also directly solved, as the initial conditions could have been quite heterogeneous. A tiny region of that heterogeneous mixture (and this tiny region was homogeneous) inflated to produce our observable, homogeneous Universe. Inflation allows for the existence of other inflated universes which occupy different domains. Theorists enjoy speculating that boundaries between these inflated domains, called domain walls, may have been present in our Universe with observable consequences.

As a consequence of predicting that space must be spatially flat, inflation then demands a Universe which is dark matter dominated. Interestingly, this provides a solution to the smoothness problem. If dark matter dominates the potential of structures which have trapped baryons, we can allow it a form that does not interact with the radiation field in the early Universe. This permits density fluctuations to begin growing at very early times. Thus, a galaxy size density fluctuation could grow from an initial dark matter density enhancement by a factor of $\approx 10^{30}$ if we start the growth at the time of electroweak symmetry breaking. A Universe at this time which was smooth to one part in 10^{30} is quite consistent with the CMB observations.

Although the inflationary paradigm operates via some unknown but clearly fundamental physics, it provides some elegant solutions to the problems encountered in the standard Big Bang model, which is dominated by baryon mass. Inflation has lately come under fire because most observations do not indicate that $\Omega_0 = 1$, as predicted. However, inflation only predicts a spatially flat Universe and the curvature of the Universe is determined by both Ω and Λ. Hence, if we believe the observations that $\Omega \leq 1$ and that the inflationary paradigm must hold, we are again driven to considerations of a nonzero Λ. And there are some esoteric but opaque "inflationary" models that do predict significant spatial curvature. Since we don't understand them, they are not considered here (but see references in Kolb and Turner).

4.6 Dark matter candidates

The preceding sections have established that there is some evidence for dark matter in the Universe but whether 90% or 99% of the Universe is made up in this form is still unresolved. The order of magnitude difference between these numbers has important implications with respect to the nature of the dark matter. There are two broad classes of dark matter to consider, baryonic and nonbaryonic. In addition to important physical differences there may be an important philosophical difference. Astronomers want to build big telescopes to probe vast distances and solve cosmological problems. But astronomers can only detect and measure baryons. If the mass of the Universe is mostly nonbaryonic, its fundamental nature will not be revealed through telescopic observation, but by some high energy experiment in an accelerator on Earth. It is thus highly regrettable that continued funding of the Superconducting Super Collider was denied, as this might have been the grandest cosmological experiment ever done.

4.6.1 Normal baryonic dark matter

An important feature concerning baryonic dark matter is that it wasn't necessarily always dark. A stellar mass black hole is the prime example, as it was once a star radiating energy. Here are some candidates for baryonic dark matter. The local space density of these candidates will be calculated by assuming that the halo of our galaxy is $10^{12}M_\odot$ and has a radius of $\approx$ 40 kpc. Since the Zaritsky *et al.* (1997) results suggest a halo size considerably larger than this, our estimates could easily be too high by factors of 10–100.

- *Bricks*: A brick is an excellent candidate for dark matter as it has very high M/L. Assuming a mass of 1 kg per brick, the required space density is 10^{28} pc^{-3}. The average spacing between bricks is $\approx$ 60 000 km and we would expect Earth to be impacted from time to time by a brick. The other problem with bricks is that they are made of heavy elements which are synthesized inside stars. Production of the dark matter brick population is necessarily preceded by a luminous phase of heavy element production in stars. The cosmic abundance of heavy elements provides a measure of all the light that has been produced in all the stars.
- *Small balls of hydrogen*: A Jupiter mass ball of hydrogen would have high M/L in the optical bands. The mass of Jupiter is $\approx 10^{-3}M_\odot$ and the required space density is $\approx$ 10 pc^{-3}, which only contributes $0.01M_\odot$ pc^{-3} of mass density in the solar neighborhood. The nearest of these objects would be $\approx$ 0.4 pc away, which would place it near the Oort cloud. Over the history of the Solar System, one might expect significant interactions between these interstellar Jupiters and the material in the Oort cloud. Moreover, although Jupiter will never be star, it does radiate significant amounts of energy in the 2–10 μm region of the spectrum. If this spectral behavior is similar for all Jupiter mass balls of hydrogen, then for distances less than 1 pc, that flux would be easily detectable from Earth-based observations. Existing 2 μm surveys of the sky as well as the *IRAS* 12 μm survey have detected no candidate interstellar Jupiters. This is consistent with the microlensing surveys described below, which have now placed stringent limits on the space density of these objects.
- *Stellar remnants*: Since our galactic disk is not yet old enough for white dwarfs to cool down to temperatures below a few hundred kelvins, or pulsars to slow down and cease pulsing, the only viable high M/L remnant that would escape local detection is stellar mass black holes. Assuming a mean black hole mass of $3M_\odot$ requires a space density of 10^{-2} pc^{-3}, which is pretty low. However, the seed population of these black holes are stars with masses $\geq 10M_\odot$. Those stars are the principal metal-producing stars in our galaxy. If they left behind $10^{12}M_\odot$ of black hole remnants then those remnants would have had to take their metals with them in order to avoid the interstellar medium being riddled with, say, gold. In fact, the general problem with appealing to stellar remnants as dominating the mass of galaxies is that, in the past, the remnants were bright and galaxies should have much more substantial luminosity evolution than is actually observed (Lilly *et al.* 1995). Moreover, in this context it is clear that the dark matter problem is

linked directly to the chemical evolution of galaxies. This forms an expectation that metal-rich galaxies have more stellar remnants and dark matter than metal-poor galaxies.

4.6.2 Exotic baryonic dark matter

- *Big stellar remnants*: From time to time there has been speculation that a population of stars must have existed in our galactic halo prior to the formation of globular clusters (Population II objects). This population is called Population III (Carr, Bond, and Arnett 1984) and is theorized to have provided the initial compliment of heavy elements that are observed to exist in Population II objects. The candidate Population III objects are known as very massive objects (VMOs), which are stars of mass $10^3 M_\odot$ to $10^6 M_\odot$. These objects would be incredibly luminous for short periods of time and would therefore have to turn on at very high redshift in order to escape detection. They would produce a substantial yield of metals and quickly collapse into a very large stellar remnant. Clusters of these remnants might gravitationally coalesce into one large mass black hole, perhaps forming the central engine to power a QSO at high redshift. If we assume an average mass of these coalesced stellar remnants of $10^6 M_\odot$, then the required space density is 10^{-8} pc^{-3}. As these objects are in the halo, their orbits carry them in and out of the galactic plane and their velocity dispersion is set by the halo mass. By our adopted halo parameters, the expected velocity dispersion would be ≈ 300 km s^{-1} and on average each massive remnant would pass through the galactic plane every 10^8 years. As 10^6 of them are required to form the halo mass, we expect a crossing of the galactic plane every 100 years or so. The passage of a $10^6 M_\odot$ black hole through the gaseous plane of our disk would probably not go unnoticed.
- *Quantum black holes*: An intersection between general relativity and the precepts of quantum mechanics allows for the existence of a very unusual particle – a mini black hole. A mini black hole has a mass equivalent to a large terrestrial mountain, about 10^{12} kg and a radius of 10^{-13} cm. Such an object could only be created by tremendous compressional forces, which might have been present in the very early Universe. As the radius of a mini black hole is like that of a nucleon, it is a quantum mechanical system. As there are no energy barriers in a quantum mechanical system, tunneling will allow energy to leak out from the event horizon of the mini black hole, causing the system to shrink and increasing the rate of energy leakage. Mini black holes are then destined to evaporate, with the last stage being a sudden release of high energy photons (gamma rays). For a mass of 10^{12} kg the evaporation time scale is 10 billion years. Although the existence of mini black holes may appear to be preposterous, as well as a desperate attempt to understand the dark matter problem, it is no more unbelievable than some of the particle schemes proposed below. At least in this case, there is a prediction. If mini black holes of mass 10^{12} kg exist, the present Universe

should exhibit gamma ray bursters which are isotropically distributed. Amazingly, this population has now been observed (Chapter 6), although no one really believes the population is due to evaporating mini black holes.

4.6.3 Nucleosynthesis constraints on baryonic matter

There appears to be no single baryonic candidate that can dominate the halo mass of our galaxy and therefore escape local detection. Undoubtedly, baryonic dark matter exists in the form of cool white dwarfs and very low mass stars (e.g., brown dwarfs) but the space density of both populations is unknown. Available data is consistent with there being as much mass density in these objects as existing stars, but certainly not 10 times more. There are also significant constraints on the total number of baryons in the Universe that come from the observed abundances of the light elements ^{3}He, ^{4}He and ^{7}Li. There are two competing theories for the production of light elements in Big Bang nucleosynthesis:

1. *The homogeneous model* assumes a homogeneous Universe with rather small lepton number. This is important as the early exchange of neutrons to protons and vice versa is mediated by neutrinos (leptons).
2. *The inhomogeneous model* assumes that during the quark–hadron phase transition there were induced density inhomogeneities that produced regions of excess neutrons. The early Universe was sufficiently small and quarks were close enough together to overcome the mediating force of any gluons; hence quarks were not found in hadrons but instead were free. As the Universe expands and cools, these quarks must get bound inside hadrons.

The inhomogeneous model was studied in much detail in the late 1980s because, with a sufficient density contrast and a short enough length scale for neutron diffusion, some models suggested that the observed baryon abundance could be made consistent with an $\Omega = 1$ **baryon-dominated** Universe. If these models were correct, they would strongly motivate the general acceptance of the inflationary paradigm but at the same time they would beg the question, Where are all these baryons? And this, coincidentally, is the subject of Chapter 6. Inhomogeneous nucleosynthesis models became a small cottage industry of the late 1980s (Alcock, Fuller, and Mathews 1987; Kurki-Suonio *et al.* 1990) and each model had its own unique diffusion process between the neutron-rich and proton-rich sites. This cottage industry was effectively shut down in the early 1990s as better-developed theory for the quark–hadron phase transition showed it to be a second-order transition instead of a first-order transition, leading to much lower levels of inhomogeneity (Olive 1991; Goyal *et al.* 1995).

Under the homogeneous model, the abundance of light elements depends primarily on the baryon-to-photon ratio η. This ratio remains unchanged after the final set of e^+e^- annihilation in the first 10 seconds. The abundance of light elements depends on η since the formation of ^{3}He requires a seed population of deuterium (^{2}H). If η is sufficiently high then ^{2}H will be photodissociated before it can fuse with another proton to form ^{3}He. In contrast, if η is too low, the expanding Universe will not contain a sufficient density of protons for the newly formed ^{2}H to find, and there will be very little helium production. We define the baryon density as ρ_b and its contribution to the total mass density of the Universe as $\Omega_b = \rho_b/\rho_c$. Recall that $\rho_c = 3H^2/8\pi G$; hence

$$\rho_b = \Omega_b \rho_c = \frac{8\pi G}{3}\Omega_b H^2 \tag{4.25}$$

Determination of the abundances of light elements has been a subject of much research. The best astrophysical locations which have been investigated are the atmospheres of metal-poor stars and meteoritic material. It is also possible to directly detect interstellar or intergalactic deuterium in the ultraviolet, although the strength of deuterium$_\alpha$ is $\approx 10^{-5}$ that of Lyman$_\alpha$. A possible detection of deuterium$_\alpha$ towards a distant QSO was reported in some HST observations by Rugers and Hogan (1995), although distinguishing that line from the myriad of QSO Lyman$_\alpha$ lines at different redshifts is difficult, so this detection is not secure (Tytler, Fan, and Burles 1997). The most recent constraint or limit on Ω_b comes from a large and comprehensive analysis by Walker *et al.* (1991), who derive

$$\Omega_b \sim (0.015 \pm 0.005)h^2 \tag{4.26}$$

For $h = 1$ ($H_0 = 100$) the value of Ω_b lies in the range 0.01–0.02. Recall that Ω for luminous stars was ≈ 0.005; hence equation (4.26) leaves ample room for baryonic dark matter at a level up to four times the mass contained in luminous stars. For $h = 0.5$ ($H_0 = 50$) the value of Ω_b could be as large as 0.08. As mentioned in Chapter 3, there is good evidence that we also have a "missing" baryon problem. This is discussed in more detail in Chapter 6.

4.6.4 Nonbaryonic dark matter

By any measure, if $\Omega = 1$ then the Universe is dominated by nonbaryonic dark matter. Furthermore, that dark matter must be distributed more smoothly than the light. Nonbaryonic dark matter comes in two basic forms, hot and cold. These terms refer to the ability of this dark matter to cool to nonrelativistic velocities and clump into smaller units. A cold dark matter (CDM) Universe gives rise to structure formation modes very different than a hot dark matter (HDM) Universe. This is the subject of the next chapter; for now we will just take an inventory of the available candidates.

CDM consists of weakly interacting massive particles (WIMPs) that become nonrelativistic at temperatures well above 10^4 K. As such, they are excellent candidates for producing small-scale structure, since before recombination they would have easily clumped together. This requires the rest mass of most CDM WIMPs to be extraordinarily high (up to 10^{16} GeV). Indeed, some WIMPs could have been created very early on, through quantum fluctuations, and if they did not immediately annihilate with their respective anti-WIMPS, they could survive as the dominant relic mass today. Many of these particles are naturally created in the supersymmetric theories (SUSY) of particle physics. SUSY makes use of a conserved quantum number called R parity. $R = +1$ for particles and -1 for their SUSY partner. R parity can be linked to baryon number B and lepton number L conservation through the spin S as

$$R = (-1)^{(3B+L+2S)} \tag{4.27}$$

The conservation of R parity has three important implications:

1. SUSY particles are always produced in pairs (hence the name of the theory).
2. Heavy SUSY particles decay into lighter SUSY particles.
3. The lightest SUSY particle produced by this decay process is stable because there are no further decay modes that exist without violating R parity.

Current potential dark matter SUSY particles are the neutralino, gravitino, photino and higgsino.

Another favorite CDM particle is known as an axion. In contrast to other CDM particles, the axion is relatively light. The axion is predicted to exist as a result of a symmetry breaking associated with the strong-CP problem in quantum chromodynamics. Although the axion mass is arbitrary over the range 10^{-12} eV to 1 MeV, the symmetry breaking occurs at a high energy scale, and hence early in the Universe. Although axions were created in the very early Universe, when it was quite hot, they have very small momenta and are born cold. Axion freeze-out is also mediated through pion-to-axion conversion with nucleons acting as a catalyst. Since nucleons only come into existence after the quark–hadron transitions (energy scale 200 MeV), they are necessarily nonrelativistic, and so are the axions created by this thermal process. However, the cosmic abundance of any thermally produced axions is orders of magnitude lower than the production associated with symmetry breaking.

But HDM consists of particles that remain relativistic for significantly longer times. This requires their masses to be less than $\approx$ 100 eV. Unlike CDM, which has no experimentally verified candidates, HDM has a definite candidate – the neutrino. Once the total number of neutrino species is known, the density of neutrinos (as well as their cosmic temperature) can be determined. Studies of the Z_0 resonances in the LEP e^+e^- collider at CERN strongly fix the number of light (e.g., $m_\nu \ll m_Z/2$) neutrino species at three (electron, muon and tau). This leads to a cosmic background neutrino density of $\approx 100\ \text{cm}^{-3}$. Any neutrino mass above 1 eV would represent a significant contribution to the overall cosmological mass density. Values of 30–100 eV are required to yield $\Omega = 1$, depending on the value of H_0. Current experimental limits on the electron neutrino are ≤ 7 eV, the muon neutrino ≤ 270 keV and the tau neutrino ≤ 31 MeV. The muon and tau neutrinos therefore have the potential to close the Universe and there is some experimental evidence for a nonzero neutrino rest mass (see below).

4.6.5 Direct detection of dark matter

By definition, dark matter can only be detected indirectly through its gravitational effects. In a broad sense, the very existence of galaxies may signify a detection of dark matter, as baryons by themselves seem unlikely candidates for producing galaxy-size potentials. The amplitude of the large-scale deviations from Hubble flows is quite sensitive to the variation of $\delta\rho/\rho$ over different scale sizes and their existence (still determined at a somewhat low signal-to-noise level) also suggests a large-scale dark matter component to the mass distribution. This discovery of gravitational lenses represents another manner in which to detect dark matter, as it's the total lensing mass which governs the observed degree of distortion to background images. If gravitational lensing by distant clusters of galaxies could be detected (e.g., Squires *et al.* 1996; Fischer *et al.* 1997), it might be possible to search locally for gravitational microlensing and therefore constrain the local space density of dark masses than can act as minilenses. This idea forms the basis for the MACHO (massive compact halo objects) survey.

4.6.6 The microlensing surveys

Deriving the equations to describe gravitational lensing is well outside the scope of the book, but the fundamentals can be gleaned from a study of Fresnel's equations for an optical lens. Microlensing occurs whenever a point mass passes very close to the line of sight for a distant star. This case is different than a standard lens which has a finite area. For microlensing, one can use the point source–point deflector approximation to describe the behavior. There are two important predictions that form the basis for the observational survey described below:

1. As the point deflector crosses the line of sight to the point source, there will be a fairly rapid amplification of the light from the point source that the observer sees. This amplification results from the requirement that the flux per solid angle seen by a detector (this can be called the surface brightness) is conserved. The lensing event creates additional optical pathways for the light from the point source to reach the detector. Hence the solid angle has increased and consequently the apparent flux must increase as well. The multiple pathways arise due to general relativistic effects associated with the sudden change in geometry of spacetime between the source and the observer. Without the point deflector, spacetime is flat between the source and the observer. The amplification factor depends on the degree of curvature and hence it depends on the mass of the point deflector; amplification also depends on the degree of alignment between the point source and the detector. For a reasonable range of baryonic point deflector candidates and degrees of alignment, amplification factors of 1–100 are expected.
2. Gravitational lensing is achromatic. There is no equivalent to the index of refraction for a normal lens, as the behavior of light passing through curved space time is independent of its wavelength. This is the unique signature of gravitational lensing as the cause of variability in the light output of a distant star. Intrinsic stellar variability (novae, binary stars, pulsating stars) always shows wavelength-dependent behavior.

If a significant fraction of the dark matter in the halo of our galaxy is composed of objects like brown dwarfs or planets, there will be occasional amplification of

Figure 4.7 Example of a MACHO field in the LMC that contains approximately one million stars. Courtesy of Dr Christopher Stubbs, University of Washington.

the light from extragalactic stars by the microlensing effect just described. But these microlensing events are extremely rare. Estimates suggest an optical depth to lensing events of order 10^{-5} to 10^{-6}. Hence to see a few events would require imaging a background field of stars that contained almost one million stars. Two such lines of sight are available for observations made in the southern hemisphere: the Large Magellanic Cloud and the galactic bulge. Out of this possibility was born the MACHO project.

The MACHO project is a collaboration between scientists at the Mt Stromlo and Siding Spring Observatories and the Center for Particle Astrophysics run jointly by the University of California and the Lawrence Livermore National Laboratory. The project leadership is provided by Charles Alcock, a physicist who works at LLNL. This collaboration represents the first successful union of astronomers and particle physicists in a quest to unravel the profound cosmological challenge posed by the dark matter problem. Using a 50 in telescope and a stunningly clever and efficient detector system, the team acquires several gigabytes of data per night on the photometric brightnesses of stars in LMC and bulge fields. An example LMC field containing about 1 million stars down to $v = 22$ mag is shown in Figure 4.7.

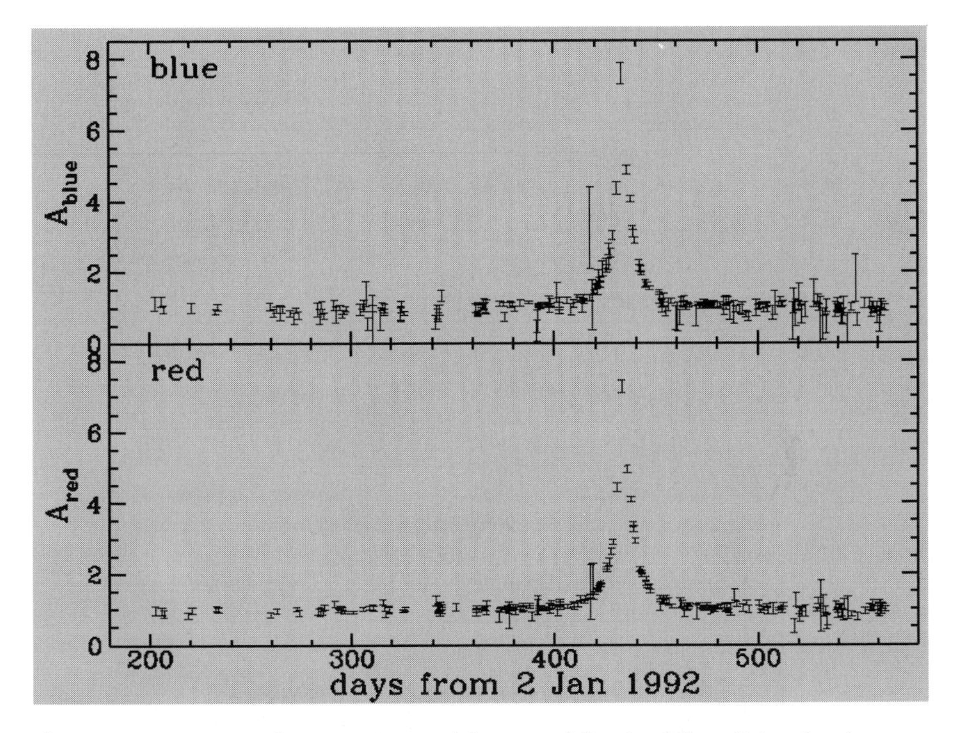

Figure 4.8 Signature for a gravitational lens amplification of starlight: the degree of amplification and the shape of the light curve are identical in the red and blue filters. Courtesy of Dr Christopher Stubbs, University of Washington.

To date, approximately 8 million stars have been observed in two colors in the LMC fields and 10 million in the bulge fields. Although there is much photometric variability of stars in these fields due to instrumental, atmospheric and intrinsic stellar variability, the unique signature of a lens event is its achromicity. Thus, the two color observations (blue and red) are the crucial components of the survey design.

The system has been on the air since January 1992, and despite the enormous amount of data that requires reduction, very likely lensing events have been identified. Figure 4.8 shows what an event looks like as a plot of light output over time. In both the red and the blue channels, the light curve is clearly the same and conforms to the expected signature of a microlensing event. To date, there have been eight candidate lens events towards the LMC and ≈ 60 toward the galactic bulge. Figure 4.9 shows the eight LMC lensing events (Sutherland *et al.* 1996). As the line of sight to the galactic bulge passes through much more of our galaxy than the line of sight to the LMC, this large difference in lensing events is to be expected. In the case of the bulge, the lensing could be occurring from point deflectors located either in the solar neighborhood, the thick disk, the halo or the bulge itself. For the case of the LMC, it is most likely that the point deflectors are members of the halo population.

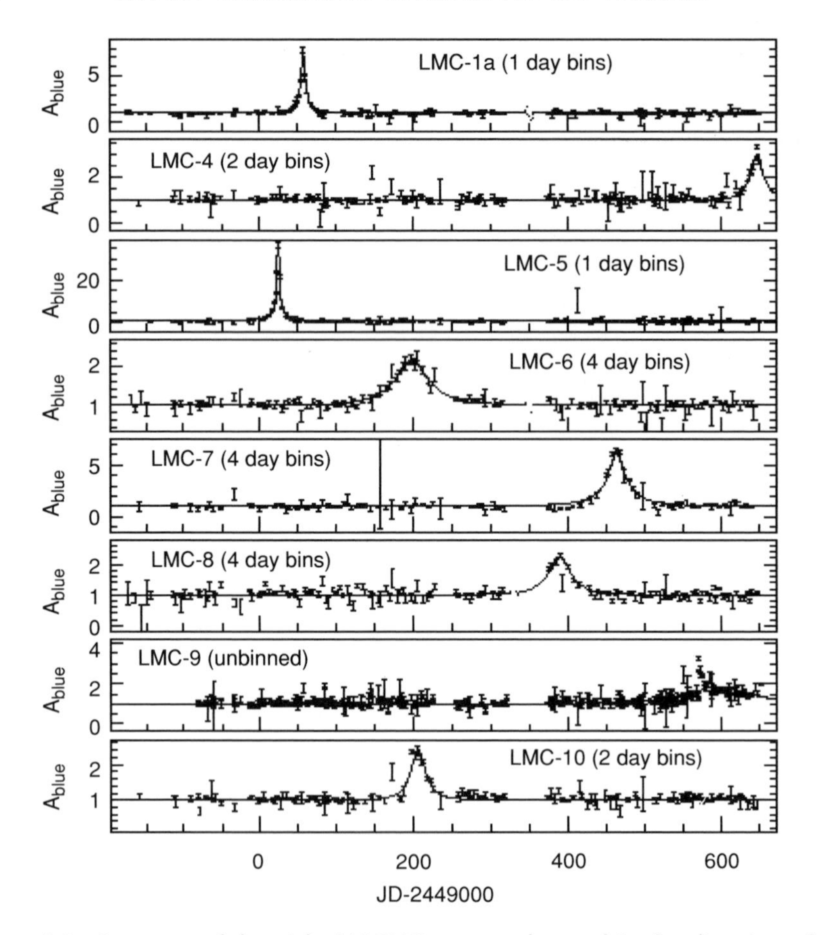

Figure 4.9 Summary of the eight MACHO events observed in the direction of the LMC: notice the wide range of amplification factors. Reproduced from Sutherland *et al.* (1996) courtesy of Dr Christopher Stubbs, University of Washington.

The relevant cosmological question to ask is: What is the range of point deflector masses that are consistent with the observations? The answer to this is nontrivial as a full understanding of the detection efficiency must be achieved. A convenient parameterization (Alcock *et al.* 1996) of the microlensing timescale is

$$\tau_l = 130\sqrt{m/M_\odot}\ \text{days} \tag{4.28}$$

Thus a Jupiter mass object will have a microlensing timescale of 4 days. If it is cloudy at the telescope for 4 days in a row, the sensitivity of the experiment to this lensing event is pretty low. Thus the survey, due to unavoidable circumstances, is not sensitive to short duration events. Similarly, the survey is not sensitive to long duration (≥ 200 days) events, hence point deflector masses larger than $3M_\odot$. Although the efficiency of the MACHO experiment is still a matter of some debate, a calibration using Monte Carlo techniques by Alcock *et al.* (1995) shows that greater than 10% efficiency is achieved for lensing timescales of 10–200 days. The peak efficiency is 30% in the range 50–100 days. These efficiencies are for cases where the amplification factor is larger than 1.35. However, since the number of lensing events increases with decreasing lensing mass, even efficiencies less than 10% at the low mass end contain information if no short duration event is ever observed.

The recent analysis of two years' worth of MACHO data (Sutherland *et al.* 1996) in the LMC direction is quite encouraging. Eight unambiguous lensing events were detected and the more events there are, the more constraining power the method has. For instance, the three events that were seen in the first year's worth of data are consistent with lensing masses in the range (0.02–0.17) $M_{\odot}$ and the fraction of halo mass which could be in this form ranges from 6 to 53% (Alcock *et al.* 1996). Much more data will be needed to distinguish between lensing caused by low mass, main sequence stars versus true brown dwarf objects. It is also worth noting that analysis of the microlensing events that happen towards the galactic bulge are providing some indication that our models for the thin disk, thick disk and halo might be incorrect.

The combined eight events for two years' worth of data has altered this picture somewhat. In particular, the distribution of lensing timescale now suggests a mean lensing mass of $\approx 0.4M_{\odot}$. This raises the serious possibility that white dwarf stars, which could be fairly numerous in the halo (Flynn, Gould, and Bahcall 1996), are responsible for the observed lensing events. Since no event with a duration of less than 20 days has been observed, Sutherland *et al.* (1996) are able to set a 95% confidence interval that less than 20% of the total halo mass is in objects with $M \leq 0.02M_{\odot}$. The data is also consistent with a zero population of these objects in the halo. This result is made more robust by a simple model in which the halo mass is dominated by these objects. If that were the case, the two year data set would have been expected to detect ≈ 45 events, when in fact no short duration events were detected. Sutherland *et al.* (1996) also give a slight revision of the Alcock *et al.* most probable lensing mass. They conclude that the most probable lens mass is $(0.5 \pm 0.25)M_{\odot}$, essentially suggesting it's the range $(0.25\text{–}0.75)M_{\odot}$ which contributes. Main sequence stars in that mass range would be easily detectable, implying that the typical microlensing event is being produced by a stellar remnant. These stellar remnants are most likely old, cool white dwarfs. However, one has to worry somewhat about the progenitor population and the brightening of halos in distant galaxies.

To date, the MACHO data set is providing interesting information about galactic structure. And it appears that lensing events associated with Jupiter mass objects are strongly ruled out. The current result, that the lensing is probing the presence of low mass main sequence stars and/or white dwarfs in the galactic halo, was the a priori expected result based on what we know about stellar populations in our galaxy. At the very least, we know that the halo mass of our galaxy is not dominated by extremely low mass stars and/or brown dwarfs. The beauty of the experiment, however, is that the constraints will only get better with time.

4.6.7 The solar neutrino experiment

Neutrinos are by-products of nuclear reactions which occur in the center of the Sun. The core neutrino flux is directly proportional to the nuclear reaction rate, which depends sensitively on the core temperature. Neutrinos created at the center of the Sun pass directly through it. Detection of these neutrinos on the Earth is difficult, but not impossible. By the early 1990s four solar neutrino telescopes were in operation. Three of these four experiments are relatively new and employ a more advanced detector design than the original solar neutrino telescope, essentially a vat of cleaning fluid, in the Homestake Mine in South Dakota. These three newer experiments give fairly consistent results and indicate a total neutrino deficit of about 35%. Specifically, the combined SAGE and Gallex results indicate a measured neutrino flux of 77 ± 10 SNUs (solar neutrino units). Two respective standard solar models give predictions of 132 ± 7 and 123 ± 7 SNUs. Each experiment also has a possible ± 6 SNU systematic error in the calibration. The level of statistical significance is between 3.3σ and 5.1σ. The neutrinos detected from the Sun are not the multitude that are generated by the standard proton–proton chain; they have insufficient energy. Instead they come from the rare process of beryllium and boron production, but the production rate is quite sensitive to the core temperature.

The significant deficiency of observed solar neutrinos has three possible resolutions:

1. The standard model of the Sun is wrong and the temperature is cooler than we think. This reduces the overall neutrino flux and the rate of beryllium and boron production. A cooler Sun seriously impacts its theoretical age, and by extrapolation the ages of globular cluster stars. This is regarded as an unlikely solution because stellar evolutionary theory has been so successful at predicting the observed properties of stars in old stellar clusters. Furthermore, Vignaud (1995) strongly concludes that even nonstandard solar models cannot be tweaked enough to explain the observed deficit.
2. The detector efficiencies are not well understood and the observed fluxes are actually consistent with theory. This is extremely unlikely to be the resolution. Unlike the Homestake Mine experiment, where understanding detector efficiency is crucial, the SAGE and Gallex experiments use a different detection scheme in which the instrumental calibration is fairly well understood. The low systematic error (10% of the observed flux) bears this out.
3. The current experiments are only sensitive to the presence of electron neutrinos. Muon and tau neutrinos pass completely through the detector without registration. Thus, if some physical process operated to change the flavor of the electron neutrino in the 8.5 minutes of travel time between the Sun and the detector, an observed deficit of electron neutrinos would result. However, since some electron neutrinos are detected, complete conversion cannot occur. A process known as the Mikheyev–Smirnov–Wolfenstein effect (Wolfenstein and Beier 1989) may come into play here. The MSW effect states that any coherent forward scattering of electron neutrinos in electronic matter results in a density-dependent effective mass. This means that electron neutrinos which travel through an inhomogeneous

medium (like the density gradients in the Sun or the Earth) have some probability of changing their effective mass and therefore their flavor. The MSW effect thus requires the neutrino to have some mass, but the effect itself is only sensitive to the square of the mass difference and cannot provide a good estimate of actual neutrino mass. The current observations suggest a mass difference between species of only $\delta m^2 \approx 10^{-5}$ eV^2.

The MSW resolution to the solar neutrino problem gives some confidence that the neutrino has a mass. The search for this mass is ongoing at several facilities but the history of claimed neutrino mass detections is not good. In particular, in early 1992 it was announced that several detectors had discovered evidence for a neutrino mass of 17 keV. Such a large mass would have caused the Universe to collapse long ago; hence these massive neutrinos, if they are real, must have decayed into other products. One possible decay channel is the photon, and there has been speculation that decay photons have sufficient energy to be relevant to the reionization of the Universe after recombination. This possibility is more fully discussed in Chapter 6. About a year later, a detector response problem was discovered which fully explained the 17 keV neutrino mass as an artifact of the detector. This ended the flurry of theoretical papers on the decay of this kind of neutrino, but the general idea that a massive particle could produce ionizing photons in its decay process has been retained, as it is of some cosmological interest.

An ongoing experiment (LSND) at Los Alamos is attempting to measure the mixing between electron and muon neutrinos. Early experimental results suggest δm^2 of order 1 eV^2, completely inconsistent with the value required to explain the solar neutrino deficit via the MSW effect. However, if one considers the effects of muon–tau neutrino mixing (Wolfenstein 1995) it is possible to reconcile LSDN with SAGE/Gallex by endowing two of the three neutrino species with a degenerate mass of ≈ 2.5 eV. This mass range has very interesting cosmological consequences as it could contribute to the large-scale features seen in the power spectrum of the galaxy distribution, without producing $\Omega = 1$. This possibility has also opened up another class of dark matter models, hybrid CDM–HDM models, known as MDM for mixed dark matter (Chapter 5).

4.7 Further implications of dark matter

Variations in M/L ratios for individual galaxies, in addition to being relevant to the overall census of cosmologically important material, also affects the determination of the extragalactic distance scale and the ability to measure large-scale flows. If dark matter dominates the potential wells of galaxies, there are two intriguing possibilities that would cause small calibration errors in the derivation of relative (and absolute) distances that would produce a false peculiar velocity signal and an erroneous value for H_0.

1. The dark matter constituent of galaxies is related to their chemical evolution if the dark matter content is stellar remnant dominated. The luminosities of stars

at different wavelengths are strongly dependent on chemical composition as atmospheric opacity often depends on the free electron density. Metals are the primary donors of free electrons. Metals are synthesized from massive stars that eventually become the dominant remnant source, so metal-rich galaxies may have more remnant mass per unit luminosity than metal-poor galaxies. This would lead to slightly higher *M*/*L* values in their luminous parts, which affects the calibration of both the TF and D_n–σ relations. There is weak evidence that the residuals from the D_n–σ relation correlate with metal line strength (Gregg 1995). Curvature in the H-band TF relation, first noticed by Aaronson *et al.* (1982) and more fully discussed by Mould, Han, and Bothun (1989), may also result from the stars in low mass metal-poor galaxies having lower H-band luminosity per unit mass (Bothun *et al.* 1984).

2. The more intriguing possibility, first suggested by Silk (1989) and Doroskevich and Klypin (1988), is that there are small-amplitude but large-scale variations in the dark matter content of the Universe. Let's imagine that, whatever it is, the dark matter is laid down in waves of wavelength 50–100 kpc with peak-to-valley amplitude of 10%. Doroskevich and Klypin show that such a distribution is very consistent with the observed large-scale flows. If we can imagine sprinkling galaxies (or the seeds of galaxy formation) down on this network of dark matter waves, then some galaxies will be located at peaks, whereas other galaxies will be located in valleys. Hence the amount of dark matter mixed in with galaxies as they formed will have ± 10% RMS variations. This means that systematic errors in the calibration of relative distances based on the TF and D_n–σ relations could occur when sampling along different lines of sight.

Trying to measure small differences in *M*/*L* is difficult at best, so there is no direct evidence for either of these phenomena. However, even systematic differences of 10% are important. Given that no cosmological structure formation scenario can account for the LP result (e.g., Strauss and Willick 1995), one needs to ask seriously if the flows are peculiar or the galaxies are peculiar. Since we do not understand the process of galaxy formation or the distribution of dark matter prior to galaxy formation, there may well be slight differences between how much dark matter is mixed into Local Group galaxies, compared to galaxies 50–100 Mpc away. This would produce a zeropoint shift in the calibration of the distance indicators, which in turn would produce a false peculiar velocity signal due to systematic errors in the determination of relative distance.

4.8 Summary

This chapter has contained a lot of information and after reading it one might well be left with lingering suspicions about the validity of the inflationary paradigm as well as the nature of the dark matter. The blunt truth is that the inflationary paradigm is so theoretically attractive that one is loathe to discard it simply on the basis that few, if any, observations support the $\Omega = 1$ prediction. In fact, as discussed in

Chapter 5, inflation makes a very specific prediction about the nature of the perturbation spectrum which produces the large-scale structure, and this prediction is verified by observation. So there is ample reason to retain the inflationary model but perhaps it requires a nonzero Λ in order for the Universe to be spatially flat. Moreover, discarding of this model in favor of a Hot Big Bang baryonic model then forces very special initial conditions that are somehow imprinted on the epoch when observations are being made.

To solidify the discussion, consider the following statements that are consistent with the current data:

- The best evidence for the existence of dark matter as the dominant mass in some potential is provided by the flat rotation curves of galaxies, as measured by their gas content out to scales $\geq 5r_h$. It is unclear if these specific results can be generalized to all spiral galaxies.
- Elliptical galaxies need to have a large-scale halo of dark matter if they are to retain their interstellar gas which has been heated and ejected by supernovae. The observations of X-ray halos around elliptical galaxies support this statement.
- Full consideration of substructure effects in clusters argues that the use of cluster dynamics does not yield reliable measures of M/L on the 1–2 Mpc scale. It remains unclear if cluster potentials have 5 or 50 times as much dark matter as an individual galaxy potential. Recent results from gravitational lensing studies (Squires *et al.* 1996), X-ray spectroscopy (Markevitch 1996), and profile fitting of X-ray clusters (Evrard, Metzler, and Navarro 1996) generally indicate M/L to be in the range 300–500.
- Determination of the cosmological parameter Ω_0 has not yet been done convincingly by any group. The unknown biasing between light and mass as a function of scale provides an additional complication. Values of $\Omega_0 = 0.2$–1.0 are consistent with someone's data set, although most dynamical determinations using large-scale flows suggest that Ω_0 lies in the range 0.1–0.3 (Strauss and Willick 1995). Possible systematic large-scale variations in individual galaxy M/L may give rise to false peculiar velocity signals, thus reducing the credibility of these dynamical determinations. The crucial issue of biasing between the galaxy distribution and the mass distribution has not been adequately resolved.
- There is encouraging evidence from the particle physics side of cosmology that the neutrino has a nonzero rest mass. Ongoing searches for CDM particles remain negative.
- If $\Omega_0 = 0.1$ then it is very likely that most of the dark matter is baryonic and in the form of stellar remnants and brown dwarfs. The suggestion that the halo potential of our galaxy is dominated by Jupiter mass objects has been ruled out by the microlensing experiment.
- If $\Omega_0 = 1.0$ then the Universe is clearly dominated by an unknown particle that does not have a distribution like the clustered distribution of light. The only observed candidate for this particle is the neutrino.

- As in the reconciliation of the ages of globular clusters with H_0^{-1}, the introduction of nonzero Λ can reconcile many of the above issues. If we let the neutrino have a small mass (1–3 eV), and allow Λ to be the dominant term in producing a spatially flat Universe, then we can accommodate both inflation and large-scale structure. Strangely, there is even data that supports this view (Chapter 5).

References

AARONSON, M., HUCHRA, J., MOULD, J., SCHECHTER, P., and TULLY, B. 1982 *Astrophysical Journal* **258**, 64

ALCOCK, C., FULLER, G., and MATHEWS, G. 1987 *Astrophysical Journal* **320**, 439

ALCOCK, C. *et al.* 1995 *Astrophysical Journal* **449**, 28

ALCOCK, C. *et al.* 1996 *Astrophysical Journal* **471**, 774

BERTSCHINGER, E., DEKEL, A., FABER, S., DRESSLER, A., and BURSTEIN, D. 1990 *Astrophysical Journal* **364**, 370

BOTHUN, G. GELLER, M., BEERS, T., and HUCHRA, J. 1983 *Astrophysical Journal* **268**, 47

BOTHUN, G., ROMANISHIN, W., STROM, S., and STROM, K. 1984 *Astronomical Journal* **89**, 1300

BOTHUN, G. *et al.* 1991 *Astronomical Journal* **101**, 2220

BRIEL, U., HENRY, P., and BOEHRINGER, H. 1992 *Astronomy and Astrophysics* **259**, L31

BUSHOUSE, H., MELOTT, A., CENTRALLA, J., and GALLAGHER, J. 1985 *Monthly Notices of the Royal Astronomical Society* **217**, 7p

BUTLER, P., MARCY, G., WILLIAMS, E., HAUSER, H., and SHIRTS, P. 1997 *Astrophysical Journal Letters* **474**, L115

CARR, B., BOND, J., and ARNETT, D. 1984 *Astrophysical Journal* **277**, 445

DEKEL, A. 1994 *Annual Reviews of Astronomy and Astrophysics* **32**, 371

DEKEL, A., BERTSCHINGER, E., YAHIL, A., STRAUSS, M., DAVIS, M., and HUCHRA, J. 1990 *Astrophysical Journal* **412**, 1

DOROSKEVICH, A. and KLYPIN, A. 1988 *Monthly Notices of the Royal Astronomical Society* **235**, 865

EVRARD, G., METZLER, C., and NAVARRO, J. 1996 *Astrophysical Journal* **469**, 494

FEIGELSON, E., WOOD, P., SCHREIER, E., HARRIS, D., and REID, M. 1987 *Astrophysical Journal* **312**, 101

FISCHER, P. and TYSON, A. 1996 *Bulletin of the American Astronomical Society* **189**, 73.03

FISCHER, P., BERNSTEIN, G., RHEE, G., and TYSON, A. 1997 *Astronomical Journal* **113**, 521

FISHER, K., DAVIS, M., STRAUSS, M., YAHIL, A., and HUCHRA, J. 1994 *Monthly Notices of the Royal Astronomical Society* **267**, 927

FLORES, R. and PRIMACK, J. 1996 *Astrophysical Journal Letters* **457**, L5

FLYNN, C., GOULD, A., and BAHCALL, J., 1996 *Astrophysical Journal Letters* **466**, L55

FORMAN, W., JONES, C., and TUCKER, W. 1985 *Astrophysical Journal* **293**, 102

GILMORE, G., WYSE, R., and KUIJKEN, K. 1989 *Annual Reviews of Astronomy and Astrophysics* **27**, 555

GOYAL, A., PATHAK, S., GUPTA, V., and ANAND, J. 1995 *Astrophysical Journal* **452**, 501

GREGG, M. 1995 *Astrophysical Journal* **443**, 527

GUTH, A. 1981 *Physical Review D* **23**, 347

HOGG, D., QUINLAN, G., and TREMAINE, S. 1991 *Astronomical Journal* **101**, 2274

HUANG, Z. and SARAZIN, C. 1996 *Astrophysical Journal* **461**, 622

KUIJKEN, K. and GILMORE, G. 1991 *Astrophysical Journal Letters* **367**, L7
KURKI-SUONIO, H., MATZNER, R., OLIVE, K., and SCHRAMM, D. 1990 *Astrophysical Journal* **353**, 406
LILLY, S., TRESSE, L., HAMMER, F., CRAMPTON, D., and LEFEVRE, O. 1995 *Astrophysical Journal* **455**, 108
MARKEVITCH, M. 1996 *Astrophysical Journal Letters* **465**, L1
MATSUMOTO, H. *et al.* 1997 *Astrophysical Journal* **482**, 133
MIRALDA-ESCUDE, J. and BABUL, A. 1995 *Astrophysical Journal* **449**, 18
MOULD, J., HAN, M., and BOTHUN, G. 1989 *Astrophysical Journal* **347**, 112
MOULD, J., OKE, B., and NEMEC, J. 1987 *Astronomical Journal* **93**, 53
NAKAJIMA, T., OPPENHEIMER, B., KULKARNI, S., GOLIMOWSKI, D., MATTHEWS, K., and DURRANCE, S. 1995 *Nature* **378**, 463
NEWBURY, P. and FAHLMAN, G. 1996 *Bulletin of the American Astronomical Society* **189**, 122.19
OLIVE, K. 1991 *Science* **251**, 1194
RUGERS, M. and HOGAN, C. 1996 *Astrophysical Journal Letters* **459**, L1
SACKETT, P. 1997 *Astrophysical Journal* **483**, 103
SANDAGE, A. 1982 *Astrophysical Journal* **252**, 553
SCHOMMER, R., BOTHUN, G., WILLIAMS, T., and MOULD, J. 1993 *Astronomical Journal* **105**, 97
SIGAD, Y., DEKEL, A., STRAUSS, M.S., and YAHIL, A. 1997 *Astrophysical Journal*, in press
SILK, J. 1989 *Astrophysical Journal Letters* **345**, L1
SQUIRES, G., KAISER, N., FAHLMAN, G., BABUL, A., and WOODS, D. 1996 *Astrophysical Journal* **469**, 73
STRAUSS, M. and WILLICK, J. 1995 *Physics Reports* **261**, 271
STRAUSS, M., CEN, R., OSTRIKER, J., LAUER, T. and POSTMAN, M. 1995 *Astrophysical Journal* **444**, 507
STRAUSS, M., DAVIS, M., YAHIL, A., and HUCHRA, J. 1992 *Astrophysical Journal* **361**, 49
SUTHERLAND, W. *et al.* 1996 in *Proceedings of the Workshop on Identification of Dark Matter* Sheffield, September 1996
TIFT, W., JEWSBURY, C., and SARGENT, T. 1973 *Astrophysical Journal* **185**, 115
TONRY, J. and DAVIS, M. 1981 *Astrophysical Journal* **241**, 666
TULLY, B. and SHAYA, E. 1984 *Astrophysical Journal* **281**, 31
TYTLER, D., FAN, X., and BURLES, S. 1996 *Nature* **381**, 207
VIGNAUD, D. 1995 preprint
VON HIPPEL, T. and BOTHUN, G. 1993 *Astrophysical Journal* **407**, 115
WALKER, T., STEIGMAN, G., SCHRAMM, D., OLIVE, K., and KANG, J. 1991 *Astrophysical Journal* **376**, 51
WEST, M. and BOTHUN, G. 1990 *Astrophysical Journal* **350**, 36
WEST, M. and RICHSTONE, D. 1988 *Astrophysical Journal* **335**, 532
WEST, M., JONES, C., and FORMAN, W. 1995 *Astrophysical Journal Letters* **451**, L5
WOLFENSTEIN, L. 1995 Los Alamos Preprint Archive Hep-ph/9506352
WOLFENSTEIN, L. and BEIER, E. 1989 *Physics Today* **42**, 28
ZARITSKY, D., SMITH, R., FRENK, C., and WHITE, S. 1997 *Astrophysical Journal* **478**, 39
ZINN, R. and WEST, M. 1984 *Astrophysical Journal Supplements* **55**, 45

CHAPTER FIVE

Structure Formation Scenarios and Observational Constraints

5.1 Gravitational instability

No cosmological model is complete without a credible theory for the formation of the structure that is observed in the Universe. This structure, fully described in Chapter 3, ranges from very small scales (e.g., globular clusters and dwarf galaxies), through bigger scales (e.g., galaxies and clusters of galaxies) to the largest scale (e.g., large voids, superclusters, clusters of superclusters, great walls). Although a wide variety of structure formation models can and have been considered, the basic idea that has persisted since the time of Newton is gravitational instability, which amplifies the growth of density fluctuations. Other scenarios are possible and will be considered here briefly, but gravitational instability has two great virtues: (1) it is the only known long-range force or process than can aggregate matter and (2) it is physically well understood. Gravitational instability is therefore the dominant paradigm for understanding structure formation, and this section is devoted to its detailed consideration. Subsequent sections focus on statistical methods of characterizing the distribution of structure, possible scenarios for structure formation, and the available observational constraints on competing cosmogenic scenarios.

The most viable framework for the formation of galaxies is still a self-gravitating fluid which experiences a critical instability. There are three basic equations from classical physics that govern the behavior of this self-gravitating fluid. We assume the fluid to be neutral and nonturbulent. In reality the role of magnetic fields and turbulence, as it relates to galaxy formation, is completely unknown but may be pivotal. Unfortunately, we lack the framework for incorporating these effects. The three basic equations we consider in identifying the process of gravitational instability and galaxy formation are, the equation of continuity, Euler's equation and Poisson's equation. These set the foundation for linear perturbation theory, as first applied by Jeans, to define various criteria against which the fluid is unstable to collapse.

5.1.1 The continuity equation

We define a small volume element v which is fixed in space in an assumed nonrelativistic fluid. Motion within the fluid will cause material to flow in and out of this fixed volume element, which will cause its density to fluctuate with time. For simplicity we assume the fluid flow to be in the x-direction. These density fluctuations are directly correlated with the rate of fluid flow through the volume element. The mass flow is ρv. Since the total mass within the fluid is assumed to be a constant, the divergence of the outward mass flow across the boundaries of the fixed volume element must be equal to the rate of decrease of the density within the volume element. This specifies the condition for the equation of continuity:

$$\frac{\partial \rho}{\partial t} + \nabla_x \cdot (\rho v) = 0 \tag{5.1}$$

5.1.2 Euler's equation

Since we assume there is no external perturbation on the fluid and no magnetic fields, the only two forces which are operative within the fluid are due to pressure and gravity. Pressure forces arise directly from any gradient in the fluid density. Gravitational forces arise from gradients in the gravitational field itself. These two forces oppose one another. Where gravity acts to collapse a fluid element, its internal pressure opposes this collapse. If these forces are not in equilibrium, there will be a net force which causes an acceleration in the fluid. This change in fluid velocity, as it responds to this net force, is specified by Newton's second law. For the fluid, we express this as

$$\rho \frac{dv}{dt} = -\nabla_x P - \rho \nabla_x \phi \tag{5.2}$$

where ∇P represents the pressure source and $\rho \nabla \phi$ is the gravitational source.

The effects of this net force on the fluid can be described in two different ways. We can describe the change in velocity at some fixed point in space or we can identify a particular volume element within the fluid and describe its change in velocity. The former treatment is historically called **Eulerian**, where the time derivatives are determined at this fixed point in space. The latter treatment is called **Lagrangian**, where the calculations follow a particular fluid element. In general, Newtonian mechanics is based on the Lagrangian formulation, but for our purposes, it is more useful to describe the behavior of the fluid in Eulerian terms. These two forms can be related by

$$\frac{d}{dt} = \frac{\partial}{\partial t} + (v \cdot \nabla) \tag{5.3}$$

The second term on the right-hand side represents changes which are caused by the bulk motion of the fluid. It is this bulk motion that needs to be accounted for if equation (5.2) is to become the fluid analog of momentum conservation. Substituting equation (5.3) into equation (5.2) yields

$$\frac{\partial v}{\partial t} + (v \cdot \nabla_x)v = -\frac{1}{\rho}\nabla_x P - \nabla_x \phi \tag{5.4}$$

This equation was first derived in 1755 and is now called **Euler's equation**. The equation describes the net change in velocity of a fluid element due to an imbalance between pressure and gravitational forces. The gravitational force, per unit mass, that acts on a fluid element is a reflection of the gradient in the gravitational potential generated by the distribution of the surrounding matter. This is fully specified by **Poisson's equation**

$$\nabla_x^2 \phi = 4\pi G \rho \tag{5.5}$$

5.1.3 Stability criterion in a static fluid

Our goal is to turn this self-gravitating fluid into a galaxy and to do this in an expanding Universe. Let's first consider the situation in a static medium. We assume the fluid is initially in a state of equilibrium that satisfies our three equations, namely

$$\frac{\partial \rho_0}{\partial t} + \nabla_x \cdot (\rho_0 v_0) = 0$$

$$\frac{\partial v_0}{\partial t} + (v_0 \cdot \nabla_x)v_0 = -\frac{1}{\rho_0}\nabla_x P_0 - \nabla_x \phi_0$$

$$\nabla_x^2 \phi_0 = 4\pi G \rho_0$$

We give a small perturbation to this fluid which causes it to deviate from the equilibrium state. If this fluctuation merely damps out, the equilibrium is restored and nothing interesting happens. On the other hand, if the perturbation causes the nonequilibrium state to amplify with time, the initial perturbation continues to grow without limit. If the pressure, density and velocity in the fluid are such that it is stable, then all perturbations will eventually decay to zero. The technique of linear perturbation analysis allows each instant of the fluid's evolution to be described as the superposition of two separable components – one corresponds to the equilibrium state and the other corresponds to the perturbed state. The key simplification is to consider arbitrarily small perturbations such that the perturbed state is very close to the equilibrium state. This allows us to write the density, pressure, velocity and gravitational potential of the perturbed state as follows:

$$\rho = \rho_0 + \rho', \qquad P = P_0 + P', \qquad v = v_0 + v', \qquad \phi = \phi_0 + \phi' \tag{5.6}$$

In all cases the perturbed quantities (denoted by $'$) are very much smaller than the initial quantity. To further simplify the physics, we apply the reasonable demand that all perturbations in the fluid involve no energy gain or loss, but only changes in density, pressure and velocity. Thus, we assume these perturbations to be adiabatic.

We now substitute our expression for the perturbed density into the equation of continuity to yield

$$\frac{\partial(\rho_0 + \rho')}{\partial t} + \nabla_x \cdot ((\rho_0 + \rho')(v_0 + v') = 0 \tag{5.7}$$

Expanding out the terms yields

$$\frac{\partial \rho_0}{\partial t} + \frac{\partial \rho'}{\partial t} + \nabla_x \cdot (\rho_0 v_0) + \nabla_x \cdot (\rho' v_0) + \nabla_x \cdot (\rho_0 v') + \nabla_x \cdot (\rho' v') = 0$$

Since the $'$ quantities are very small, the product $\rho' v'$ is vanishingly small and can be dropped. The purely equilibrium terms cancel and we are left with

$$\frac{\partial \rho'}{\partial t} + \nabla_x \cdot (\rho' v_0) + \nabla_x \cdot (\rho_0 v') = 0 \tag{5.8}$$

Linear perturbation theory thus produces an equation which is linear in all the surviving terms. We can apply the same technique to the Euler and Poisson equations to yield

$$\frac{\partial v'}{\partial t} + (v_0 \cdot \nabla_x) v' + (v' \cdot \nabla_x) v_0 = -\frac{\rho'}{\rho_0^2} \nabla_x P_0 - \frac{1}{\rho_0} \nabla_x P' - \nabla_x \phi' \tag{5.9}$$

$$\nabla_x^2 \phi' = 4\pi G \rho' \tag{5.10}$$

The equations presented here were first derived by Jeans. His solution was to assume a uniform and static fluid in which $v = v_0$ at all times; hence ρ and P are constants. However, Euler's equation indicates this condition can only be satisfied if $\nabla\phi_0 = 0$; a condition that, by Poisson's equation, requires $\rho_0 = 0$. Hence, for the case of a static fluid, it is only possible to satisfy the equilibrium equations if we have the unphysical situation that $\rho_0 = 0$. This means that no equilibrium state is even possible for a static, self-gravitating fluid.

Yet it is clear that perturbed fluids can only be understood in this framework if we have the condition $\nabla\phi_0 = 0$. Jeans' resolution to this dilemma was to apply some physical insight to simplify the mathematics by assuming that the gravitational field which originates from the unperturbed state can be ignored; hence the self-gravity of the fluid is determined only by the perturbed component. This assumption, known informally as the Jeans swindle, has most validity in the case of an infinitely long, uniform fluid. In this case, any dynamic evolution in the fluid is uniquely determined by the self-gravity associated with the perturbed state. If we now set $v_0 = 0$, $\rho_0 =$ constant, $P_0 =$ constant and $\nabla\phi_0 = 0$ the linearly perturbed fluid dynamics equations become

$$\frac{\partial \rho'}{\partial t} + \nabla_x \cdot (\rho_0 v') = 0 \tag{5.11a}$$

$$\frac{\partial v'}{\partial t} = -\frac{1}{\rho_0} \nabla_x P' - \nabla_x \phi' \tag{5.11b}$$

$$\nabla_x^2 \phi' = 4\pi G \rho' \tag{5.11c}$$

The time derivative of equation (5.11a) is

$$\frac{\partial}{\partial t}\left[\frac{\partial \rho'}{\partial t} + \rho_0 \nabla_x \cdot v'\right] = \frac{\partial^2 \rho'}{\partial t^2} + \rho_0 \nabla_x \cdot \frac{\partial v'}{\partial t} = 0 \tag{5.12a}$$

and the divergence of equation (5.11b) is

$$\nabla_x \cdot \frac{\partial v'}{\partial t} = \frac{-1}{\rho_0} \nabla_x^2 P' - \nabla_x^2 \phi' \tag{5.12b}$$

This gives us the term $\nabla^2 \phi$, which appears in Poisson's equation. Equations (5.12a) and (5.12b) can now be combined to yield

$$\frac{\partial^2 \rho'}{\partial t^2} - \nabla_x^2 P' - 4\pi G \rho_0 \rho' = 0 \tag{5.13}$$

which fully describes the time rate of change of the small density perturbation ρ'. This time rate of change competes with the contribution of pressure gradients and changes in the gravitational potential. Hydrostatic equilibrium ($\partial^2 \rho' / \partial t^2 = 0$) is recovered from this framework if the pressure and gravity terms exactly cancel each other. But the physical importance of equation (5.13) lies in the fact that density perturbations (e.g., ρ') require increasingly larger pressure gradients to stabilize them against further amplification. Equation (5.13) specifies the conditions for an effective runaway process that allows continued growth of the density perturbation. But amplification of the density perturbation requires that the perturbation travels through the fluid medium at some characteristic speed. We can easily derive this characteristic speed by ignoring the effects of gravity. In this case

$$\frac{\partial^2 \rho'}{\partial t^2} - \nabla_x^2 P' = 0 \tag{5.14}$$

This equation has two unknowns (ρ' and P') which are related to each other via the equation of state of the fluid. Since our fluid is really hydrogen gas in the early Universe, it is reasonable to use the ideal gas law

$$P = \frac{kT}{m} \rho^\gamma \tag{5.15}$$

where k is Boltzmann's constant, T is the temperature of the gas, m is the mass of an individual particle in the gas and $\gamma = 1$ for an isothermal gas or 5/3 for an adiabatic gas. In this formulation any density increases lead directly to increases in pressure, which push outward to smooth any inhomogeneities in the density. This pressure exerts a force on regions that surround our fluid element, causing their density and pressure to increase. In this manner a "pressure wave" propagates through the fluid as individual fluid elements are coupled to one another. This pressure or acoustic wave propagates at a characteristic speed, the sound speed c_s. The sound speed is directly related to the rate of change of pressure with density at constant entropy. For an ideal gas law, the sound speed is defined as $c_s^2 \equiv \partial P/\partial\rho$. From equation (5.15) we have

$$c_s^2 = \gamma\frac{P}{\rho} \tag{5.16}$$

We substitute this into equation (5.14) to yield what looks very much like a classical wave equation:

$$\frac{\partial^2\rho'}{\partial t^2} - \frac{c_s^2}{\gamma}\nabla_x^2\rho' = 0 \tag{5.17}$$

In the absence of gravity, small density perturbations will lead to small pressure perturbations that propagate through the fluid like a wave. In essence this leads to variations in density through the medium that also propagate in the x-direction like a wave and can be described as

$$\rho'(x, t) = A\cos(kx - \omega t + \phi)$$

where A is the amplitude of the wave, ω is the frequency and k is the wavenumber of the oscillation where k is defined as $2\pi/\lambda$, and ϕ is the phase of the wave. In general, A can be identified with the initial density enhancement and ϕ can be set to zero by the proper choice of coordinates. Furthermore, c_s is related to k and ω via the familiar dispersion relation

$$\omega^2 = c_s^2k^2$$

If we consider the effects of gravity via the same plane wave analogy, an additional term appears in the dispersion relation:

$$\omega^2 = c_s^2k^2 - 4\pi G\rho_0 \tag{5.18}$$

This equation now specifies the behavior of the fluid in different physical limits. At very high wavenumbers (very short wavelengths) gravity can be effectively ignored and the physics reduces to the acoustic wave behavior we obtained previously. In this case the perturbation oscillates without energy dissipation until it is damped out by viscosity and/or friction. In the very early Universe, radiation and matter are coupled such that photons become a significant source of viscosity, effectively preventing the growth of short wavelength perturbations. This would lead to a characteristic length scale below which structure should not form.

Gravity, however, does dominate in the limit of very long wavelength perturbations, in which case the pressure terms become unimportant. Then ω^2 becomes negative and requires a solution that is a complex number. The simplest form of this solution is $\omega^2 = -b$. For this solution the density perturbation has a time dependence which goes as

$$\rho' \propto e^{\pm b^{1/2}t} \quad \text{or} \quad e^{\pm t/\tau} \tag{5.19}$$

where τ is the e-folding time over which the perturbation grows or decays:

$$\tau = \frac{1}{\sqrt{4\pi G\rho_0}} \tag{5.20}$$

We have now derived the general result that very long wavelength density perturbations in the fluid will have an amplitude that grows exponentially in time and the growth rate is inversely proportional to the square root of the initial fluid density. This result should be intuitive. Denser perturbations have more self-gravity associated with them; hence they can drive the instability at a faster rate. It then follows there must be some critical length scale λ_j which defines the boundary between perturbations that damp out from those that amplify exponentially. This critical value is known as the **Jeans length**, defined as

$$\lambda_j \equiv \frac{2\pi}{K_j} = c_s\left(\frac{\pi}{G\rho_0}\right)^{1/2} \tag{5.21}$$

And for a spherical perturbation of diameter λ_j the **Jeans mass** is

$$M_j = \frac{\pi}{6}\rho_0(\lambda_j)^3 = \frac{\pi}{6}\rho_0\left(\frac{\pi c_s^2}{G\rho_0}\right)^{3/2}$$

The only physical parameters in this characterization are c_s and ρ_0. Since c_s is essentially a measure of the pressure in the fluid, the Jeans criterion for gravitational collapse is a direct competition between internal pressure and external self-gravity. And since gravity is largely a volume effect, this analysis implies that on larger scales the self-gravity will ultimately exceed the pressure. This is the physical premise behind star formation in molecular clouds. As we will see shortly, the main physical difference with respect to galaxy formation is how this criterion must be established in an expanding Universe where the density is constantly dropping. Since pressure is proportional to temperature, a larger Jeans mass is required to overcome the internal pressure for high temperatures. This essentially precludes any possibility of structure formation at early times when the temperature is high.

High temperature also means very high radiation pressure. As the matter is coupled to this radiation, the matter experiences radiation drag and is redistributed in a manner which is the same as the distribution of the radiation. The CMB observations show that the radiation is distributed nearly homogeneously. If the matter distribution ends up completely homogeneous, there will be no net gravity and no structures will form. Hence density inhomogeneities must be maintained throughout the radiation-dominated era in order that structure can form in the matter-dominated era. These density enhancements in the matter produce the Sachs–Wolfe effect discussed earlier. *COBE* has now measured the overall amplitude to be $\approx 1.5 \times 10^{-5}$. Can such a minute density enhancement actually be amplified to produce structure? The answer is, easily, because of the exponentially growing nature of the fluctuation.

Let's consider the case of a modest galaxy with total mass $\approx 10^{11} M_\odot$. If we assume this object is initially composed only of hydrogen gas then there are 10^{68} atoms involved. Purely random fluctuations (which go as $1/\sqrt{N}$) in the initial distribution of atoms would then lead to a random density fluctuation of $\rho_1/\rho_0 \approx 10^{-34}$.

After 80 *e*-folding times, this almost negligible density perturbation would have grown to $\rho_1 \approx \rho_0$. By 100 *e*-folding times, this perturbation would have grown to $\rho_1 \approx 10^4\rho_0$, which is about the current ratio of the average density of a galaxy to the density of the Universe. Suppose we start growing this perturbation after recombination has occurred and the Universe is no longer ionized. At this redshift ($z \approx 1100$) the average density of the Universe is $\approx 10^{-18}$ g cm^{-3}, leading to an *e*-folding time of $\tau \approx 10^4$ years, ample time for the perturbation to grow. The fallacy of this argument is that the Universe is expanding, so ρ is steadily decreasing, causing τ to increase. This means the growth rate of perturbations in an expanding Universe is considerably slower than for a static medium.

There is an additional complication. Galaxies are not mildly overdense structures but have a density contrast of 10^4 to 10^5. They are therefore highly nonlinear structures, which means that linear perturbation theory breaks down long before the process of galaxy formation is complete. This nonlinearity occurs on larger scales as well. For instance, a typical cluster of galaxies has a density contrast of ≈ 200 within $R \approx 1$ Mpc. Even a supercluster of galaxies, in scales of 5–10 Mpc, is overdense by factors of 10–20. As the observed structure in the Universe is strongly nonlinear, this means that growing gravitational perturbations rapidly cross into the nonlinear regime, thus nullifying the original premise of *small* perturbations.

To determine the physics of what happens when this boundary is crossed requires an understanding of the very complex theory of nonlinear perturbations. A chief feature of nonlinear perturbations involves nonlinear partial differential equations, which contain important cross terms that relate to the coupling of perturbations at different wavelengths. Recall that in linear perturbation theory there is no coupling. This mode coupling likely plays a crucial, but largely unknown role in the advanced physical development of these instabilities. Currently, these effects are best studied via large N numerical simulations. Analytical descriptions of the process remain fairly opaque and certainly beyond the level of this book. This doesn't mean that linear theory is without merit, but a necessary condition for the growth of perturbations via nonlinear processes is clearly the existence on some scale of perturbations having $\rho_1/\rho_0 \geq 1$. Hence linear perturbation theory should continue to be an adequate description of the initial growth and development of the perturbations that have evolved to produce the observed structure. Our difficulty with nonlinear effects suggests that the formation of smaller-scale structure (e.g., galaxies) may be much more difficult to describe than the formation of the largest-scale, lowest density features that we observe.

5.1.4 Gravitational instability in an expanding Universe

The expanding medium means that, for any small density perturbation, there will be competition between self-gravity, attempting to increase the density, and the general expansion of the Universe, which decreases the density. The expansion rate of the Universe was initially very high, which makes it difficult for any perturbation to grow by continually increasing its density. Nevertheless, if we restrict ourselves to a specific physical regime, we can apply the previous formalism to the problem of perturbation growth in an expanding medium. We need to make three assumptions:

- We only consider perturbations on scales smaller than the horizon size (the *COBE*-measured anisotropy is for many horizons).
- We assume nonrelativistic perturbations only.
- We assume that pressure is negligible; on the scale of the horizon, this assumption is justified after inflation ends.

In the previous treatment in a static medium, we were able to set $v_0 = 0$. But since the medium is now expanding, its initial velocity clearly can't be zero. However, we can still retain the simplifying assumptions of homogeneity and isotropy in the fluid:

$$\nabla_x \rho_0 = 0 \quad \text{and} \quad \nabla_x P_0 = 0$$

We then proceed as before but retain all terms which have v_0; we start with

$$\frac{\partial \rho'}{\partial t} + \nabla_x \cdot (\rho_0 v') + \rho' \nabla_x \cdot v_0 + v_0 \cdot \nabla_x \rho' = 0 \tag{5.22a}$$

$$\frac{\partial \rho'}{\partial t} + (v_0 \cdot \nabla_x) v' + (v' \cdot \nabla_x) v_0 = \frac{-1}{\rho_0} \nabla_x P' - \nabla_x \phi' \tag{5.22b}$$

$$\nabla_x^2 \phi' = 4\pi G \rho' \tag{5.22c}$$

Only Poisson's equation remains unchanged.

Recall that the Eulerian approach examined the behavior of a fixed point in space. In an expanding Universe, the density at this fixed point will change with time both due to the expansion of the Universe and the real growth of perturbations due to gravitational instability. If we transform the fluid dynamic equations to a reference frame that follows the general expansion, these two effects can be separated. Transforming to this frame means converting from the Eulerian description to the Lagrangian description. The relationship between the Lagrangian and Eulerian derivatives can be expressed as

$$\frac{\partial \rho'}{\partial t} = \frac{d\rho'}{dt} - (v_0 \cdot \nabla_x) \rho'$$

Substituting this into the equation of continuity and using the vector identity $\nabla_x \cdot \rho v = \rho \nabla_x \cdot v + v \cdot \nabla_x \rho$ yields

$$\frac{d\rho'}{dt} + \rho_0 \nabla_x \cdot v' = 0 \tag{5.23}$$

or

$$\frac{d}{dt}\left(\frac{\rho'}{\rho_0}\right) + \nabla_x \cdot v' = 0 \tag{5.24}$$

We now define the density contrast δ to be the dimensionless quantity ρ'/ρ, which leads to

$$\frac{d\delta}{dt} + \nabla_x \cdot v' = 0 \tag{5.25}$$

This is the Lagrangian form for the equation of continuity. Implicit in this form is the use of an absolute coordinate system. We now make use of a coordinate transformation to define a system of coordinates that are comoving with the universal expansion. The advantage of describing the behavior of the fluid in comoving coordinates is that the comoving position r remains constant for a fluid element moving with the overall expansion. Its physical coordinate x would be constantly changing. The comoving position r can be expressed as

$$r = \frac{x}{a(t)}$$

where $a(t)$ is the time evolution of the universal scale factor a, first presented in Chapter 1 as $R(t)$. Thus

$$\nabla_r = \frac{1}{a}\nabla_x \tag{5.26}$$

where ∇_r refers to derivatives that are taken with respect to comoving coordinates. In addition to a spatial coordinate transformation, we also want to transform the absolute velocity of a fluid element which is located at fixed spatial position x. The motion of this fluid will have two components: a component driven by the general expansion of the Universe, and a peculiar velocity that can arise from some external perturbation. We transform the proper velocity $v(r, t)$ via the relation

$$v(r,t) = \frac{dx}{dt} = r\frac{da}{dt} + a\frac{dr}{dt}$$

where the first term represents the normal expansion of the fluid due to the expansion of the Universe (this term is equivalent to v_0 in the previous treatment) and the second term is the peculiar velocity. In the context of linear perturbation theory, this peculiar velocity, if small, can be associated with what we have been calling v'. It is convenient to express the comoving perturbed velocity dr/dt as u, where

$$v'(r,t) = a(t)\,u$$

We can now specify the comoving Lagrangian form of the three basic fluid equations:

- *Equation of continuity*

$$\frac{d\delta}{dt} + \nabla \cdot v' = 0 \rightarrow \frac{d\delta}{dt} + \nabla_r \cdot u = 0 \tag{5.26a}$$

- *Perturbed Euler equation*

$$\frac{dv'}{dt} + (v' \cdot \nabla_x)v_0 = \frac{-1}{\rho_0}\nabla_x P' - \nabla_x \phi' \rightarrow \frac{du}{dt} + \frac{2}{a}\frac{da}{dt}u$$

$$= \frac{-\rho_0}{a^2}\nabla_r P' - \frac{1}{a^2}\nabla_r \phi' \tag{5.26b}$$

- *Perturbed Poisson equation*

$$\nabla_x^2 \phi' = 4\pi G \rho_0 \delta \rightarrow \nabla_r^2 \phi' = 4\pi G \rho_0 \delta a^2 \tag{5.26c}$$

Recall that our main objective is to determine the rate of growth of density perturbations δ in this expanding medium. We proceed as before by first taking the time derivative of the equation of continuity, now expressed in comoving form:

$$\frac{d}{dt}\left(\frac{d\delta}{dt}\nabla_r \cdot u\right) = \frac{d^2\delta}{dt^2} + \nabla_r \cdot \frac{du}{dt} = 0 \tag{5.27}$$

Next we take the divergence of Euler's equation

$$\nabla_r \cdot \frac{du}{dt} + \frac{2}{a}\frac{da}{dt}\nabla_r \cdot u = -\frac{\rho_0}{a^2}\nabla_r^2 P' - \frac{1}{a^2}\nabla_r^2 \phi' \tag{5.28}$$

Using the previous expressions we can now eliminate terms involving u, as they have been rewritten as terms involving δ, the physical quantity of interest. Finally, by making use of Poisson's equation, we can derive

$$-\frac{d^2\delta}{dt^2} - \frac{2}{a}\frac{da}{dt}\frac{d\delta}{dt} = -\frac{\rho_0}{a^2}\nabla_r^2 P' - \frac{1}{a^2}4\pi G \rho_0 \delta a^2 \tag{5.29}$$

and using the relation $c_s^2 = dp/d\rho = P'/\rho'$ we can derive

$$\frac{d^2\delta}{dt^2} + \frac{2}{a}\frac{da}{dt}\frac{d\delta}{dt} = 4\pi G \rho_0 \delta + \frac{c_s^2}{a^2}\nabla_r^2 \delta \tag{5.30}$$

We have now arrived at our full specification of the time evolution of density perturbations in an expanding universe. We can compare this "wave" equation with equation (5.17), the case of the static medium, to discern the presence of the additional term involving da/dt (or $\dot{a}$). This term, which must be positive in an expanding Universe, clearly shows that universal expansion acts to retard the growth rate of density fluctuations. Furthermore, as in the static medium, we still have the same competition between pressure and gravity. Hence the Jeans length criterion remains valid and in this formulation it can be thought of as a competition between two timescales: the characteristic timescale for gravitational growth $(G\rho_0)^{-1/2}$ and the crossing time λ_j/c_s for a pressure wave to move across the Jeans length scale. For wavelengths much longer than the Jeans length, this crossing time will be very much longer than the gravitational free-fall time, so the solutions to equation (5.30) are pressureless. In this case we solve

$$\frac{d^2\delta}{dt^2}+\frac{2}{a}\frac{da}{dt}\frac{d\delta}{dt}=4\pi G\rho_0\delta \tag{5.31}$$

in two important limiting cases, $\Omega=1$ and $\Omega \ll 1$.

5.1.5 The $\Omega = 1$ solution

As emphasized in Chapter 4, $\Omega=1$ is predicted by inflation; hence it remains the preferred model. Chapter 1 established the following relations for this model:

$$\frac{1}{a}\dot{a}=\frac{2}{3}t^{-1}$$

$$\rho=\frac{1}{6\pi G}t^{-2}$$

which we can now substitute into equation (5.31) to yield

$$\frac{d^2\delta}{dt^2}\frac{4}{3}t^{-1}\frac{d\delta}{dt}=\frac{2}{3}t^{-2}\delta$$

The solution to this partial differential equation is a power law of the form

$$\delta \propto t^{\alpha}$$

Substitution yields

$$\alpha(\alpha-1)t^{\alpha-2}+\tfrac{4}{3}\alpha t^{-1}t^{\alpha-1}=\tfrac{2}{3}t^{-2}t^{\alpha}$$

which has two solutions, $\alpha=2/3$ and $\alpha=-1$. These two solutions correspond to the growing mode (2/3) and the decaying mode (−1). The decaying mode has much faster time evolution, so it quickly becomes unimportant, leading to the result that the growing mode amplifies at the rate of

$$\delta \propto t^{2/3} \propto a \propto (1+z)^{-1}$$

In the static medium we derived an exponential growth rate, but in the expanding medium the growth rate is very much slower. In fact, such a slow growth rate caused early cosmologists (Lifshitz 1946; Bonnor 1957) to conclude that gravitational instability could not have produced the structure we observe today, because the growth rate is too slow relative to the age of the Universe. Indeed, this remains a problem for the very largest features (e.g., the Great Wall) that we see in the galaxy distribution.

5.1.6 The $\Omega \ll 1$ solution

Here the simplest thing to do is to assume $\Omega=\rho=0$. Then

$$\frac{da}{dt} = \frac{a}{t}$$

and our wave equation reduces easily to

$$\frac{d^2\delta}{dt} + 2\frac{d\delta}{dt}t^{-1} = 0$$

Using our trial solution $\delta \propto t^{\alpha}$, we can write

$$\alpha(\alpha - 1) + 2\alpha = 0$$

which has the two solutions $\alpha = -1$ (similar to before) and $\alpha = 0$. Thus, in an empty Universe, or more realistically in a very open Universe, there is no time dependence on the growth of fluctuations. Instead they maintain themselves at a constant comoving density. This is because there is simply very little matter to generate any self-gravity structure. The matter density Ω is a function of the redshift at which it is measured. Specifically, for Friedmann models, we have

$$\Omega(z) = \frac{\Omega_0 + \Omega_0 z}{(1 + \Omega_0 z)}$$

where Ω_0 is the present value. From the observations previously discussed, Ω_0 appears to be in the range 0.1–0.3. In this case, at $z = 10$, Ω was at least 0.55 and at $z = 100$ it would be nearly 1. In fact, $\Omega(z)$ is only independent of redshift in the special cases of $\Omega = 0$ or 1. When $\Omega = 0$ the Universe has always been massless and none of us are here, whereas $\Omega = 1$ is predicted from inflation. The important point is that, even in a low density Universe, the major time of perturbation growth from $z = 1100$ to $z = 10$ would have occurred in the domain of $\Omega \approx 1$, in which case the growth rate would go as $t^{2/3}$. Note that in an open Universe there will be some redshift at which Ω does begin to significantly deviate from 1, leading to a much slower growth rate. Hence structure formation in an open Universe is effectively over when $\Omega(z)$ approaches 0. This condition is satisfied by $z \approx (1/\Omega_0) - 1$. Thus if Ω_0 is 0.1 then structure formation by this process should be over at redshift $z = 9$.

5.1.7 Numerical games with the Jeans mass

Recall that the Jeans mass at any time in the Universe is a function of its temperature and density. For our case of an ideal gas, the Jeans mass for a spherical perturbation is

$$M_J = \frac{\pi}{6}\rho\left(\frac{\pi kT}{m\rho_0 G}\right)^{3/2}$$

It is interesting to calculate M_J at recombination to see if that corresponds to any known astrophysical objects. Recombination occurs at $z \approx 1100$. The temperature of the Universe scales as $(1+z)$, so if it's 2.7 K currently then it was ≈ 3000 K at recombination. The density scales as $(1+z)^3$. The current density of the Universe is in the range $\rho = 10^{-29}$ to 10^{-30} or $\rho = 10^{-20}$ to 10^{-21} at recombination. The lower range of values would correspond to a very open Universe and the upper range is the approximate density in an $\Omega = 1$ model (recall that ρ_{crit} depends on H_0). For this range of current densities, M_J is $(1\text{–}4) \times 10^5 M_\odot$. This mass is very similar to the masses of globular clusters, which are known to be the oldest collections of stars in the Universe. There are many who think this is just a coincidence, but from a physics point of view, we have the logically consistent argument that objects with mass a few times $10^5 M_\odot$ could collapse under their own self-gravity following recombination. Since the Jeans mass is a minimum mass, it is also clear that galaxy $(10^{11} M_\odot)$ and cluster $(10^{15} M_\odot)$ size perturbations are far above the Jeans criterion, so pressure can be ignored.

If we measure the present-day densities of globular clusters, galaxies and clusters of galaxies, we find a sequence of progressively decreasing densities. Since the dynamical timescale of a density enhancement goes as $(G\rho)^{-1/2}$, these dense, globular cluster mass perturbations are the first to collapse. Although it's difficult to guess the initial size and density of these perturbations, we can make a rough estimate of how fast they could have collapsed. If our galaxy-size perturbation of mass $10^{11} M_\odot$ had an initial radius of ≈ 50 kpc, the initial density would be $\approx 2 \times 10^{-25}$ g cm^{-3}, leading to a dynamical timescale of 300 million years. At $z = 10$ the Universe is approximately 3% of its present age, meaning there is time for a galaxy-size perturbation to have gravitationally collapsed by this redshift. The halo parameters deduced by Zaritsky *et al.* (1997) lead to a collapse time approximately twice as long as this case. Chapter 6 will introduce the properties of a newly discovered population of galaxies known as low surface brightness (LSB) galaxies. These galaxies appear to be up to 100 times lower in mass density than normal galaxies; hence they would represent galaxy-size perturbations which have long collapse times and late formation. In fact, it's important to emphasize that any range of matter densities in protogalaxies, at fixed mass, produces a range of collapse times and formation epochs.

At the smallest scale are globular clusters which range in mass from 10^5 to $10^6 M_\odot$ over a scale size approximately 1000 times less than a typical galaxy. This leads to a density which is $\approx 10^4$ times that of a galaxy. Hence their dynamical timescales are only a few million years at most and they easily could have formed by $z = 100$. Extremely dense globular clusters would have formed even earlier and could be subject to some interesting dynamical evolution that might lead to the formation of gamma ray burst objects (Chapter 6).

At the other size scale, a cluster of galaxies will have a significantly longer collapse time. Although these structures have 10^3 to 10^4 times more mass than a galaxy, their scale sizes are at least 100 times larger. Hence the overall density is down by two orders of magnitude, which yields to order of magnitude longer collapse times. We therefore would not expect to find clusters of galaxies at high redshift. For a 15 billion year old Universe and an average collapse time of 3 billion years, we would expect no clusters to exist beyond redshift $z \approx 2$.

5.2 Statistical characterization of structure

Structure as a function of physical scale size (wavenumber) is usually described in terms of a power spectrum

$$P(k) \equiv \langle |\delta_k|^2 \rangle = A k^n \tag{5.32}$$

where δ_k is the Fourier transform of the primeval density fluctuations which are amplified by gravity to produce the observed structure. These density fluctuations give rise to the observed CMB anisotropy, thus δ_k has some observational constraint. The spectral index n determines the relative distribution of power on various scales. Values of n which are less than 0 produce a spectrum with power on very large scales. The amplitude A depends upon which structure formation scenario is being considered. In principle, A is also subject to observational constraint. $P(k)$ itself is most correctly considered as the functional representation of the power per unit volume in k-space (Bertschinger 1992). Observations reveal the power (or correlation function) per unit volume in physical space. It is then necessary to define a framework that allows these observations to be mapped back on to $P(K)$. This greatly restricts the choices of structure formation scenarios which can be observationally constrained (Strauss and Willick 1995). As discussed in more detail below, this mapping can only be done under the hypothesis that the phases of δ_k are random. Fortunately, the random-phase hypothesis is directly predicted from inflation and, in fact, it would hold in any isotropic Universe (i.e., there is no preferred direction).

The determination of $P(k)$ only provides a statistical description of the observed distribution of density fluctuations. By itself, it provides little physical information for the actual formation of structure; instead it statistically characterizes the range and frequency of scales over which structure does form. For instance, it is possible (and in some cases easy) to construct a physical scenario that predicts a form for $P(k)$ that roughly agrees with observations, but which contains no real physics to produce structure itself. Hence one needs to be wary and to carefully differentiate between statistical structure formation models and physical models (one obviously wants a model which can do both).

This need to differentiate is quite clear when one considers structure formation via gravitational instability. The amplification of initial density fluctuations on some scale δ_r naturally produces a statistical density field. In turn, this statistical density field produces a statistical distribution of structure collapse and formation times. As structures on different scales collapse at different times, there may well be energy feedback to the entire system, which interferes with the collapse of lower density structures. This feedback is not accounted for by any statistical theory. A possible example is provided by QSOs. Suppose QSOs are the collapse of subgalaxy-scale, very dense perturbations. A massive star cluster forms out of the gas initially (perhaps being the first generation of stars to produce metals). The subsequent evolution of the massive star remnants into black holes and neutron stars is a gravitational coalescence to form a supermassive black hole. As baryonic gas continues to infall on this dense seed, it becomes the power source for the QSO, and suddenly the

Universe contains a large number of sources of ionizing radiation. If this occurs while other collections of neutral hydrogen gas inside dark matter potentials are quiescently collapsing, it will be a significant source of reheating and will further delay the formation of these systems. One could even conceive of situations where the energy feedback from material falling on these dense structures might evacuate regions around them, leading to a void.

Thus, in fairness, it is really quite unclear whether the apparent cellular pattern in the galaxy distribution – manifest in the distribution of voids, walls and clusters – can be physically produced from gravitational instability alone. Indeed, the overall local topology strongly implies that structure formation has a hydrodynamic component associated with it, and it is quite difficult to fold this into efforts to recover both $P(k)$ and real galaxy formation. Structures like the Great Wall are extremely difficult to understand from gravitational instability considerations alone, as it's likely the Universe is not old enough to have built such a large structure in this manner. Thus the profound theoretical challenge that is posed by the complexity of the observed galaxy distribution lies in achieving a physical understanding of the processes for creating a void-filled Universe, with small-scale structure apparently forming at the intersections between these voids.

Adding to this complex mix is our unfortunate situation regarding the unknown makeup and amount of dark matter in the Universe. Structure formation scenarios are almost totally driven by the assumed form of dark matter as well as its overall contribution to the mass density of the Universe. Each form of dark matter carries with it different predictions for the form of $P(k)$. In the simplest terms, CDM scenarios involve nonrelativistic particles that dissipate and clump at very early times to form **small-scale structure**. HDM involves relativistic particles that cannot clump early on; they can generate only large-scale perturbations and thus power on large scales.

5.2.1 What comes first – galaxy-size or cluster-size potentials?

The real Universe is probably some complex hybrid of HDM and CDM, but we can still consider three limiting cases of structure formation that arise under the gravitational instability paradigm. Each of these three cases assumes that a single kind of particle dominates the mass in the Universe.

The top-down scenario

Under this scenario, all fluctuations which are smaller than a horizon size are erased by the free streaming motion of relativistic particles. This scenario best applies in an HDM-dominated Universe in which the HDM particle is a neutrino. A convenient way to express the critical density in units of energy density is

$$\rho c \leq 11h^2 \text{ keV cm}^{-3} \rightarrow m_\nu \leq 110h^2 \text{ eV}$$

where $h = H_0/100$. The conversion from h to m_ν uses a neutrino density at $z = 0$ of 100 cm^{-3}. For $h = 1$ ($H_0 = 100$) $\Omega_\nu = 1$ requires $m_\nu \approx 100$ eV For $h = 1/2$ ($H_0 = 50$) $m_\nu \approx 30$ eV yields $\Omega_\nu = 1$. If, as suggested in Chapters 3 and 4, $h \approx 0.8$ and $m_\nu \approx 2.5$ eV then $\Omega_\nu = 0.03$.

The free streaming of neutrinos stops when they become nonrelativistic. This occurs through expansion and cooling of the Universe. When the Universe has cooled to the point where kT is equal to the rest mass energy of the neutrino (m_ν), the neutrinos become nonrelativistic. At this point the Universe has some horizon size r_{hor} and variations in neutrino density can only occur on scales larger than r_{hor}. The total mass within the horizon is approximately

$$r_{hor}^3 m_\nu^4$$

For $m_\nu \approx 30$ eV the horizon mass is $\approx 10^{16} M_\odot$. This mass is similar to the mass of the putative Great Attractor discussed in the last chapter. For a neutrino-dominated Universe, potentials of this mass would be the first to form. The formation of smaller-scale structure would occur from fragmentation of gas within these potentials. This scenario then predicts that all smaller-scale structures should be embedded in larger-scale structures, which is in good qualitative agreement with the observations. The greatest strength of this scenario lies in its natural ability to produce power on large scales (e.g., superclusters). Its greatest weakness (see below) lies in the supreme difficulty of producing small-scale structure early on in the Universe.

The bottom-up scenario

This is the scenario that Newton would have preferred, as structure in the Universe is built by the hierarchical gravitational clustering of subunits. The minimum mass of these subunits is set by the Jeans instability criterion. In a CDM-dominated Universe, the hypothetical CDM particles are not subject to radiation drag in the early Universe, so they can begin to clump via gravitational instability at very early times (perhaps as early as the end of the inflationary epoch). Amplification of these seeds will then produce density fluctuations which can accrete baryonic material after recombination. As this material continues to flow into the density fluctuation, it continues to grow in size, thus sweeping up more material in the vicinity. Galaxy-size objects are eventually made via this gravitational coalescence of subunits then clusters of galaxies are made later in the Universe via the continuation of this gravitational clustering hierarchy.

The greatest strength of the CDM-dominated scenario is the natural production of small-scale structure that should be embedded in a large-scale distribution of dark matter. Furthermore, galaxy formation is something that occurs early on. The greatest weakness of the CDM model lies in its inability to produce the truly large-scale structure that is observed. An interesting consequence of the bottom-up scenario is the suggestion there may be totally dark galaxies, i.e., CDM-dominated potentials which were unable to trap and confine baryonic gas that subsequently fragmented into stars to produce a luminous galaxy to mark the location of that potential. And the bottom-up scenario also predicts that cluster formation via gravitational merging of subunits is continuing through the present epoch. This would provide a natural source to generate the observed peculiar velocities. The detection of high redshift clusters would not be expected under this scenario.

The baryon-dominated scenario

This is a variant of the bottom-up scenario that would occur in a low Ω Universe dominated by baryons. Baryonic fluctuations cannot easily grow during the radiation-dominated era, so the relevant issue is the amplitude of the Jeans length at the time of recombination. Although it is quite unlikely that the Universe is baryon dominated, this scenario naturally produces old globular clusters whose formation is difficult to understand in the other two scenarios. But a particular problem with this scenario is galaxy formation. Presumably it would occur via the gravitational coalescence of globular cluster size objects. During this process, tidal forces between globular clusters should liberate a lot of the more weakly bound stars. Although this mechanism does produce a field population of old halo stars, observations of our own galaxy as well as others show that this halo of stars is extremely diffuse and fails, by two orders of magnitude, to contain enough mass to account for flat rotation curves. It also seems likely that the formation of galaxies via the gravitational clustering of Jeans mass density enhancements is not very efficient, so we would not expect to find many baryons in galaxies. Hence, a test of this scenario would involve determining the ratio of baryons in galaxies to baryons distributed throughout an intergalactic population. Although there is some evidence for an intergalactic population of baryons (Chapter 6) it is fairly clear there is not an order of magnitude more baryons outside of galactic potentials than inside them.

5.2.2 Biasing in the Universe

The concept of linear biasing between the distribution of light and mass was introduced in Chapter 3. Since the structure formation models we are considering all predict the distribution of mass, the role of biasing is pivotal when mapping the observed light distribution back into the model. One of the first indicators that biasing does exist came from comparing the galaxy–galaxy correlation function to the cluster–cluster correlation function. The observations indicate that, although both correlation functions have the same power law slope, the cluster–cluster correlation function has an amplitude about 20 times larger. If galaxies and clusters have both arisen due to gravitational instability and amplification of fluctuations in the primordial density field, then both scales should trace large-scale structure equally well. But the large difference in correlation amplitude is quite inconsistent with this expectation.

To reconcile this discrepancy, Nick Kaiser (1984) shaped the following physical argument. Very rich clusters of galaxies are obviously the most massive objects that have collapsed by the present epoch into some kind of equilibrium state. These clusters (e.g., the Coma cluster) are also extremely rare, as the number density of rich clusters is orders of magnitude smaller than the number density of galaxies. In the gravitational instability scenario, at fixed mass scale, the first objects to collapse are the densest. For simplicity we assume that the primordial spectrum of density fluctuations was Gaussian in nature. This assumption arises naturally out of our original hypothesis that the phases of δ_k are random. Furthermore, analysis of the four-year *COBE* anisotropy data shows virtually no departure from Gaussian behavior (Hinshaw *et al.* 1995, 1996). For Gaussian fluctuations it follows that these clusters rich in high density regions are rare, because their initial density was several σ above the mean value.

A general statistical property of Gaussian random fields is that the rare high-σ peaks tend to occur near other high peaks. Quantitatively, the relatively weak correlation on large scales is amplified with respect to the background matter density by a factor of ν^2 where ν equals the number of σ that corresponds to the density fluctuation. To account for the significantly stronger correlation function among clusters, relative to galaxies, requires that $\nu^2 \approx 20$ or ν approaches 4; such events in the Universe would be quite rare, accounting for the very low space density of rich clusters. This is known as statistical biasing. Clusters form at very high peaks and these peaks are near other such peaks. Density fluctuations of lower amplitude would collapse on longer timescales and be distributed in a way that is more representative of the overall mass distribution.

The same argument of statistical biasing in the formation of rich clusters can also be applied on a smaller scale to the formation of galaxies. In a landmark study, Alan Dressler (1980) presented strong, almost indisputable evidence that galaxy morphological type correlates with the local galaxy density. In particular, he found that elliptical galaxies favored regions of high local galaxy density (e.g., clusters of galaxies) and spirals favored lower density regions. With the discovery of LSB galaxies and a study of their environments, we can extend the Dressler morphology–density relation to a more general statement which says that the phase-space density within a galaxy correlates with local galaxy density. The very low density LSB galaxies are relatively isolated. If rich clusters of galaxies are formed in a biased manner and they mostly contain ellipticals, then this suggests that galaxy formation is similarly biased. Dense galaxies (e.g., ellipticals) form from rare 3σ–4σ peaks and lower density galaxies form from the more common, less biased 1σ–2σ peaks.

The idea of biased galaxy formation was at the heart of the first-generation CDM models in the mid 1980s. In the numerical simulations of Davis *et al.* (1985) a value of $b = 2.4$ was needed to explain the small-scale correlations of galaxies. This is a rather high bias factor. Since that time, a great effort has been undertaken to produce fair samples of galaxies in redshift surveys in order to measure b more robustly. This is highly relevant to any structure formation scenarios, for if it were possible to predict b from theory, then strong observational constraints would exist. Estimations of the amount of biasing have greatly changed in the last 10 years with some recent studies (e.g., Lin *et al.* 1996; Benoist *et al.* 1996) now showing that $b = 1$ (no bias) or b is slightly less than one (indicating an antibias $\rightarrow$ mass is more strongly clustered than light). In terms of our structure formation scenarios the least biased sample would correspond to those objects that formed at low density (e.g.,

1σ–2σ peaks) fluctuations. Although these are thought to be mostly spiral galaxies, it is unclear whether there exists a unique morphological signature of a galaxy which formed from a low density fluctuation, so it is difficult to a priori select the most unbiased sample.

However, this scenario for biasing explicitly assumes that galaxy formation is the result of the collapse of individual potentials whose density distribution is Gaussian. But this simplified idea of galaxy formation may be somewhat inconsistent with recent HST images of high redshift galaxies. In particular, the Hubble deep field (HDF) (Figure 5.3) provides a strong indication that galaxies are essentially assembled from smaller subunits (Pascarelle *et al.* 1996) instead of forming from the smooth collapse of gas inside a potential. If this is indeed the case, then the morphology–density relation may not reflect statistical biasing at all; instead it is produced by environmental processes that determine the rate at which galaxies are assembled. Gas removal processes during this assembly phase may then determine the kind of galaxy that ultimately forms. At the very least, the HDF data indicates that the formation of galaxies is likely to be a complex and extended process (see below).

5.3 Some possible cosmogenic scenarios

5.3.1 The power spectrum of the galaxy distribution

One of the strongest observational constraints on structure formation scenarios is provided by the observed power spectrum of the galaxy distribution. This is shown in Figure 5.1, in which power is plotted as a function of scale in log–log space. Different redshift surveys (e.g., Schuecker, Off, and Seitter 1996; Lin *et al.* 1996; Mo and Fukugita 1996; Klypin, Primack, and Holtzman 1996; Oliver *et al.* 1996) produce slightly different variants of this diagram. As such, Figure 5.1 is a good characterization and shows the general tendency for stronger clustering on smaller scales. The point at the largest scale ($\approx 200h^{-1}$ Mpc) is the most uncertain. Some redshift surveys indicate a turnover in power on this scale, whereas others have it uniformly rising. The important point, however, is that large-scale power does exist and needs to be accounted for by the model. Three models (Park *et al.* 1994) are fitted to the power spectrum in Figure 5.1:

- The standard CDM model normalized on small scales (e.g., the correlation length scale of galaxies). This model plus its normalization completely fails to account for structure on scales larger than $50h^{-1}$ Mpc.
- If we take this same model and normalize it to the *COBE* scale, we get the unfortunate consequence that we greatly overproduce small-scale structure. If this is the correct model, we have clearly missed a substantial population of nearby galaxies.
- The dashed line in Figure 5.1 is the model which fits the data best. This model can be the either an open Universe model, which is not directly allowed for by the inflationary paradigm, or a spatially flat model, which has a nonzero cosmological constant.

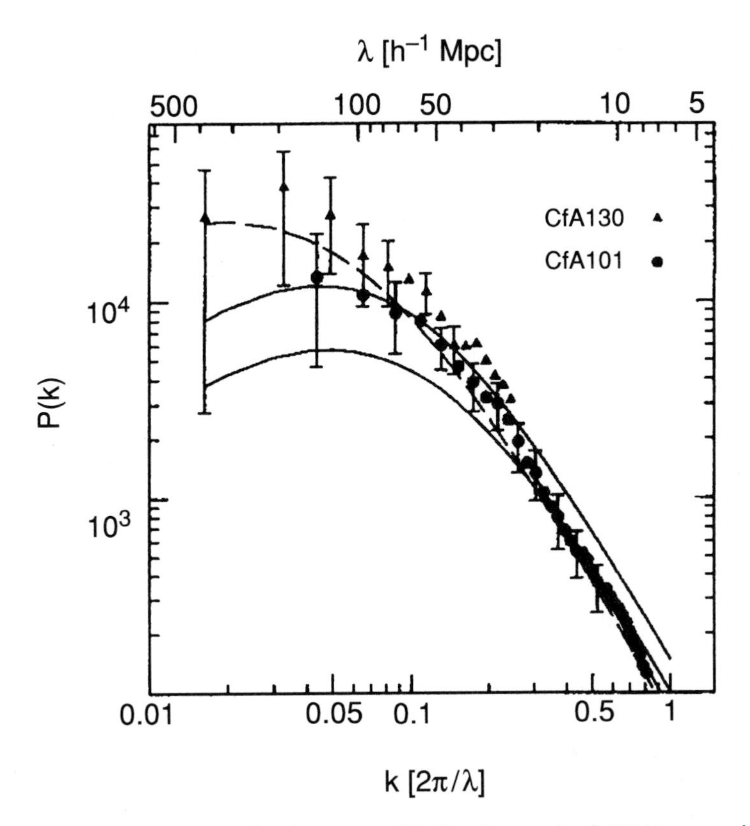

Figure 5.1 Power spectrum: the bottom solid line is standard CDM normalized to the small scale; the supper solid line is standard CDM normalized to the larger scale; the dashed line is an open Universe or λ-dominated model. Reproduced, with permission, from Park *et al.* (1994).

5.3.2 Can CDM be saved?

The basic result that is communicated in Figure 5.1 is that the CDM model cannot simultaneously fit the large-scale and small-scale powers. In this sense, we have a true cosmological crisis in that we have no viable structure formation model that readily accounts for all the scales on which structure is observed. However, the CDM model remains quite attractive as a seed model because qualitatively it has the correct spectral shape and it is a natural consequence of the inflationary paradigm. Since the shape of the CDM spectrum is essentially correct, the problem when comparing with the data is its amplitude at some spatial scale. Perhaps CDM can be can be augmented with elements of other theories to correct this.

Turner (1995a,b) has advanced five clever twists or augmentations that keep CDM somewhat viable. Here is a brief outline of these five variations of standard

CDM and the available observational constraints. In general, these variations are designed to "fix" CDM so that it produces the correct shape and normalization of the power spectrum at both large and small scales. From both observational and physical points of view, some of these modifications should best be viewed as "desperate" or at least rather complex.

- *Low Hubble constant + standard CDM*: From Chapter 1 we have that the critical density of the Universe goes as H_0^2. Lowering H_0 then significantly lowers the matter density, which in turn means it takes longer for the Universe to reach the point where the energy density in the radiation field is equal to the energy density in the matter field. This gives the Universe more time to wash out small-scale fluctuations and thus reduces the clustering on small scales. Furthermore, lowering H_0 makes the Universe older; hence there is more time available for gravitational instability to build the largest structures observed. However, for this variant to work, H_0 has to be around 30 and there is no observational evidence for a value this low.
- *Mixed dark matter*: This is a case of fine tuning where the idea is to mix in just enough HDM to allow for the observed power on large scales, while retaining enough CDM to allow for early structure formation on small scales. The required amounts range from 10 to 30% of HDM, which puts rather stringent limits on the combined mass of the various neutrino species.
- *Extra radiation + CDM*: The goal here is to delay the epoch of matter–radiation energy density equality. The Low H_0 model lets this happen by lowering the matter density. Equivalently we can simply raise the radiation density. Since the observed entropy of the Universe provides a strong constraint on the radiation in the form of CMB photons, we must look towards extra sources. One which has been proposed is an unstable relativistic particle (in particular the tau neutrino) whose main decay channel is radiation. But again, some fine tuning is necessary because if this particle decayed during the epoch of primordial nucleosynthesis, it would upset one of the more accurate predictions of Big Bang cosmology. Hence we need just the right mass range for this particle to allow for a relatively late decay.
- *Extra sources of anisotropy*: In its simplest form, inflation strongly predicts a scale-invariant spectrum of Gaussian density perturbations. The scale-invariant limit has spectral index $n \approx 1$ (equation 5.32), in excellent agreement with the *COBE* observations. But if the spectrum is not quite scale invariant and has a spectral index slightly less than 1, there will be less power on small scales. This deviation from the $n \approx 1$ case is called tilted CDM. A similar "fix" can occur if we allow gravitational radiation to be a significant source of the anisotropy observed in the CMB. In this case the overall amplitude of the density perturbations must also be lower.
- *Nonzero Λ*: The standard inflationary theory strongly predicts that the Universe has zero spatial curvature at the present day. For most models in the past, this is accomplished by letting $\Omega = 1$. However, a broader class of inflationary models

reaches zero curvature via a combination of Ω and Λ. If most of the contribution to zero curvature comes from the Λ term, then the lower Ω leads to lower matter density, as in the case of low H_0. Λ-dominated zero curvature also leads to a larger expansion age at fixed H_0, which helps to relieve some of the apparent conflict discussed in Chapter 3 between H_0^{-1} and the ages of globular clusters. This larger age also allows more time for gravitational instability and aggregation to build larger-scale structure. Hence nonzero Λ would appear to solve several problems simultaneously.

5.3.3 Beyond CDM: more exotic scenarios

For completeness, here are four other fairly nonstandard structure formation scenarios that complement and/or dominate the gravitational instability paradigm.

Primordial Baryon Isocurvature (PBI) model

This model from Peebles (1987) is quite elegant in its simplicity and directly relates to what we presently observe in the Universe. The model is an evolution of work done in the early 1970s when the cosmological parameters H_0 and Ω were thought to be known accurately. In the PBI model, the Universe consists of photons, baryons and three species of massless neutrinos, hence it is strictly baryon dominated. Initial density perturbations take the form of entropy perturbations which are fluctuations in the baryon–photon and/or baryon–neutrino number densities. Since there is no obvious mechanism to generate these entropy perturbations, they are assumed to take the form of a power law. The index n of the power law is inferred from present-day observations, hence the observed $n \approx 1$ case is perfectly consistent with PBI. Since baryons are strongly coupled to photons in the early Universe, there will be some scale over which photon diffusion erases the perturbation. Below this scale, the initial entropy fluctuations become the density perturbations that seed large-scale structure. As density perturbations above this scale cannot form, the slope of the power spectrum rises significantly and becomes quite steep ($\approx n + 4$) just below this scale. In this model, structure formation can rapidly occur right after recombination, and dense perturbations could give rise to the formation of massive star clusters and/or QSOs, which become significant sources of ionization for some time after recombination.

Topological defects

This is a very complex theory whose origin is motivated by the physically reasonable proposition that, as the very early Universe undergoes a phase transition, symmetry is spontaneously broken and gives rise to some kind of defect (e.g., a cosmic string, a domain wall, magnetic monopoles). The defect network then evolves and provides the seeds for structure formation. As the network constantly evolves (the defects can't "damp" out), density perturbations are constantly being produced in a manner that is not easily characterized by the normal random phase hypothesis that leads to Gaussian fields. Hence evidence of any significant non-Gaussian behavior in the *COBE* temperature fluctuation data would be consistent with defect-driven structure formation models. The absence of this component, however, would likely rule out this model (Bennet and Rhie 1993).

Explosions

When the first CFA slice results were presented and the cellular pattern in the galaxy distribution first became apparent, Jerry Ostriker and his coworkers at Princeton came up with a series of models that involve primordial explosions and subsequent hydrodynamic evolution of a network of expanding shells, due to the explosions, that sweep up the material. This model naturally evacuates large regions, leading to voids, and collects this material at the intersection of shells. Qualitatively, this agrees quite well with the observed galaxy distribution. These explosions occur after matter and radiation decouple, hence they serve as a nongravitational component to structure formation. Although the source of the explosions is unknown, it can plausibly be associated with the energetics of galaxy formation and the release of energy via supernova explosions. The kinetic energy carried by the supernova plows into the surrounding medium and pushes it into a shell of radius R_s. The physical parameters which determine this radius are the released kinetic energy, the timescale over which this energy is transported (if this is longer than the expansion timescale, there is no effect), and the density of the surrounding medium.

We can make an approximation that the timescale is essentially given by $(G\rho)^{-1/2}$, which leaves us with only two parameters, the timescale and the kinetic energy E. To yield units of length, the correct combination of these two parameters is

$$R_s \sim (GEt^4)^{1/5}$$

where E can be identified with the energy from supernova explosions. For typical values of E associated with the formation of a large galaxy, R_s is in the range 1–5 Mpc, comparable to the average separation between galaxies (Chapter 3). Although the explosion scenario is not likely to be correct in detail, it does offer two important points: (1) the formation of galaxies is likely to be affected by the process itself due to energy feedback (just like the formation of stars in giant molecular clouds is affected) and (2) it may be wise to consider the hydrodynamical evolution of density perturbations since we clearly live in a void-filled universe. Further work by Cen and Ostriker (1994) clearly shows the relevance of hydrodynamics in structure formation scenarios.

Primordial turbulence

This is a rather old theory whose details were worked out in the late 1960s and early 1970s by groups in Italy and the Soviet Union (e.g., Bonometto *et al.* 1974, 1975; Dallaporta and Lucchin 1973; Ozernoi *et al.* 1968; Sunyaev 1970). By most accounts, it has been discarded or forgotten today. However, this theory contained the seeds of the hydrodynamic treatment in today's models and can be considered as a reasonable precursor to the explosion scenario. In this theory, structure formation is a consequence of the initial turbulence spectrum in the early Universe. Eddy viscosity serves as a significant source of damping for perturbations on a small scale and virtually all proponents of this theory demonstrate that, in high Ω Universes, photon viscosity damps out galaxy-size perturbations. Hence, when inflation appeared in 1980, there was no more room for this theory in the consideration of cosmological models. Now, however, since we have no viable structure formation model, turbulence might as well reemerge as a contender. In this theory the density fluctuations can be expressed as

$$\frac{\delta\rho}{\rho} \propto \frac{v_t^2}{c_s^2} \tag{5.33}$$

where v_t is the turbulence velocity and c_s prior to recombination is

$$c_s^2 = 1/3c^2\left(1 + \frac{3}{4}\frac{\rho_r}{\rho_m}\right)^{-1} \tag{5.34}$$

where ρ_r and ρ_m are the radiation and matter densities. Hence random motions in the turbulent fluid that are subsonic before recombination become supersonic as ρ_r decays. The subsequent shock waves associated with this supersonic turbulence act to compress matter into high density regions that condense from the expanding background. The perturbation spectrum is set by the nonlinear transfer of energy from large scales to small scales, which is given by the familiar Kolmogorov spectrum

$$v_t(r) \propto r^{1/3}$$

with the largest scale set by the condition that the hydrodynamical interaction timescale is equal to the expansion age of the Universe.

Silk and Ames (1972) postulated that the maximum turbulence velocities were $\leq 0.1c$ else the density fluctuations would be too large and would have collapsed into relatively dense structures at early epochs. Their treatment shows that the turbulence velocity decays fairly slowly. The rapid expansion of the horizon feeds the turbulence and allows density fluctuations to grow with time. The larger-scale eddies are then dissipated by viscous decay before recombination. After recombination, supersonic turbulence then destroys any large-scale surviving fluctuations. Silk and Ames (1972) contend that only small-scale fluctuations can then survive. These fluctuations grow into galaxies through conventional gravitational instability. In direct contrast to this model, Stein (1974) argued that turbulence is a natural mechanism for producing cluster-size fluctuations instead of galaxy-size fluctuations. Stein (1974) developed a rigorous criterion for eddies to damp out as a function of their turbulence velocity. The physical criterion is defined by the familiar Reynolds number of fluid dynamics. If the Reynolds number is of the order of unity before recombination then photon viscosity will completely erase the turbulence. The turbulence velocity is assumed to arise out of some primordial chaotic velocity (V_L).

Values of $V_L/c \leq 0.4(\Omega h^2)^{13/8}$ damp out. This conforms to other models which show that the initial turbulence velocity must be quite high (e.g., $\geq 0.4c$) if turbulence is to survive until recombination. In general, large turbulence velocities are present in larger-scale eddies. Stein (1974) shows that density fluctuations that arise from turbulent eddies with a mass of $10^{14} M_\odot$ have gravitational binding energy that exceeds the turbulence energy; hence they are bound. This mass corresponds to the mass of a moderate cluster of galaxies. The predicted collapse redshift of these turbulence-produced, bound density perturbations is $z = 1.3 - 10.5$. Such structures would be predicted to have substantial angular momentum, and hence the clusters should rotate. Since cluster velocity fields are complicated by the presence of substructure, or the quadrupole anisotropy introduced by surrounding clusters, net cluster rotation which is significantly less than the internal velocity dispersion would be difficult to detect.

Finally, in addition to being intrinsically nonlinear, turbulence also has the advantage of naturally producing angular momentum in structure. In general, it is difficult to account for the observed angular momentum in galaxies. In CDM and its variants, the angular momentum is thought to arise as a result of tidal torques between neighboring protogalaxies. However, observations show that the angular momentum of a galaxy is independent of its environment.

5.4 Observational constraints

Several important observational constraints are now available, on large and small scales. The plethora of structure formation scenarios just considered can be weighed against the data to make some appear implausible. But no structure formation theory has so far been able to satisfy all the constraints. Still, significant progress is made when theories can be falsified by good data. We begin with the largest-scale constraints.

5.4.1 Large-scale constraints

COBE

Matter and radiation were coupled in the early Universe. Any density fluctuation in the matter would represent gravitational potential wells, from which the radiation (photons) would have to climb in order to escape. This effect, called the Sachs–Wolfe effect (discussed in Chapter 3), causes the photons in these potential wells to lose a small bit of energy and become redshifted with respect to photons that are not in the vicinity of a potential well. At the surface of last scattering, these small energy differences will be manifest in the CMB as small temperature anisotropies. The structure and amplitude of the temperature anisotropy map of the CMB directly reflects the spectrum of initial density fluctuations that produced the structure observed today. As there is a spectrum of density perturbations, then each successive perturbation encountered by the photon may be either of smaller or larger amplitude than the perturbation previously encountered. Thus, photons can either gain (blueshift) or lose (redshift) energy through these repeated encounters. Under inflationary cosmology, density perturbations are generated through initial quantum fluctuations in the inflation field. These are predicted to be highly Gaussian in nature. In this case the corresponding temperature fluctuations in the CMB, over sufficient angular scale, will also be Gaussian. However, even fluctuations that are non-Gaussian will again, when averaged over many horizons, produce mostly Gaussian temperature fluctuations. Hence the detection of any higher-order departure from a purely Gaussian temperature fluctuation spectrum in the *COBE* data would be highly significant. To date (Hinshaw *et al.* 1995), none have been convincingly detected and this would seem to rule out many exotic models which appeal to non-Gaussian initial density fluctuations.

For Gaussian fluctuations there is a simple way to estimate the approximate expected anisotropy. The relation between scale factor and redshift in the matter-dominated Universe is

$$z \propto a(t)^{2/3}$$

Hence $a(t)$ has grown by a factor of 30 000 from the redshift epoch $z = 1000$ (where decoupling occurs) until $z = 0$. Since the observed Universe exhibits factors of 2 over density (e.g., $\delta\rho/\rho = 1$) on large scales, then it's clear that a perturbation of amplitude 1/30 000 could have been present at the surface of last scattering. This leads to the simple expectation that the fluctuation level should be a few $\times 10^{-5}$. Based on four years of analyzed data, the observed fluctuation level is $\approx 1.5 \times 10^{-5}$ (Bennett *et al.* 1996). And the measured spectral index is $n \approx 1.2 \pm 0.3$, perfectly consistent with the $n = 1$ prediction from inflation and scale-invariant fluctuations. Overall the *COBE* data provides an accurate normalization for structure on large scales. Any structure formation scenario, regardless of its nature, must be firmly anchored to this normalization. Furthermore, the fact that anisotropy was detected, at about the expected level, really provides excellent confirmation that the gravitational instability paradigm must be basically correct.

At recombination the angular size of the horizon is approximately 2° but the *COBE* beam has a resolution of 7°; hence the measured anisotropy is beam-averaged over a few horizon scales. Within one horizon there can be considerably larger amplitude fluctuations, which correspond to some of the large-scale structure features previously discussed. Fluctuations on the degree scale are particularly important to detect, as these may well correspond with the largest-scale features in the observed power spectrum of galaxies. Detection of anisotropy on smaller angular scales (e.g., 0.5–2°) by using terrestrial techniques (e.g., balloon-borne detectors or direct observation with radio telescopes) provides yet more evidence in favor of the gravitational instability paradigm. The recent observations with the Cambridge Anisotropy Telescope (Wollack *et al.* 1997), have strongly confirmed the *COBE* results on these smaller scales. However, as pointed out by Kogut and Hinshaw (1996), there is considerable disagreement between various experiments on this angular size scale. An overall mean of the anisotropy measurements on the degree scale is approximately $(3.5 \pm 1.0) \times 10^{-5}$.

In sum, the *COBE* results rule out any model which predicts anisotropies of order 10^{-4} or greater. Because of that, these models are not discussed here. The *COBE* normalization of the power spectrum would also seem to strongly rule out the standard CDM model. The highly Gaussian nature of the temperature fluctuations is inconsistent with topological defects and similar models. Hu and Sugiyama (1996) also show that the *COBE* anisotropy is fairly inconsistent with PBI. The surviving models are the variations of CDM, all of which predict a nearly scale-invariant spectrum, as predicted by inflation and observed by *COBE*.

The value of H_0

Although many CDM-based models do survive the *COBE* test, most of them are unlikely to survive the H_0 constraint if H_0 is above ≈ 70. Low H_0 and MDM models will be ruled out. PBI is not ruled out. Furthermore, if H_0 is above 80, a positive cosmological constant is required, assuming we believe the ages of globular clusters. If this is the case, then nonzero Λ models are obviously favored.

The observed power spectrum

Redshift surveys of galaxies and the generation of the power spectrum is now something of a cottage industry. Large surveys in the northern and southern hemispheres are well underway and, as of early 1995, an all-sky "magnitude-limited" sample of $\approx 15\,000$ galaxies was available (but see Chapter 6 for the myth of magnitude-limited galaxy samples). Various data sets can be constructed over various magnitude-limited ranges and angular regions of the sky. To date, all derived power spectra agree within the errors out to a scale of $100h^{-1}$ Mpc and the power spectra keep rising. The disagreement between various surveys or samples occurs on scales of $200-400h^{-1}$ Mpc and in the particular scale the power spectrum turns over. The existence of power on scales of at least $100h^{-1}$ Mpc is no longer in dispute (Landy *et al.* 1996).

However, it's clear that not enough data exists on a large scale to have a definitive representation of the power spectrum on these length scales. One of the primary goals of the Sloan Digital Sky Survey, which expects to obtain 10^6 redshifts over five years, is to obtain this large-scale data. Hence any attempt at constraining structure formation scenarios by bridging the gap between the well-sampled region out to $100h^{-1}$ Mpc and the *COBE* scale ($\approx 1000h^{-1}$ Mpc) is premature. Rather than reproducing tons of power spectra, it suffices to say that most strongly resemble Figure 5.1. According to the most recent treatment by Lin *et al.* (1996), the following classes of model remain consistent with the power spectrum:

- Flat CDM models with $\Omega_0 \approx 0.4-0.5$, $\Lambda \approx 0.5-0.6$ and $H_0 \approx 50$. These models have very little bias (e.g., $b = 0.9$).
- Open CDM models with $\Omega_0 \leq 0.5$ and H_0 as large as 80 ($b = 0.9$).
- Flat $\Omega_0 = 1$ models which have mixed dark matter such that the contribution of neutrinos to Ω is ≈ 0.2. These models require $H_0 \approx 50$ and are mildly antibiased ($b = 0.8$).

- Flat $\Omega_0 = 1$ models with a tilted spectrum ($n \approx 0.7$). These also require $H_0 \approx 50$ and are mildly biased ($b = 1.1$–1.3).

Even though these models do survive the constraint imposed by the observed power spectrum of the galaxy distribution, most of them could also be eliminated if H_0 were greater than ≈ 70.

The clear result is that CDM models, if they are to be salvaged, require exotic modifications such as nonzero Λ, an open Universe, a mixture of HDM or a tilted spectrum. The standard CDM model is no longer viable. These conclusions are consistent with another way to characterize large-scale structure, namely by the observed topology. Topology is a measure of the connectedness of high and low density regions in the Universe. In the most recent treatment that incorporates all the available data, Gott, Cen, and Ostriker (1996) conclude that the standard CDM model is also ruled out, whereas the nonzero Λ and/or CDM open Universe models are consistent with the data.

Clusters of galaxies

The number density of rich clusters, their baryonic mass fractions, the amount of substructure they contain, the cluster–cluster correlation function and the epoch of virialization for clusters are all probes of Ω and structure formation scenarios. In general, most of the structure formation models under consideration do not overproduce rich clusters and are consistent with the cluster–cluster correlation function. Cluster baryonic mass fractions, however, have become a recent concern. Since most of the baryons in a cluster are not in the member galaxies, but in the hot intracluster medium (ICM), this requires accurate cluster masses as inferred from X-ray observations.

Simon White and collaborators (White *et al.* 1993) have shown that the ratio Ω_b/Ω_0 measured for a cluster should not be significantly different than the universal value. The baryonic mass in clusters consists of two forms, a visible component (e.g., luminous galaxies and cluster X-ray emission) denoted by f_b and a dark component (e.g., stellar remnants, low mass stars). The total baryonic density Ω_b is inferred from primordial nucleosynthesis as previously discussed. Hence, if f_b can be determined for clusters then Ω_0 can be inferred from the relation $\Omega_0 = \Omega_b/f_b$. Current observations indicate that $f_b \geq 0.04h^{-3/2}$. When combined with the nucleosynthesis limits ($\Omega_b \sim 0.015h^{-2}$; Walker *et al.* 1991), this leads to $\Omega_0 \leq 0.3h^{-1/2}$. To reconcile this with $\Omega = 1$ models requires either $H_0 \leq 30$ or the possibility that total cluster masses have been systematically underestimated. The latter possibility does not appear to be the case (Evrard, Metzler, and Navarro 1996); hence the measured values of f_b in clusters appear quite inconsistent with $\Omega = 1$.

But a low value of Ω appears to be inconsistent with the substructure arguments that suggest the formation of clusters is still ongoing (or at least terminated rather recently). Late cluster formation requires high Ω. Identifying the formation epoch of clusters is also a strong constraint, as in any CDM model the clusters form after galaxies have formed; hence we would not expect much clustering at high redshift.

Recent data obtained with the Hubble Space Telescope is beginning to suggest that galaxies are somewhat clustered at redshifts $z = 2-3$, consistent with deep ground-based images of fields around high redshift QSOs that show they are often surrounded by other galaxies. Although this is not strongly inconsistent with the gravitational instability paradigm for the formation of clusters of galaxies, it does serve as a reminder that other important nonlinear effects may come in to play that effectively speed up the formation process. One of these processes might be the statistical biasing discussed earlier.

Identifying cluster formation, however, is a very ambiguous problem. Substructure in nearby clusters, for instance, indicates they are still accreting material. This kind of cluster augmentation provides clear evidence of merger processes involving smaller structures which have shorter collapse times. If there is sufficient power on large scales so that these smaller units are available to infall at later times (as appears to be the case), then cluster formation is a process that occurs on a much longer timescale than galaxy formation. Still, there are examples of distant clusters ($z \approx 0.9$) which look very much like the core of the Coma cluster (Figure 3.3) looks now (Postman *et al.* 1996). This brings up what may be a very powerful test of cluster formation and virialization first proposed by Perrenrod and Henry (1981).

Most nearby clusters with strongly virialized cores are also sources of X-ray emission. The X-ray emission is a consequence of intracluster gas being heated by the cluster potential to its virial temperature. For typical clusters, the virial temperature is a few million degrees, leading to strong X-ray emission by the gas in the energy range 0.5–5 keV. Equilibrium should occur approximately on a dynamical timescale. The origin of the intracluster gas is unclear, although likely sources are (1) tidally liberated gas caused by interactions between protogalaxies as they form in the overall cluster potential, (2) gas which has been driven out of galaxy potential wells into the cluster potential by energetic internal processes such as star formation and supernova heating and (3) leftover gas that didn't get incorporated into any galaxy-size potentials. For most clusters, the X-ray gas contains strong lines of ionized iron, indicative of metal abundances which are near solar. This indicates processing of the gas in galaxies and subsequent expulsion (via supernova heating).

Observations of the evolution of the X-ray luminosity function of clusters as a function of redshift may reveal the epoch of cluster core virialization. To date, the sensitivity of various X-ray satellites (e.g., *Einstein*, *ROSAT*, *ASCA*) has allowed the detection of X-ray emission in clusters of galaxies out to $z \approx 1$ (Hattori *et al.* 1997). To date, the most distant cluster detected in X-rays has $z = 1.0$ and was detected on the basis of an emission line at 3.35 keV, which is thought to be the redshifted 6.7 keV iron line (Hattori *et al.* 1997). There is not yet enough data to see whether there is a characteristic redshift at which most clusters "turn on." However, with future increases in X-ray satellite sensitivity (e.g., AXAF) it may be possible to detect or to strongly constrain this turn-on epoch to redshifts less than some value. In fact, although the existence of some clustering at high redshift may not be too surprising, the detection of substantial X-ray emission originating from a virialized cluster core at redshifts $z \geq 2$ would seem to strongly rule out the formation of these cores via gravitational instability, or to indicate a new population of significantly denser clusters than currently are known.

Other aspects of clusters of galaxies can also act as a constraint on structure formation models. Zabludoff and Geller (1994) use kinematic observations of the densest clusters of galaxies to show that models which match the power on large scales do not match the observed distribution of velocity dispersions. Moreover, biased models predict too few high velocity dispersion clusters compared to the number of low velocity dispersion clusters. In fact, they conclude that no model matches both the statistics of the galaxy distribution on large scales and the small-scale velocity dispersion characteristics of clusters of galaxies. Crone and Geller (1995) consider the effects of merging on the evolution of cluster velocity dispersions. Their models show that the abundance of clusters with $\sigma_v \geq 1200$ km s^{-1} increases with time, whereas the number of groups decreases with time. The particular evolutionary rates depend upon choice of cosmogenic scenario. The models which match the data best are either $\Omega_0 = 0.2$ or biased $\Omega = 1$. All models, however, predict fewer low velocity dispersion systems than is actually observed (Zabludoff *et al.* 1993). Finally, Dell'Antonio, Geller, and Fabricant (1995) show that the baryonic fraction in low velocity dispersion clusters is approximately one-half that observed for higher velocity dispersions clusters (Evrard, Metzler, and Navarro 1996). This is a curious result as it suggests baryonic and dark matter may be segregated in different ways, depending upon the overall depth of the potential well. This conclusion is strongly at odds with the good match between the X-ray and optical distributions for both high and low σ_v clusters.

QSO absorption lines

A possible strong constraint on structure formation timescale comes from from observations of QSO absorption lines. As baryonic gas inside a dark matter potential collapses and forms stars, any massive stars rapidly evolve and feed back heavy elements into the gas. The production of the first heavy elements can be equated to the "epoch of galaxy formation." To detect these metals, the line of sight to a QSO must pass through one of these "forming galaxies." The probability of this occurring is directly proportional to the size of the protogalaxy and the number density of QSOs at high redshift. To date, metal lines, specifically those of carbon (e.g., carbon IV) can be identified in QSO spectra back to a redshift of ≈ 4 (Steidel 1992). The distribution of metal line strengths with redshifts (Figure 5.2) indicates the following:

- Up to $z = 2.5$–3 the metal line strength is fairly low, indicating little processing.
- A major episode of metal production, which can plausibly be identified with the formation of the disk components of galaxies, seems to be occurring at $z = 1.5 - 3$. During this time the mean metallicity appears to increase from 0.01 solar to ≈ 0.1 solar.

There is also a class of QSO absorption lines which are called damped Lyman$_\alpha$ systems. Here the absorption is through a sufficient column of H I ($N_{HI} > 10^{20}$ cm^{-2})

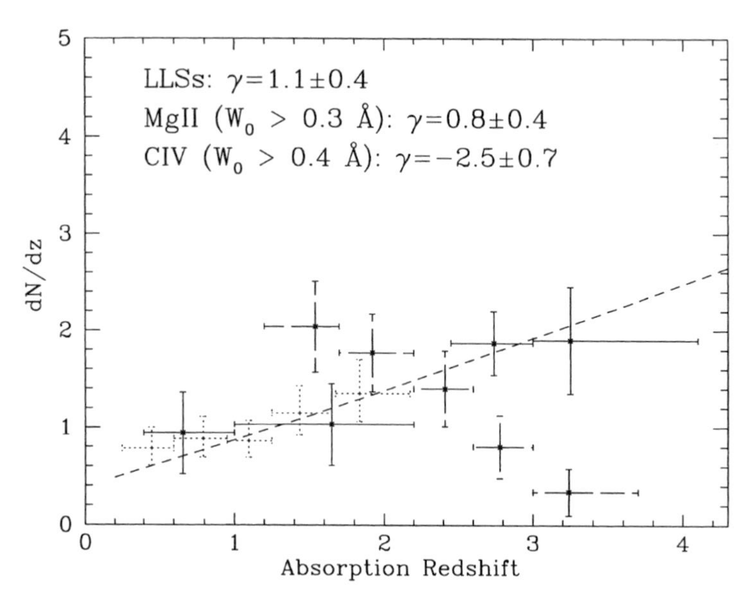

Figure 5.2 Distribution of the number density of various QSO absorption line systems. The solid crosses are for pure hydrogen absorption systems whose number density, as expected, increases with increasing redshift. Of interest here is the behavior of the carbon IV systems (dashed crosses), whose number density evolution shows a strong monotonic decrease with increasing redshift, approaching zero by redshift 3.5. Courtesy of Charles Steidel (1992).

to produce noticeable damping wings on the Lyman$_\alpha$ line. These systems can be plausibly identified with H I disks. At the moment, the highest redshift, damped Lyman$_\alpha$ system is at $z = 4.38$ towards a $z = 4.7$ QSO (Lu *et al.* 1996). The number of these damped systems per unit redshift interval monotonically increases from $z = 0$ to $z = 4$ with no obvious peak in the distribution. However, the amount of matter contained in these systems does seem to show a peak at $z \approx 3$ (Storrie-Lombardi, McMahon, and Irwin 1996). This suggests that protogaseous disks are in place by $z = 3$ and that high column density gas arranged in a disk configuration occurred within a couple of billion years since recombination. The presence of neutral hydrogen at $z \approx 4$ also demonstrates that the Universe can't have been completely reionized (by QSOs) at this redshift.

Gravitational lensing

In principle the number density of gravitational lenses as a function of redshift provides a strong constraint on Ω and Λ. This is because, in either a low Ω or a Λ-dominated Universe, the amount of volume increases with unit redshift interval compared with an $\Omega = 1$ Universe. An unbiased survey for gravitational lensing would thus have significant cosmological value. Unfortunately, surveys for gravitational lensing have a variety of selection effects associated with them, the most serious of which is the difficulty of finding splittings on angular scales larger than 5 arcseconds (Kochanek 1995). Such large splittings are naively expected. For instance, a singular isothermal sphere with velocity dispersion of 1000 km s^{-1} (e.g., a typical cluster) produces lensing with average separations of 28 arcseconds. This splitting can be reduced if the cluster has a strong central mass concentration.

Currently there are two confirmed lenses with separations larger than 3 arcseconds. These are Q 0957 + 061 (Walsh, Carswell, and Weymann 1979) and Q 2016 + 112 (Lawrence *et al.* 1984). As pointed out in Kochanek (1995), there are four unconfirmed candidates as well, the largest of which has an angular splitting of 7.3 arcseconds. Via an elaborate model of selection effects, both in detecting lenses and the current QSO catalogs, Kochanek (1995) demonstrates that the observations are inconsistent with standard CDM, as normalized by the *COBE* data. To reconcile the observations with this model again appeals to the earlier fixes, e.g., tilt with $n \sim 0.3$–0.7, low H_0 ($H_0 \leq 30$), low Ω or Λ-dominated spatial flatness. Hence the galaxy power spectrum and the observed number of large angular separation gravitational lens systems point to the same general required modifications of standard CDM.

However, this consistency check is not particular strong as the results are quite model dependent for lenses. For instance, if clusters with $\sigma_v \approx 1500$ km s^{-1} have core radii in excess of $40h^{-1}$ kpc, the number of expected lenses is reduced by a factor of 10 and the constraints on the CDM model are invalid (Kochanek 1995). Finally, some (e.g., Kochanek 1995; Maoz and Rix 1993) have argued that the current data on gravitational lensing systems already rules out models in which Λ exceeds 0.6. However, this constraint is also not very strong, as it's highly dependent on the form of the assumed selection effects that are operative in current surveys for gravitational lenses (e.g., Sugiyama, Silk, and Vittorio 1993). Eventually, over the next decade, more lens systems will be discovered and the selection effects will be better quantified. I believe it's too premature to argue that the current detection of gravitational lensing as a function of redshift is able to constrain the value of Λ. But eventually that constraining power will come.

5.4.2 Small-scale constraints

Dwarf galaxies and massive halos

Although the HDM scenario is very attractive for giving the Universe its observed power on large scales and for having an identified candidate particle (e.g., the neutrino), there are two small-scale constraints that essentially rule out the theory completely. One constraint is provided by the maximum phase space density that neutrinos can have. This was first pointed out by Tremaine and Gunn (1979) and works as follows.

If neutrinos have a mass, then at some point in their cosmic evolution they must become nonrelativistic, and like baryons they become trapped in a galactic potential. If the neutrinos are sufficiently massive (e.g., ≥ 30 eV) they can dominate this halo mass. For a spherical halo, the escape velocity is given from Newtonian dynamics as

$$V_{esc} = \sqrt{\frac{2GM}{R}}$$

For galaxies like the Milky Way, which have $M \sim 10^{11}$ to 10^{12} and $R \sim$ 10–30 kpc, V_{esc} ranges from 300 to 500 km s^{-1}. If the halo is dominated by neutrinos, then an important exclusion principle comes into play.

Neutrinos are fermions, and only one fermion can occupy a unit volume of phase space. The maximum positional space is the volume of the halo and the maximum momentum space is $m_\nu V_{esc}$. Hence the total phase space volume, which by the exclusion principle is equivalent to the total number of neutrinos that exist in this volume, is given by

$$\frac{4\pi}{3} R^3 \left(\frac{4\pi}{3}\right) (m_\nu V_{esc})^3 \tag{5.35}$$

or, using the expression for escape velocity, we obtain

$$\frac{16\pi^2}{9} (m_\nu^3)(2GMR)^{3/2} \tag{5.36}$$

An upper bound on the total mass of neutrinos in this halo is then

$$M \leq \frac{16\pi^2}{9} (m_\nu^4)(2GMR)^{3/2} \tag{5.37}$$

where M on the left-hand side is really the total number of neutrinos (N_ν) times m_ν. Thus we can now write a lower bound

$$m_\nu \geq (G^3 R^3 M)^{-1/8} \tag{5.38}$$

For normal galaxies the constraint is relatively uninteresting as it leads to a lower bound on neutrino mass of a few electronvolts. However, there is some evidence that the dwarf satellite companions to the Milky Way (e.g., Draco, Ursa Minor) are gravitationally bound systems instead of expanding systems due to tidal encounters associated with their low perigalactic orbits (Piatek and Pryor 1995). These systems are characterized by $M \sim 10^7$ and $R \sim 0.1$ kpc. For these objects, equation (5.38) then gives a lower bound on m_ν of 500 eV. A stable neutrino species with a mass this large can easily be ruled out on cosmological grounds, as the Universe would have collapsed long ago.

High redshift galaxies

Another strong constraint comes from the existence of galaxies at high redshift. Top-down scenarios require fairly long times for large-scale instabilities to fragment down to smaller scales. For a neutrino-dominated Universe, the initial mass scale is 10^4 times that of even a large galaxy. A rough timescale for this process to occur would be a dynamical timescale. At $z = 0$ a typical supercluster has a radius of 5–10 Mpc and a velocity dispersion of 500 km s^{-1}. These structures themselves are not virialized but they do contain one or more virialized cores, which we identify with smaller-scale clusters of galaxies of radius ≈ 1 Mpc. From these parameters we derive a crossing time of $\approx 10^{10}$ years, an appreciable fraction of the Hubble time. Thus we expect late galaxy condensation and formation in this top-down scenario. This is clearly *not* observed and is regarded by most as conclusive evidence that we do not live in a neutrino-dominated Universe.

The epoch of galaxy formation

The best constraint on galaxy formation will clearly come when we actually observe the process and identify at what redshift galaxies began to form. Recent ground-based observations have now detected galaxies at redshift $z \approx 3$. Steidel *et al.* (1996) have unambiguously detected star-forming galaxies at this redshift. The amount of star formation present at this redshift appears to be 5–10 times less than at redshift $z = 1$–1.5. At redshift $z = 4$ the overall star formation is down by a factor of 5 relative to $z = 3$. Furthermore, in these high redshift objects, the star formation seems to be confined to much smaller spatial scales, centered on the galaxy, than at lower redshift. Since the production of metals is strongly correlated with the star formation per unit volume at some epoch z, the rise in this rate should correlate with the rise in the metal abundance of QSO absorption lines. So far the data is consistent with this expectation and indicates a rough peak in star formation per unit volume at $z = 1.5$–2. In addition to the high redshift galaxies detected by Steidel *et al.* (1996), Hu, McMahon, and Egami (1996) have detected Lyman$_\alpha$ emitting galaxies near QSOs at $z \approx 4.5$. Although the exact nature of these galaxies is unclear, they do conform to the simplest expectation that the initial epoch of star formation at high redshift in galaxies should give rise to Lyman$_\alpha$ recombination radiation.

A rich data set for further investigating the properties of high redshift galaxies is the Hubble deep field (HDF) data, obtained in December 1995. Figure 5.3 (see plate section) shows the data and it is quite striking. Many of the galaxies in this field look as though they are in the process of formation; they are composed of multiple condensations which may be in the process of merging together to form one large galaxy. The redshifts of these interesting objects are yet to be determined. Attempts to infer their redshifts from their colors combined with stellar population models are highly uncertain at best. A recent analysis by Lanzetta, Yahil, and Fernandez-Soto (1996) argues that the reddest objects in the HDF have $z \geq 6$. Clearly this requires spectroscopic confirmation, which will be difficult as the candidate galaxies are faint.

In summary, the available data on the properties of high redshift galaxies and QSO absorption lines suggest the following:

- The initial stage of galaxy formation, defined as when the first generation of stars is formed, occurred prior to $z = 3$ and is best identified with the formation of spheroids (either elliptical galaxies or spiral bulges). The formation of extended disks clearly takes a longer time and was apparently very active between $z = 1$ and

$z = 2$. Vogt *et al.* (1996) show convincing evidence that objects with normal disk kinematics are in place by $z = 1$. The presence of these high redshift structures severely limits the amount of matter that can be obtained in any HDM model.

- At $z = 5$ the Universe is 7% of its present age or 0.7–1.4 billion years. QSOs have been detected at this redshift, so we know that small-scale structure formation can occur on the 1 Gyr timescale. It's possible that these distant QSOs are the manifestation of galaxy formation and the formation of the first generation of stars. To generate the QSO activity requires the presence of a massive black hole. Possibly it is these massive black holes that have acted as the seeds to attract additional baryonic material. In fact, the origin of these massive black holes, 1 billion years after the birth of the Universe, is really quite interesting. If they are the evolved remnants of massive star clusters, they obviously formed much earlier than $z = 5$.
- The simple idea that a protogalaxy would form the bulk of its stars during the initial collapse is probably incorrect. Over a dynamical timescale (a few $\times 10^8$ years for galactic potentials), if most of the gas turns into stars, then a star formation rate of $(100–1000)M_\odot$ per year would result. Although such a large star formation rate has been observed in some ultraluminous *IRAS* galaxies (Sanders *et al.* 1988), which are most likely the merger of two well-formed galaxies, no objects at high redshift have yet been identified that exhibit this behavior. This is a strong argument that galaxy formation is not a quick process, marked by a very large star formation rate (and a very large supernova and metal enrichment rate), but perhaps it is a far more quiescent and longer process. Indeed, detailed studies of elliptical galaxies at $z = 0$ now strongly suggest there is a range of ages in their stellar populations and their full formation occurred over several billion years (Rose *et al.* 1994).
- The role of feedback to the galaxy formation process, either through a supernova or the formation of QSOs, is not yet well understood. If the Universe has been completely reionized by QSOs, the observations indicate this occurred at $z \geq 5$. Possibly this event served further to delay the general process of galaxy formation.
- The observations of Steidel *et al.* that star formation in galaxies was well in place by $z = 3.5$ is difficult to understand in CDM models as it implies there was already small-scale power by this redshfit. Mo and Fukugita (1996) demonstrate that the presence of small-scale power at this redshift is greatly aided by nonzero Λ as the time per unit redshift interval is greater in this case.
- The morphology of objects in the HDF gives the strong visual impression that galaxy formation is occurring via an assembly line process in which small subunits are being accreted into a larger entity. However, these subunits are already composed of gas and stars, so some process had to produce them at a much earlier epoch. Possibly, this process is the one physical process that we understand – simple Jeans mass collapse at high redshift. These (baryonic) subunits then produce galaxies, via merging, as they respond to the underlying mass distribution, which is dominated by dark matter. This is a potentially complex physical process that will challenge our understanding.

Pairwise and peculiar velocities

The final small-scale constraint which can be considered is the average velocity and/or spatial separation between two random galaxies. Peculiar velocities that might arise from gravitational interactions between galaxies or between a galaxy and an overdense region, such as a cluster, cause deviations from Hubble flow but do not alter the position of the galaxy on the plane of the sky. Thus spatial correlation functions performed in physical space, functions which may be isotropic, become anisotropic when mapped onto redshift space (Kaiser 1987). The amount of anisotropy in redshift space can be measured through the lower-order moments of the peculiar velocity distribution. For galaxy pairs, the first moment of the distribution, v_{12}, is sensitive to the growth of the spatial or two-point correlation function. The second moment σ_{12} provides a direct measurement of the kinetic energy of any random motions. In the equilibrium gas, σ_{12} balances the gravitational potential, hence it can be used to measure the effective mass. This is the situation in a cluster of galaxies in hydrostatic equilibrium.

For standard CDM, normalized to give the observed power on small scales, σ_{12} is predicted to be $\approx 1000\,\mathrm{km\,s^{-1}}$. Open models in which $\Omega_0 \approx 0.2$ predict $\sigma_{12} \approx 500\,\mathrm{km\,s^{-1}}$. The most recent determination of σ_{12} is based on a sample of 12 800 galaxies that comprise a well-defined subset of the Northern and Southern Sky Redshift surveys. The results (Marzke *et al.* 1995) of this analysis are unfortunately ambiguous:

- The measured σ_{12} is $540 \pm 180\,\mathrm{km\,s^{-1}}$. Although this is larger than the 1983 measurement of $340 \pm 40\,\mathrm{km\,s^{-1}}$ (Davis and Peebles 1983), it still does not effectively discriminate between open and closed CDM models.
- The samples are "contaminated" by the presence of rich clusters where σ_{12} reflects the cluster velocity dispersion, which is significantly higher than σ_{12} for field galaxies. This "contamination" is severe. When galaxies which are thought to be members of rich clusters are removed from the sample, σ_{12} lowers significantly to $295 \pm 100\,\mathrm{km\,s^{-1}}$. In essence this removal is accounting for the most nonlinear structures that are present, and these aren't necessarily a good probe of CDM structure formation scenarios. In this case it would seem that the open Universe CDM models are strongly favored.
- The amount of "contamination" depends on the volume of the redshift survey. Local samples are biased against selecting galaxies that are members of rich clusters, hence σ_{12} is biased to low values. This explains the low value originally measured by Davis and Peebles. If one uses the observed distribution of cluster velocity dispersions (Zabludoff *et al.* 1993), it is possible to estimate how big a volume must be obtained in order for this "contamination" not to be a dominant effect in the sample. Marzke *et al.* (1995) estimate that the required volume exceeds the volume of the existing redshift sample; therefore no fair sample yet exists to properly measure σ_{12}. Nevertheless, the indications are that σ_{12} is relatively low and the small-scale velocity field is therefore mostly quiescent.

This quiesence would appear to rule out most of the explosion models and the large-scale hydrodynamic models of Cen and Ostriker (1994). Since those models introduce a nongravitational component to the peculiar velocity, they necessarily produce high σ_{12}. However, one way to reduce σ_{12} is via galaxy–galaxy interactions and dynamical friction. Accounting for the possible role of mergers appears to make the σ_{12} measurements consistent with the predictions of high resolution hydrodynamic simulations such as those of Zurek *et al.* (1994). But if this is true, the small-scale clustering and dynamical properties of galaxies would then be probing the evolution of galaxy merging more than structure formation scenarios. A recent analysis of the Canadian deep redshift survey finds strong evidence for increased merger activity out to $z \approx 0.3$ and derives a merging rate that goes as $(1+z)^{2.9 \pm 0.9}$, consistent with the expected $(1+z)^3$ dependence (Patton *et al.* 1997).

In fact, the effects of merging, which mean that the number density of galaxies as a function of redshift is not conserved, have serious implications on the use of small-scale structure to constrain structure formation scenarios. This is because merging greatly modifies what is observed on small scales and leads to an overall decrease in the galaxy density on small scales if numerous small galaxies merge into one larger galaxy. Hence the pairwise velocity dispersion, the two-point correlation function and the amount of power on small scales could all be modified, by merging, from their original values. Such a signature, however, would be clearly detected as a strong redshift evolution of these quantities. At present, insufficient data exists to search for this signature.

5.5 Summary

This chapter has discussed the gravitational instability paradigm for the formation of structure in the Universe. The detection of the Sachs–Wolfe effect by *COBE*, at close to the a priori predicted level provides the strongest observational support for this paradigm. But the observed cellular distribution of galaxies argues for some kind of hydrodynamic augmentation of the gravitational instability process. This augmentation may be the result of energy feedback from the process of galaxy formation itself. Unfortunately, this process remains unobserved.

At present, the main challenge in structure formation scenarios is to simultaneously match the large-scale power with the observed clustering on small scales. Formulations of the galaxy power spectrum show that power does exist on very large scales and this simply can't be reproduced by standard CDM when it is normalized to small scales. Hence somewhat exotic augmentations of CDM that act to suppress the formation of small-scale structure must be present. Current data cannot distinguish between these various augmentations, but most can be ruled out if H_0 is greater than 70. New classes of nonzero Λ models, however, can successfully reproduce many of the features of the power spectrum. Currently there are no strong observational constraints against nonzero values for Λ.

References

Bennett, C. *et al.* 1996 *Astrophysical Journal Letters* **464**, L1
Bennett, D. and Rhie, S. 1993 *Astrophysical Journal Letters* **406**, L7

BENOIST, C., MAUROGORDATO, S., DA COSTA L., CAPPI, A., and SCHAEFFER, R. 1996 *Astrophysical Journal* **472**, 452

BERTSCHINGER, E. 1992 in *New Insights into the Universe*, eds V. MARTINEZ, M. PORTILLA, and D. SAEZ, New York: Springer-Verlag, p. 65

BINNEY, J. and TREMAINE, S. 1987 *Galactic Dynamics*, Princeton NJ: Princeton University Press

BONNOR, W. 1957 *Monthly Notices of the Royal Astronomical Society* **117**, 104

BONOMETTO, S. *et al.* 1974 *Astronomy and Astrophysics* **35**, 267

BONOMETTO, S. *et al.* 1975 *Astronomy and Astrophysics* **41**, 55

CEN, R. and OSTRIKER, J. 1994 *Astrophysical Journal* **431**, 451

CRONE, M. and GELLER, M. 1995 *Astronomical Journal* **110**, 21

DALLAPORTA, A. and LUCCHIN, F. 1973 *Astronomy and Astrophysics* **26**, 325

DAVIS, M. and PEEBLES, P.J.E. 1983 *Astrophysical Journal* **267**, 465

DAVIS, M., EFSTATHIOU, G., FRENK, C., and WHITE, S. 1985 *Astrophysical Journal* **292**, 371

DELL'ANTONIO, I., GELLER, M., and FABRICANT, D. 1995 *Astronomical Journal* **110**, 502

DRESSLER, A. 1980 *Astrophysical Journal* **236**, 351

EVRARD, G., METZLER, C., and NAVARRO, J. 1996 *Astrophysical Journal* **469**, 494

GOTT, R., CEN, R., and OSTRIKER, J. 1996 *Astrophysical Journal* **465**, 499

HATTORI, M. *et al.* 1997 preprint

HINSHAW, G., BANDAY, A., BENNETT, C., GORSKI, K., and KOGUT, A. 1995 *Astrophysical Journal Letters* **446**, L67

HINSHAW, G., BANDAY, A., BENNETT, C., GORSKI, K., and KOGUT, A. 1996 *Astrophysical Journal* **464**, L29

HU, E., MCMAHON, R., and EGAMI, E. 1996 *Bulletin of the American Astronomical Society* **188**, 2107

HU, W. and SUGIYAMA, N. 1996 *Astrophysical Journal* **471**, 542

KAISER, N. 1984 *Astrophysical Journal Letters* **284**, L9

KAISER, N. 1987 *Monthly Notices of the Royal Astronomical Society* **227**, 1

KLYPIN, A., PRIMACK, J., and HOLTZMAN, J. 1996 *Astrophysical Journal* **466**, 13

KOCHANEK, C. 1995 *Astrophysical Journal* **453**, 545

KOGUT, A. and HINSHAW, G. 1996 *Astrophysical Journal Letters* **464**, L39

LANDY, S. *et al.* 1996 *Astrophysical Journal Letters* **456**, L1

LANZETTA, K., YAHIL, A., and FERNANDEZ-SOTO, A. 1996 *Nature* **381**, 759

LAWRENCE, C. *et al.* 1984 *Science* **223**, 46

LIFSHITZ, E. 1946 *Journal of Physics USSR* **10**, 116

LIN, H. *et al.* 1996 *Astrophysical Journal* **471**, 617

LU, L., SARGENT, W., WOMBLE, D., and TAKADA-HIDA, M. 1996 *Astrophysical Journal* **472**, 509

MAOZ, D. and RIX, H. 1993 *Astrophysical Journal* **416**, 425

MARZKE, R., GELLER, M., DA COSTA, L., and HUCHRA, J., 1995 *Astronomical Journal* **110**, 477

MO, H. and FUKUGITA, M. 1996 *Astrophysical Journal Letters* **467**, L9

OLIVER, S. *et al.* 1996 *Monthly Notices of the Royal Astronomical Society* **280**, 673

OZERNOI, L. *et al.* 1968 SOVIET ASTRONOMY JOURNAL **11**, 907

PARK, C., VOGELEY, M., GELLER, M., and HUCHRA, J. 1994 *Astrophysical Journal* **431**, 569

PATTON, D., PRITCHET, C., YEE, H., ELLINGSON, E., and CARLBERG, R. 1997 *Astrophysical Journal* **475**, 29

PASCARELLE, S., WINDHORST, R., KEEL, W., and ODEWAHN, S. 1996 *Nature* **383**, 45

PEEBLES, P.J.E. 1987 *Astrophysical Journal Letters* **315**, L73

PERRENOD, S. and HENRY, P. 1981 *Astrophysical Journal Letters* **247**, L1
PIATEK, S. and PRYOR, C. 1995 *Astronomical Journal* **109**, 1071
POSTMAN, M. *et al.* 1996 *Astronomical Journal* **111**, 615
ROSE, J. *et al.* 1994 *Astronomical Journal* **108**, 2054
SANDERS, D., SOIFER, T., ELIAS, J., NEUGEBAUER, G., and MATTHEWS, K. 1988 *Astrophysical Journal Letters* **328**, L35
SCHUECKER, P., OTT, H., and SEITTER, W. 1996 *Astrophysical Journal* **472**, 485
SILK, J. and AMES, M. 1972 *Astrophysical Journal* **178**, 77
STEIDEL, C. 1992 *Publications of the Astronomical Society of the Pacific* **104**, 843
STEIDEL, C., GIAVALISCO, M., PETTINI, M., DICKINSON, M., and ADELBERGER, K. 1996 *Astrophysical Journal Letters* **462**, L17
STEIN, R. 1974 *Astronomy and Astrophysics* **35**, 17
STORRIE-LOMBARDI, L., MCMAHON, R., and IRWIN, M. 1996 *Monthly Notices of the Royal Astronomical Society* **283**, L79
STRAUSS, M. and WILLICK, J. 1995 *Physics Reports* **261**, 271
SUGIYAMA, N., SILK, J., and VITTORIO, N. 1993 *Astrophysical Journal Letters* **419**, L1
SUNYAEV, N. 1970 *Astronomy and Astrophysics* **121**, 90
TREMAINE, S. and GUNN, J. 1979 *Physical Review Letters* **42**, 407
TURNER, M. 1995a The Hot Big Bang and Beyond, in IAU **168**: *Diffuse Background Radiation*, International Astronomical Union
TURNER, M. 1995b *Annals of the New York Academy of Science* **759**, 153
VOGT, N. *et al.* 1996 *Astrophysical Journal Letters* **465**, L15
WALKER, T., STEIGMAN, G., SCHRAMM, D., OLIVE, K., and KANG, J. 1991 *Astrophysical Journal* **376**, 51
WALSH, D., CARSWELL, R., and WEYMANN, R. 1979 *Nature* **279**, 381
WHITE, S., NAVARRO, J., EVRARD, A., and FRENK, C. 1993 *Nature* **366**, 429
WOLLACK, E., DEVLIN, M., JAROSIK, N., NETTERFIELD, C., PAGE, L., and WILKINSON, D. 1997 *Astrophysical Journal* **476**, 440
ZABLUDOFF, A. and GELLER, M. 1994 *Astronomical Journal* **107**, 1929
ZABLUDOFF, A., GELLER, M., HUCHRA, J., and RAMELLA, M. 1993 *Astronomical Journal* **106**, 1301
ZARITSKY, D., SMITH, R., FRENK, C., and WHITE, S. 1997 *Astrophysical Journal* **478**, 39
ZUREK, W., QUINN, P., SALMON, J., and WARREN, M. 1994 *Astrophysical Journal* **431**, 559

CHAPTER SIX

The Distribution of Baryons in the Universe

6.1 Extragalactic backgrounds

The final matter of cosmological interest to consider is one that is often neglected. We have earlier addressed the issue of the general baryonic content of the Universe, as measured by primordial nucleosynthesis and other techniques, but we have not addressed the issue of where these baryons are located. In general, there are two choices: baryons are either trapped inside a potential well, hence located in a galaxy, or they are distributed outside of galaxies. This is determined by the efficiency of galaxy formation. What percentage of the primordial hydrogen and helium gas managed to creep into some preexisting gravitational potential to ultimately fragment, forming stars inside of galaxies? Theory, regardless of the particular structure formation scenario, does not make a reliable prediction of this efficiency, so it has to be determined observationally. This chapter will consider the evidence regarding three possible repositories for baryonic material: (1) easy-to-detect galaxies, (2) hard-to-detect galaxies and (3) a distributed background that is detected via recombination radiation at some particular wavelength. Observationally, this entails determining the galaxy luminosity function (GLF) and searching for evolution with look-back time as well as trying to detect various radiation processes that are tracers of intergalactic atoms. At the very least, we wish to ascertain whether there are more baryons located inside galaxies compared to outside. And this section begins with a summary of various searches for diffuse extragalactic backgrounds that might arise from a significant intergalactic population of baryons.

With the exception of the CMB, establishing the existence of electromagnetic backgrounds at other wavelengths has been a difficult chore. In order of increasing wavelength, here is a summary of the current evidence for different backgrounds and their possible interpretations. In general, these backgrounds have been discovered via satellite observations. These observations are of rather low resolution and therefore cannot effectively distinguish between backgrounds that represent a uniform distribution of hot or cold gas from those that are due to an aggregate of discrete sources. This is much like making a naked-eye observation of the plane of the Milky Way and concluding there is a diffuse, continuous band of light that runs across the sky. Higher resolution observations reveal the source of the radiation to be discrete stars. Finally much care must be taken to isolate extragalactic backgrounds from galactic backgrounds. This requires an all-sky survey and subsequent analysis of the data in galactic coordinates.

6.1.1 The radio background

In the centimeter to decameter range there is a well-known background of radiation due to galactic synchrotron and thermal processes. This was discovered when the first radio telescope was built (Jansky 1933; Reber 1940). There is also a very large population of extragalactic radio sources, usually active galaxies. Deep radio surveys can be used like deep optical surveys (described below) to construct the source count vs. flux relation in an attempt to measure the geometry of the Universe (e.g., Ω). This is one of the classical cosmological tests. We have not discussed tests like this because they require an accurate understanding of the evolution of the sources (e.g., the radio or optical luminosity evolution of galaxies). Understanding the luminosity evolution of galaxies is extremely model dependent, hence the use of source counts as a function of apparent flux (or better still, redshift) currently has little constraining power.

At shorter wavelengths the spectrum is completely dominated by the CMB. This spectrum has now been established by *COBE* to be that of a blackbody to very high precision (Figure 1.5). There is no longer any room for the extra components that were postulated in the late 1980s when balloon measurements hinted at a possible submillimeter excess of radiation in the total spectrum (Matsumoto *et al.* 1988). The first hints of these components prompted theorists to construct fairly elaborate models of high redshift dust (mostly for conference proceedings). The dust was heated by QSOs or the first generation of massive stars in protogalaxies (Bernstein *et al.* 1989; Lahav, Loeb, and McKee 1990; Ostriker and Strassler 1989; Heisler and Ostriker 1988) to temperatures of 20–40 K and its thermal spectrum was greatly redshifted to appear as the submillimeter excess. If real, this submillimeter excess would indicate a possibly large population of intergalactic baryons in the form of dust. That population would have clear cosmological implications. The spectrum determined by *COBE* now completely precludes this possibility and serves as a reminder that it's very important to properly characterize any suspected cosmological background before attempting to model it.

6.1.2 The infrared background

A cosmological background in this waveband has been difficult to establish unambiguously. In the range 10–30 μm there is a strong source of radiation from heated

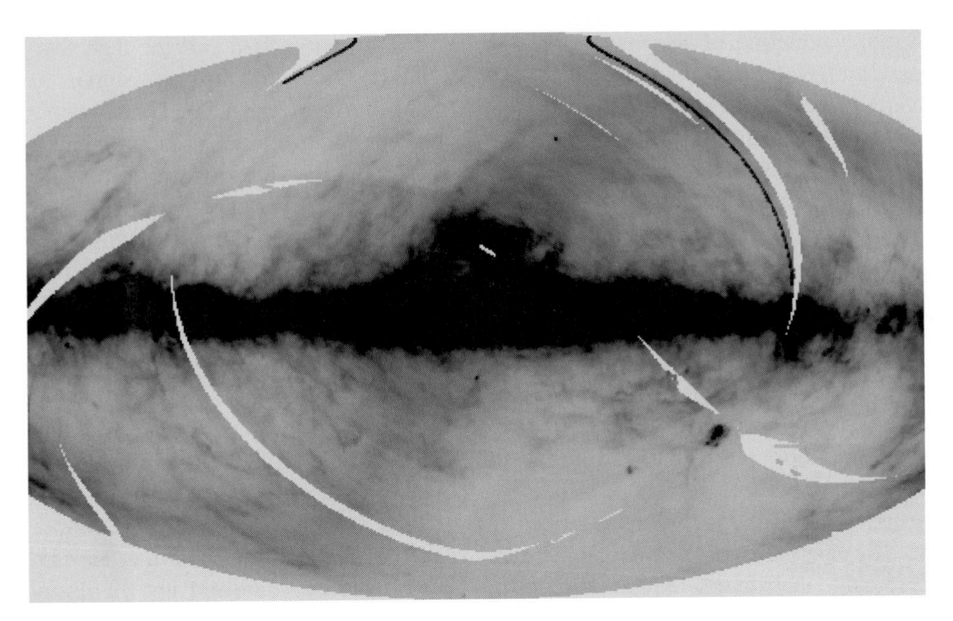

Figure 6.1 *IRAS* 100 μm image of our galaxy showing emission (cirrus) at high latitudes. Courtesy of IPAC and Joe Mazzarella.

interplanetary dust (which is why deep far-infrared surveys are done at the ecliptic poles). At wavelengths, ≈ 30–150 μm heating of the dust in regions of star formation and/or by the general interstellar radiation field produces a very strong galactic background even at relatively high galactic latitude. The *IRAS* satellite, launched in 1984, was sensitive to source emission in the range 12–100 μm. The all-sky map (Figure 6.1) produced at 100 μm by *IRAS* clearly shows the presence of large regions of emission at high galactic latitude. These sources have been named Galactic Cirrus as their structure is fairly cloud-like. The heating sources of this high latitude cirrus aren't completely clear, but most of the heating probably comes from stars in the galactic plane, where the optical and UV radiation can escape through regions of low opacity and penetrate to high latitude. The typical dust temperature of a cirrus structure is 20–30 K. *IRAS* observations of nearby large galaxies (e.g., M31, M33) have shown this cirrus component to be ubiquitous (Walterbos and Schwering 1987).

Since heated dust has been determined as pervasive in galaxies, the aggregate of all the galaxies in the Universe should produce a redshift-smeared background over the range 10–400 μm. In general, the spectrum of dust emission is that of a blackbody convolved with the emissivity of the dust grains. The total energy emitted by the dust scales as T_{dust}^{n+4} where n is the emissivity index ν^n. Large grains, which dominate the emission at long wavelengths, have an emissivity which goes as λ^{-1}, hence the total energy goes as T^5. This is an important point of energy conservation and balance in galaxies. Small differences in dust temperature between galaxies, or between regions in the same galaxy, reflect very large differences in energy input.

Simple modeling of the interaction between the radiation field and the dust temperature (Bothun, Lonsdale, and Rice 1989) suggests that the dust temperature is a good diagnostic of the nature of the heating sources (e.g., UV radiation from newly formed stars vs. the ambient light from older stars). Since the extragalactic background represents the sum of sources at different wavelengths, it will not be characterized by a simple blackbody of some given temperature. However, it is possible there might be a "feature" in the spectrum that would represent high star formation rates at high redshift. This assumes that dust, which must come from previous generations of massive star formation, is already in place in these galaxies at high redshift.

Detection of a possible cosmological infrared background (CIB) means detecting an isotropic signal of unknown spectral signature against the strong signal of our galaxy, itself nearly isotropic due to high latitude cirrus emission. Further difficulty arises when trying to calibrate the absolute strength of the CIB, as all the strong foreground sources need to be properly removed. To unambiguously detect the CIB requires very good modeling of the the known Solar System and galactic foregrounds to examine whether there are significant sources of residual emission that are isotropically distributed. Attempts to do this with the *IRAS* data did not yield any strong results. However, one of the instruments onboard *COBE* was the diffuse infrared background experiment (DIRBE), which made measurements over the range 1.25–240 μm (significantly longer than *IRAS*). The DIRBE measurements are potentially quite sensitive to the existence of the CIB.

Estimated minimum strengths of this background can be obtained by using the existing far-infrared (FIR) luminosity function (LF) of galaxies, as determined by *IRAS*. Over the range 100–300 μm, these minimum strengths are in the range $(2–4) \times 10^{-9}$ W m^{-2} sr^{-1} when the FIR LF is integrated out to $z = 3$. Any luminosity evolution in the sources and/or increase in space density with redshift would make the real background potentially much higher than this minimum estimate. The current status of the modeling, coupled with the DIRBE observations, shows residuals of $(20–50) \times 10^{-9}$ W m^{-2} sr^{-1} at high latitude in this wavelength range.

Although this is a possible detection of the CIB, at a level at least 10 times the predicted minimum strength, it could equally as well be an indication that the current modeling of foreground sources is inadequate. Furthermore, FIR missions, such as ISO and SIRTF, will help to better determine the CIB if it exists. Moreover, even if the CIB were unambiguously established, its origin would still be difficult to interpret. The low angular resolution at FIR means that many discrete sources (e.g., galaxies) would fill a single beam. The expectation, of course, is that the CIB is produced by galaxies, but at high redshift, intergalactic dust heated by quasistellar objects (QSOs) could also produce a CIB (Heisler and Ostriker 1988).

6.1.3 The UV background

As in the case of the observed infrared background, the observed diffuse ultraviolet background is dominated by dust scattering and interstellar emission in our own galaxy. However, when the galaxy is subtracted, the result is a nearly isotropic residual with a photon amplitude of $I_\lambda \sim 100\ \mathrm{s^{-1}\ cm^{-2}\ sr^{-1}\ \AA^{-1}}$ over the range 1300–2500 Å (Henry and Murthy 1993). Unfortunately, the calibration of this residual is very difficult and the quoted amplitude is uncertain by a factor of 2–3. There are three plausible extragalactic sources for the observed residual: (1) diffuse thermal emission that would arise from material between the galaxies, the intergalactic medium; (2) the aggregate light from discrete sources such as UV bright QSOs and star-forming galaxies; (3) radiation associated with the decay of some relatively long-lived massive particle.

Although the primordial nucleosynthesis constraints on Ω_b are fairly stringent, it is nevertheless important to observationally constrain the amount of material that might be contained in a hot intergalactic medium (IGM). The presence of a relatively warm (e.g., 10^4 to 10^6 K) dense IGM would manifest itself in the UV through the redshift blending of individual emission line features, principally Lyman$_\alpha$ from hydrogen and He II λ 304. To date, the best UV probes of the nature of any "hot" IGM come from the myriad of QSO absorption line studies. This gives information on both the density of the IGM along various lines of sight out to redshift $z \approx 4$ and the state of its ionization. From these studies it is clearly known that the IGM contains some baryonic matter, mostly in the form of the ubiquitous "Lyman$_\alpha$ forest" clouds first characterized by Sargent *et al.* (1980). The source of ionization of these structures is most likely the QSOs themselves which, possibly together with forming galaxies, produce the metagalactic UV flux.

In fact, the contribution of QSOs to the metagalactic UV flux can be directly estimated from a phenomenon known as the "proximity effect." Imagine there is a string of Lyman$_\alpha$ forest clouds at various redshifts along the line of sight toward some distant QSOs. If one of these clouds is sufficiently close to the QSO itself, that cloud will see a source of ionizing flux which is above the metagalactic flux, owing to its proximity to a single QSO. This extra UV flux from the QSO itself raises the overall state of ionization in the proximate clouds, hence it reduces their ability to produce absorption lines. Since the UV flux from the QSO itself is directly observed, the metagalactic UV flux can be inferred by comparing the UV absorption line strengths from those clouds that are near and far from the QSO. Over the redshift range 1.5–4, the level of this metagalactic UV flux is $I_{\nu H} \sim 10^{-21}\ \mathrm{erg\ s^{-1}\ cm^{-2}\ sr^{-1}\ Hz^{-1}}$.

This flux serves to ionize any intergalactic hydrogen or helium. These ionized gases cool via recombination line emission. The two dominant channels are the Lyman$_\alpha$ and He II λ 304 recombination lines. In this way, the photoionized gas acts as a photon energy downconverter (e.g., continuum ionizing photons of wavelength less than 912 Å are downconverted into 1216 Å Lyman$_\alpha$ photons). Since the intensity of the recombination flux can never exceed the intensity of the input ionizing flux, individual line contributions to the total UV background can be estimated from the observed or inferred metagalactic UV flux over some redshift interval. In numerical terms, this can be expressed as

$$I_\lambda(\lambda) \le \frac{I_{\nu H}(z)}{2\pi\hbar\lambda_H(1+z)^4} \qquad (6.1)$$

where λ_h refers to Lyman limit photons. The $(1+z)^4$ term accounts for photon dilution effects as the radiation is redshifted to $z=0$. Redshift-smeared Lyman$_\alpha$ recombination radiation from sources in the redshift range 0–0.6 would contribute to the 1200–2000 Å UV background. There are rather few QSOs in this redshift range and the available ionizing flux is down by ≈ 2 orders of magnitude relative to that quoted above for the redshift range 1.5–4. From equation (6.1) this yields an upper limit on the photon amplitude of 2 s^{-1} cm^{-2} sr^{-1} Å^{-1}, which is only 2% of the observed extragalactic flux. The corresponding case of redshift-smeared He II λ 304 recombination radiation, which comes from the redshift range $z=3$–5, suffers from more severe photon dilution. Using the nominal flux of $I_{\nu H} \sim 10^{-21}$ J s^{-1} cm^{-2} sr^{-1} Hz^{-1} in this redshift range yields a photon amplitude of less than 1 s^{-1} cm^{-2} sr^{-1} Å^{-1}.

These simple calculations effectively demonstrate that recombination radiation from a dense IGM of hydrogen and helium does not effectively contribute to the observed extragalactic UV flux at $z=0$. However, there is an alternate possibility that the IGM is shock heated; hence hydrogen and helium are collisionally ionized. This can be completely ruled out in the case of hydrogen as the total amount of neutral hydrogen in the IGM is strongly constrained by the lack of H I absorption troughs seen in the spectra of distant QSOs – historically this is known as the Gunn–Peterson test (Giallongo *et al.* 1994; Fang and Crotts 1995). For the case of helium, absorption troughs have now been detected toward at least two distant QSOs, indicating the existence of some neutral intergalactic helium at redshifts $z \approx 3$ (Davidsen, Kriss, and Zheng 1996). However, any λ 304 emission associated with the collisional ionization of this helium IGM would be rapidly damped out via cumulative absorption by the "Lyman limit" hydrogen systems. We can thus confidently conclude that thermal emission from any IGM is a negligible component of the observed UV background. The most viable candidate for producing the background is therefore the integrated light from QSOs and star-forming galaxies as we dismiss the decaying particle hypothesis below.

6.1.4 The X-ray background

An isotropic distribution of X-rays was originally discovered in 1962 by Ricarrdo Giacconi and colleagues. As it predated discovery of the CMB, an X-ray background (XRB) was initially somewhat puzzling. Over the spectral range 3–40 keV, the XRB spectrum is very well described by a thermal bremsstrahlung model with a characteristic temperature of 40 keV. This led to a series of models in the 1980s in which the spectrum was produced by a rather hot, diffuse IGM (Subrahmanyan and Cowski 1989; Anderson and Margon 1987; Giacconi 1987). The heating sources of this hot ($\approx 10^8$) gas, however, are rather unclear. Care must be taken not to violate any Big Bang constraints, and several investigations (Taylor and Wright 1989) have shown that a hot IGM requires the gas to be reheated in the redshift range $z=3$–5, which means the rest-frame temperature of the IGM is of order 150–250 keV. To obtain the observed flux in the XRB also requires a baryon density of $\Omega_b \geq 0.2$. Thus, the confirmation of the XRB as being due to a hot diffuse IGM has major significance for the question of how the baryons are distributed. This confirmation would also seriously call into question the limits on Ω_b obtained from primordial nucleosynthesis arguments.

A clear prediction of the existence of a hot IGM would be the production of Compton distortions in the CMB, of the kind described in Chapter 2 regarding the Sunyaev–Zeldovich effect. These distortions would have been detected in just the first few minutes of observation with the *COBE* satellite and they were one of the first things that could be ruled out (Mather *et al.* 1990). With more *COBE* data, it is now possible to constrain the contribution of a smoothly distributed hot, diffuse IGM to less than 1 part in 10^4 to the observed XRB (Wright *et al.* 1995). Although a cooler, more strongly clumped IGM can circumvent some of this constraint, in order to be consistent with the observed *COBE* anisotropy, it requires a very large number of clumps that are essentially galaxy sized but nevertheless are sources of X-ray emission. The gas that is bound in galaxy halos by dark matter can be a source of X-ray emission, but its characteristic temperature, which reflects the depth of the potential, is much too low to account for the observed XRB. These considerations make it plain that the only feasible contribution to the XRB comes from discrete sources.

Galaxies, galaxy clusters and QSOs/AGNs are all potential discrete source contributors to the XRB. Sorting out the potential contribution of all these sources became a serious endeavor in the mid 1980s and early 1990s. Now, with the *ROSAT* all-sky survey in soft X-rays as well the *GINGA*, *EXOSAT* and *ASCA* missions, there is a great deal of observational data which can be used to constrain the contributions from each of these sources. Of these sources, the overall spectral shape of the XRB now strongly favors QSOs/AGNs as being the dominant contributors (Comastri *et al.* 1995); essentially there is no room for additional classes of objects and certainly no room for a hot, diffuse IGM. The key feature of the model which allows for this conclusion is the incorporation of absorbed sources as a function of neutral hydrogen column density. It is now realized that most active galactic nuclei (AGNs) are surrounded by a torus of absorbing material (e.g., gas and dust) and the source–torus–observer geometry determines the observational attributes of a particular AGN. Thus, some fraction of X-ray AGNs have their emission strongly self-absorbed and/or scattered by this torus. Incorporating this feature into the model produces a very good fit to the XRB data, particularly that determined by *ROSAT* (Hasinger *et al.* 1993).

6.1.5 The gamma ray background

Diffuse gamma ray radiation was first detected in 1972 by the *OSO-3* satellite (Clark 1972) and has been confirmed by a number of other satellite missions. In 1991, the *Compton Gamma Ray Observatory* (*CGRO*) was launched. On board was the Energetic Gamma Ray Experiment Telescope (EGRET), which had an order of magnitude more sensitivity and lower instrumental background than any previous gamma ray detectors launched into orbit. The primary scientific objective of EGRET was to perform an all-sky survey for gamma radiation with photon energies greater than 30 MeV. Prior to *CGRO* there was a well-established correlation between the observed diffuse emission and the column density of interstellar gas in our galaxy. However, there remained hints of residual emission along lines of sight that didn't intersect any galactic gas, thus hinting at an extragalactic origin. These hints, however, were unable to be verified by other missions, due to pre-mature instrumental failure or just lack of instrumental sensitivity.

To date, EGRET has detected very strong diffuse emission from the galactic plane. At higher galactic latitudes, this diffuse emission is considerably weaker, thus facilitating the detection of point sources. A total of 36 points sources had been identified by late 1995 with a class of active galactic nuclei known as blazars (Impey 1996). Blazars are strong radio sources and often show very strong polarized optical emission. Some blazars exhibit superluminal motion (see below) and most have high time variability of their emission (at all wavelengths). The identification of blazars as gamma ray sources then opens up the possibility that their redshift-smeared emission can produce an extragalactic background. However, this background has not yet been detected by *CGRO* as it's a rather difficult measurement. The intensity of the background is low and there is no predicted spectral signature to look for. Furthermore, the instrumental background is still rather high and little can be done to correct this, as the background is produced by interactions between cosmic rays and the material the detector is made from. In addition, a secure detection of a diffuse gamma ray background of cosmological origin requires a very good model of the galactic emission, which needs to be subtracted.

The current model of gamma ray emission from our galaxy (Bertsch *et al.* 1996) is in very good agreement with the EGRET data, so it becomes a matter of understanding the instrumental response of EGRET to high precision. A preliminary analysis of many high galactic latitude fields indicates that some residual extragalactic background gamma rays are present. The overall spectrum is a power law with a slope of -2.2 ± 0.2 but the normalization and calibration of this spectrum is not yet available from the data. In addition, EGRET confirms the presence of one spectral feature noted earlier – a significant enhancement of flux, above the power law fit, at energies ≈ 3 MeV (known as the 3 MeV bump). If we therefore assume that the gamma ray background represents the sum of discrete sources, such as active galactic nuclei (AGNs) or quasistellar objects (QSOs), then the slope of the power law and the presence of the 3 MeV bump need to be accounted for.

As a starting point, one can use the observation that discrete gamma ray sources tend to be associated with radio-loud quasars. Dermer and Schlickeiser (1992) have used a linear scaling between radio flux and gamma ray flux and have integrated over the radio source counts to reach the conclusion that radio sources cannot be the sole contributors to the gamma ray background (keeping in mind that the overall background level is still uncertain). Furthermore, detailed studies of a few gamma ray AGNs have shown them to have the wrong spectral slope. Thus it seems that additional sources are required. Perhaps these sources are the more numerous radio-quiet QSOs and/or lower luminosity AGNs. Still, the possible

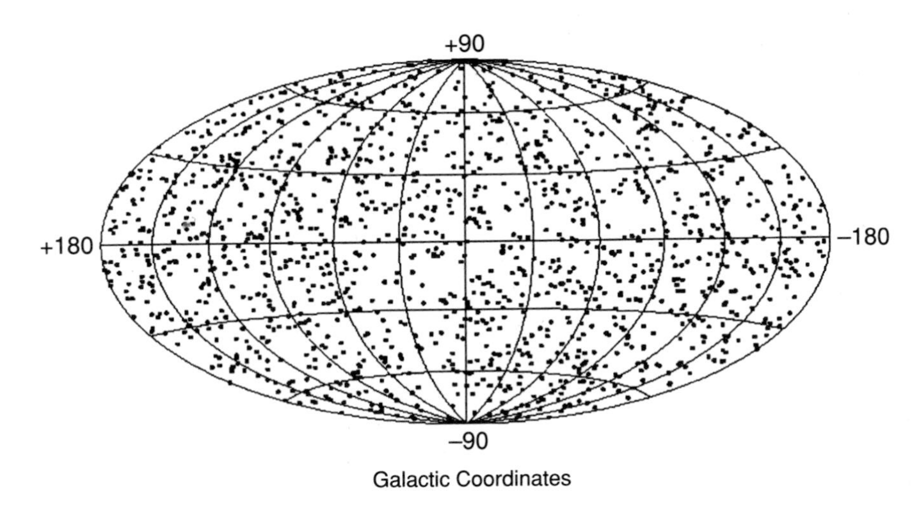

Figure 6.2 Spatial distribution of detected gamma ray bursters as of June 1997: the isotropic distribution strongly implies a cosmological origin. Courtesy of NASA and the Compton Observatory Science Support Center.

existence of a sourceless gamma ray background can't yet be ruled out. One model which would produce this appeals to late matter–antimatter annihilation occurring in the range $z = 10$–100. In this model (Stecker and Salamon 1996), the 3 MeV bump is produced by the redshifted 70 MeV annihilation line associated with the decay of relativistic pions. Hence it is important to see whether the data can yield a firm distinction between a discrete source background or a truly diffuse background that comes from matter–antimatter annihilation (the diffuse background would be difficult or impossible to understand from our current cosmological models).

The mystery of gamma ray bursters

Another observation which points to the importance of discrete sources is the profound mystery of gamma ray bursters (GRBs). Although they have been known for 20 years, *CGRO* has discovered several hundred new GRBs using the BATSE detector. In general, they have burst durations that range from a few seconds to a few minutes. In a couple of cases, burst durations of less than one second have been detected. On March 1, 1994 a very bright burst was detected. A search for a radio counterpart of that burst was made for ≈ 100 postburst days with no detection recorded (see below). To date, no GRB has yet had an observed counterpart at other wavelengths. Part of this is due to the low positional accuracy (only about 10°) that BATSE is capable of providing.

The real mystery of the GRBs is revealed in Figure 6.2, which shows the distribution of ≈ 1000 GRBs plotted in galactic coordinates. Notice that the distribution of sources is isotropic. Even though our galaxy is known to be a strong source of gamma ray emission, the burst phenomenon does not appear to be associated with these processes. To explain their isotropic nature, three classes of models have been suggested: (1) the Oort cloud model in which GRBs are actually objects spherically distributed around the Solar System; (2) they are distributed in a very extended spherical halo around our galaxy; (3) they are of cosmological origin. This failure even to know the distance scale that corresponds to the GRBs by 12 orders of magnitude qualifies them as being the one astrophysical phenomenon that we know least about!

The Oort cloud model suffers from the lack of any known reasonable physical object which might produce gamma rays (evaporating mini black holes have been suggested but too high a space density seems to be required). The extended halo model appeals to collisions of objects with neutron stars to generate the gamma rays: (1) it requires a velocity dispersion of the halo which is considerably larger than observed and (2) it suggests an anisotropy in the source counts should be observed towards nearby galaxies like M31, as a similar process would be occurring in its halo. Thus we are left with a cosmological origin as the most *reasonable* of the hypotheses!

Consistent with a cosmological origin is the observation that, in addition to their isotropic distribution, there are more bright bursts than faint bursts. This suggests that the distribution of these objects is bounded by some limit and that we are near its center. The lack of fainter bursts argues that the more numerous and more distant objects yield fluxes that are below the detection threshold of the instrument. A cosmological origin would suggest that galaxies are the likely hosts of GRBs. In this case, many models predict that postburst flux densities at centimeter wavelengths of a few tens of millijanskys should appear in a few weeks following the initial burst. So far, no positive detections have occurred, including the best case of the March 1, 1994 outburst, whose field was observed at radio wavelengths for 100 postburst days (Dessenne *et al.* 1996; Koranyi *et al.* 1995; Frail *et al.* 1994).

Because of the poor angular resolution of the BATSE detector and the subsequent positional uncertainty of a few degrees, deep imaging studies that look for the optical counterpart are not practical, as there will be literally millions of candidates to choose from over the field of view corresponding to the GRB position. To optimize the detection of GRB host galaxies requires accurate burst localizations. These locations can be obtained by using a widely separated network of satellites that are sensitive to gamma ray bursts (in addition to *CGRO* there are several military satellites available that use this signature to search for nuclear test-ban violations) using precise time-of-arrival measurements. Through triangulation, positional error boxes of a few arcminutes can be realized. Larson, McClean, and Becklin (1996) have made deep near-infrared measurements of the six GRB error boxes which have the lowest positional uncertainties as determined in this manner. Each of the six fields does contain an obvious galaxy whose angular size and flux level is consistent with it being located at high redshift, making it the candidate host galaxy of the GRB. In early March 1997 the *BeppoSAX* X-ray satellite, which was designed to find GRBs, detected a new GRB, and after 8 hours of observation it narrowed its probable position down to a circle of radius 50 arcsec. Subsequent optical observations in that area have revealed a candidate galaxy, at redshift $z \sim 1$, as the possible host of this event. At this distance scale, the energy of the event is comparable to the total energy (optical luminosity + mechanical energy) of 2–4 type I supernovae. This new observation would seem to confirm that GRBs are generated in galactic cores. If galaxies are the hosts of GRBs, what then is the physical source of the radiation?

There seem to be two plausible physical mechanisms that could generate GRBs. The first appeals to collisions between neutron stars (NSs). The burst is a manifestation of a thermonuclear runaway on the surface of the neutron star. Alternatively, a burst could be generated as the signature of the release of large amounts of gravitational binding energy. Either kind of event should produce optical and soft X-ray photons as the higher energy radiation produced by the burst interacts with the surrounding medium and is absorbed and reradiated as lower energy photons. This provides the motivation for searching for optical or soft X-ray counterparts. The main question about this method of producing GRBs revolves around the environmental requirements that a galaxy must have to facilitate NS–NS collision (or a collision of another star with an NS).

The second source of GRB appeals to a mechanism that is already known to be operative. There exists a class of about 20 objects discovered to date which are known as superluminal radio sources. These objects are galaxies with active nuclei that emit bipolar jets of material. From our observational perspective, the time evolution of the angular separation of the jets on the plane of the sky apparently requires faster-than-light motion. This paradox can be understood if are looking down essentially a beam pipe of relativistic outflowing material, as explained in Chapter 2. To date, practically all of the known superluminal sources have been detected by EGRET; the nondetections are thought to be the result of "bent" jets (von Montigny *et al.* 1995). So we know that relativistic beaming and the associated Lorentz boosting of photon energies can produce gamma rays. In addition, this relativistic jet of material should contain electrons and positrons which will annihilate and produce the 0.511 MeV annihilation line. In the observer's frame, this radiation can be greatly blueshifted through the motion of the jet, then later redshifted via universal expansion. The superposition of many of these beamed jets could produce the observed 3 MeV bump in the gamma ray background. Since the acceleration mechanism of these jets is still unknown (one plausible model appeals to radiation pressure that accelerates a plasma to a bulk relativistic velocity), it is difficult to understand how this process might produce an actual burst of gamma rays that is significantly above their continuous production. More to the point, if relativistic beaming by distant extragalactic sources is the main producer of GRBs, the intrinsic number of GRBs in the Universe must be quite large, as only a fraction of these relativistic beam pipes will point at the Earth. One then has to wonder about the effect of a large space density of these "plasma guns" on galaxies that are trying to form at high redshift and which might be located near a gun.

At present, there are over 100 theories (see the suggested further reading) that purport to explain GRBs (obviously theorists love unexplained phenomena), and to these I add one more. The principle requirement of these theories is to produce an isotropic background. Within the framework of the gravitational instability paradigm there may exist a natural system for the production of GRBs, a system associated with structure formation. Recall that the Jeans mass at the time of recombination is similar to the mass of a globular cluster. Let's consider globular clusters with an order of magnitude more density than those presently observed, and assume they formed at $z \approx 10$–50. What would be the dynamical fate of these clusters? Such clusters would have very short dynamical timescales and they would collapse and form stars quickly. The key is to have a system with a relaxation timescale which is approximately the same as the stellar evolutionary timescale. This would be $\approx 10^7$ years. Spitzer and Hart (1971) parameterize the relaxation time for globular clusters as

$$\tau_{relax} = \left(\frac{6.5 \times 10^8 \text{ yr}}{\ln 0.4N}\right)\left(\frac{M}{10^5 M_\odot}\right)\left(\frac{M_\odot}{m}\right)\left(\frac{r_h}{1 \text{ pc}}\right)^{3/2} \tag{6.2}$$

An initial cluster of $N \approx 10^5$ and $M \approx 10^5 M_\odot$ which is confined to a radius of $r_h \approx 0.5$ pc (corresponding to $\rho \approx 10^{-20}$ g cm^{-3}) gives a relaxation time of $\approx 10^7$ years. This is the timescale over which equipartition of energy occurs, the massive stars sink to the center of the cluster and the lower mass stars are heated, possibly to escape velocity. As the core collapses due to relaxation (and dynamical friction), the NS remnants naturally sink to the center, thus greatly increasing the probability of an NS–NS binary (one of these binaries is known to exist in our galaxy). Energy loss from gravitational radiation of the NS–NS binary system will eventually cause the two stars to merge and possibly produce a GRB. The rest of the cluster stars would have evaporated away by this time. Hence the equations of stellar dynamics tell us that, if a sufficiently dense perturbation is present to form a dense stellar cluster, the resulting match between the dynamical relaxation timescale and the stellar evolutionary timescale produces a natural breeding ground and a confinement mechanism for NS–NS binaries. As this is expected to occur at very high redshift, before any clustering sets in, the distribution of these objects would be isotropic and there would be no surviving remnant today, because the lighter stars would have evaporated. Pretty cool, huh?

6.1.6 A possible decaying particle background

During the period 1990–1995, the British cosmologist Dennis Sciama introduced an important new idea of cosmological significance. Sciama surmised that the current level of ionization in our galaxy seemed to be higher than could be accounted for by the known contribution of young, ionizing stars in the galactic disk. In particular, the free electron scale height in the galaxy is observed to be approximately 900 pc and it's difficult to account for ionizing radiation that would make it this high above the thin (≤ 100 pc) plane defined by young stars. And it has always been unclear whether the combined ionizing flux of QSOs was sufficient to produce the partial ionization states of the Lyman$_\alpha$ forest clouds and/or the metallic line systems. To account for a possible extra source of ionization in the Universe, Sciama hypothesized the existence of a neutrino with a nonzero rest mass whose principle decay channel was an ionizing photon. The foundation of this idea, that some massive particle might experience late decay though photon channels, does have significant cosmological consequences, so it's important to test this idea rigorously.

The basis of the idea is well grounded in particle physics. One of the main ingredients of the standard model for particle physics is the existence of three families of neutrinos, the electron, muon and tau neutrinos. Suppose the tau neutrino were unstable: it could decay along a channel

$$\nu_\tau \rightarrow \nu_\mu \, (\text{or } \nu_e) + \gamma \tag{6.3}$$

The decay lifetime could be anything, depending upon unknown details in the standard model of particle physics. In order to conserve energy and momentum, the energy of the decay photon is given by

$$E_\gamma = \frac{1}{2} m_1 (1 - \frac{m_1^2}{m_2^2}) \tag{6.4}$$

where m_1 refers to the heavier neutrino species (e.g., ν_τ) and m_2 refers to the lighter species (ν_μ or ν_e). To be consistent with the solar neutrino experiment, one requires $m_1 \gg m_2$ so that

$$E_\gamma \sim \tfrac{1}{2} m_1 \tag{6.5}$$

Since we require E_γ to be larger than the ionization energy of hydrogen (13.6 eV), this leads to a lower limit on the mass of the tau neutrino of 27.2 eV. This is a cosmologically interesting mass because it allows theses neutrinos to have a significant contribution to Ω. However, the decay time of these neutrinos must be quite long, of the order 10^{23} seconds (Sciama 1990). Recall that the expansion age of the Universe is of the order 10^{17} seconds. This long decay time is set by the requirement that too short a decay time produces far too much UV background, but too long a decay time means there are not enough ionization photons at $z = 0$ to account for the large free electron scale height (900 pc; Lyne *et al.* 1990) in our galaxy, as inferred from dispersion measures to globular clusters. Indeed, it is difficult to understand this large scale height if the ionization is solely due to young OB stars in the galactic disk which have a scale height ≤ 100 pc.

The main aesthetic complaint to this particular theory is that two very disconnected physical properties of the Universe, i.e., the ionization potential of hydrogen and Ω are now strongly coupled, leaving only a very small range of neutrino masses than can satisfy both conditions. When one considers recent observational data on the metagalactic ionizing UV radiation, this small range becomes even more narrow. Vogel *et al.* (1995) report on observations of a large intergalactic H I cloud discovered originally by Giovanelli and Haynes (1989); see also Impey *et al.* (1990),

Salzer *et al.* (1991) and Chengalur, Giovanelli, and Haynes (1995). This cloud, which may be a form of LSB galaxy (see below), represents an ideal laboratory for determining the metagalactic UV flux because it has no identifiable internal sources of ionization. The limits on H_α recombination from the Vogel *et al.* study constrain the photon flux of ionizing radiation to be $\leq 1.6 \times 10^5$ cm^{-2} s^{-1}. The flux of decaying neutrinos is given by

$$\frac{N_\nu}{\tau}\frac{c}{H_0} \tag{6.6}$$

where N_ν is the number density of neutrinos at $z = 0$ (≈ 100 cm^{-3}) and τ is the decay lifetime. From Chapter 1 we have that c/H_0 is the horizon scale or the radius of the observable Universe. Since the decaying neutrinos have some redshift distribution associated with them (due to a distribution of decay times) then

$$E_\gamma = 13.6 + \varepsilon \text{ eV} \tag{6.7}$$

where ε represents the fractional volume of the Universe over which the neutrinos can decay and still have 13.6 eV of energy at $z = 0$. We thus have the firm observational constraint that

$$F \leq 1.6 \times 10^5 = \frac{N_\nu}{\tau}\frac{c}{H_0}\frac{\varepsilon}{13.6} \tag{6.8}$$

which leads to ε in the range 0.2–0.4 eV. Thus, if this theory is correct, observations have fixed the mass of the tau neutrino to be 27.6–28.0 eV. We have either solved cosmology or dismissed the decaying neutrino hypothesis (Sciama 1995).

There are other observational constraints that are also inconsistent with this hypothesis. The first of them involves the ionization of nitrogen, which requires photons of energy greater than 14.5 eV. Under the current constraints, the decaying neutrino hypothesis would not result in the ionization of nitrogen. In our galaxy, nitrogen is observed to be partially ionized when there are no apparent local sources of ionization. And as discussed above, the observed UV background flux is quite consistent with the integrated contribution of galaxies and it doesn't appear to require extra sources. For the decay parameters presented here, decay photons would end up providing approximately 70% of the extragalactic background at 1500 Å (Sciama 1995). Finally, if decaying massive neutrinos contribute most of the binding mass to clusters of galaxies, then very massive clusters (which have a high density of neutrinos) should be sources of weak UV emission at the specific wavelength of $\lambda_0 = \lambda_e(1 + z)$ where λ_e corresponds to a photon of energy 13.8–14.0 eV. Observations of A665 using the Hopkins Ultraviolet Telescope (HUT), at $z = 0.18$, failed to detect any emission at the predicted wavelength (Davidsen *et al.* 1991). To explain this nondetection requires a longer neutrino decay time. However, a longer neutrino decay time will not supply the needed ionizing photons to account for the 900 pc free electron scale height in our galaxy. On balance, this intriguing idea does not appear to be viable.

6.2 The space density of galaxies

The preceding discussion has established that there is no strong evidence for a significant amount of hydrogen or helium distributed in a diffuse background. Similarly, there does not seem to be solid evidence for a decaying particle background. Hence we are driven to the conclusion that, other than the QSO absorption line systems, it would seem that most of the baryons in the Universe are confined to galactic potentials. There is no significant population of intergalactic baryons. If this is indeed the case, then our baryon census is equivalent to determining the luminosity function of galaxies. Recall, however, that the luminous component of easily visible galaxies contributes only 0.5% to the closure density. This is well below the limits on Ω_b derived from primordial nucleosynthesis and suggests that many of the baryons in galactic potential wells are "dark" or there may exist "dark" galaxies that have not yet been properly accounted for in the census of nearby galaxies. Hence we really have a "missing" baryon problem to deal with.

There are basically two approaches to determining the space density of galaxies. The first is to construct the galaxy luminosity function (GLF) from what is considered to be a fair sample of redshifts. The determination of the GLF is one of the fundamental cosmological observations that can be made. With the GLF one can estimate the mean luminosity density of the Universe and the mean M/L ratio of galaxies. This in turn provides a direct indication of the contribution of nonluminous matter in galaxies to Ω. The shape of the luminosity function can also provide an important constraint for structure formation theories. As a consequence of its fundamental nature, the GLF has been worked on by many, many groups. Until recently, most determinations of the GLF have been similar but recent advances in galaxy sampling and redshift survey data suggest that a proper determination of the GLF still remains.

6.2.1 The luminosity function of galaxies

For most purposes, the GLF is assumed to have the functional representation first suggested by Schechter (1976):

$$\Phi(l)dL = (\Phi_0 L^{-\alpha} e^{-L/L_*})dL \tag{6.9}$$

where $\Phi(l)$ specifies the space density of galaxies over some luminosity interval dL. This function is schematically illustrated in Figure 6.3. There are three free parameters that must be observationally determined:

- $\Phi(0)$ provides the overall normalization (at $z = 0$) of the GLF. An accurate measurement of $\Phi(0)$ requires a fair, volume-limited sample of galaxies.
- L_* is the normalization of the sharp exponential cutoff term in equation (6.9). Galaxies with luminosity in excess of L_* have quite low space density. There have even been attempts (e.g., Trevese, Cirimele, and Appodia 1996; Oegerle and Hoessel 1989; Gudehus 1989) to construct the extragalactic distance scale by assuming that L_* is universal and independent of environment. Observations

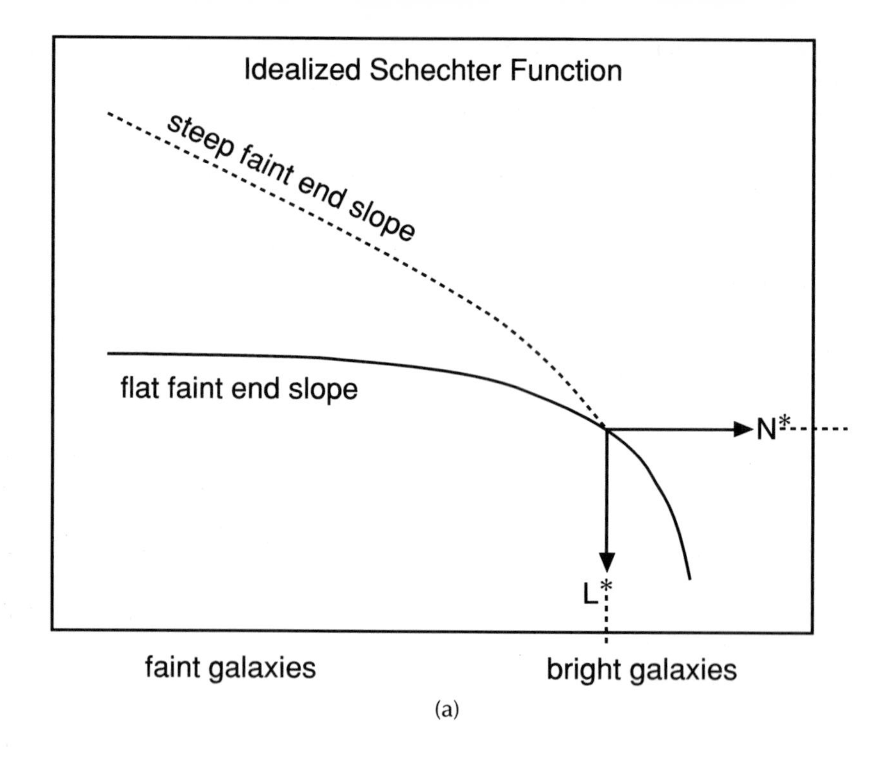
Idealized Schechter Function
steep faint end slope
flat faint end slope
N*
L*
faint galaxies
bright galaxies

(a)

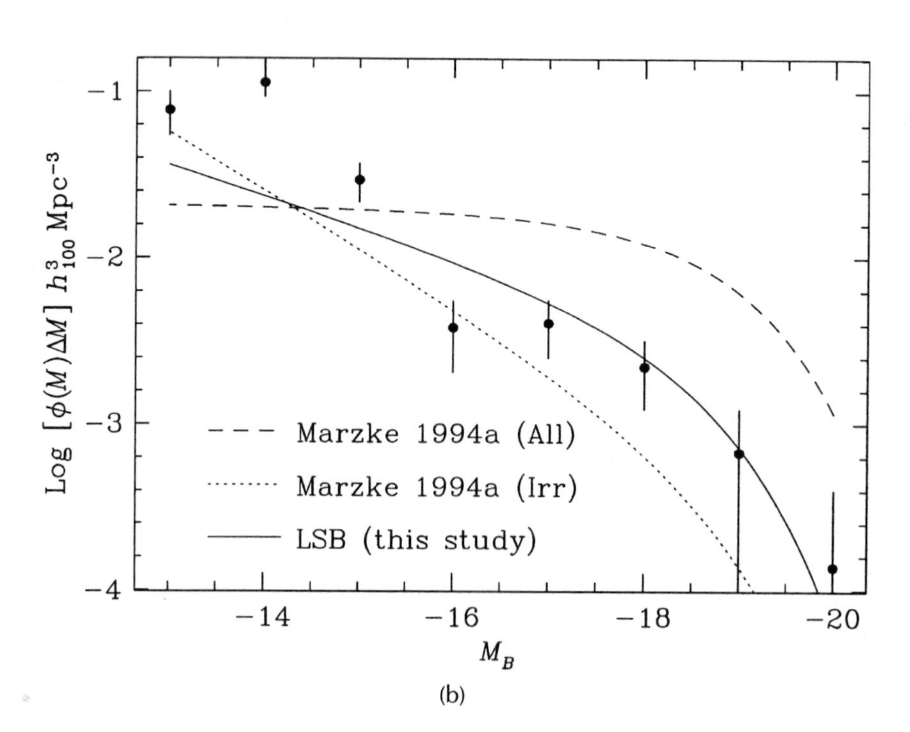
Log [φ(M)ΔM] h^3_{100} Mpc^{-3}
M_B
Marzke 1994a (All)
Marzke 1994a (Irr)
LSB (this study)
−1
−2
−3
−4
−14
−16
−18
−20

(b)

on clusters of galaxies reveal that L_* can be used to obtain relative cluster distances. We do not yet have a secure determination of the overall GLF, hence using its "features" to determine extragalactic distances is likely to contain unknown systematic errors (see below). However, it is true that most studies of nearby redshift samples yield approximately the same value of L_*. For $H_0 = 100$ this value is 10^{10}*blue* $L_\odot$.

- α is the faint end slope of the GLF. Of the three free parameters, this has the most cosmological significance as it determines the amount of mass that can potentially be locked up in low luminosity and low mass galaxies. For many years α was thought to have a slope of ≈ -1.25 for field galaxies and -1–1.1 for galaxies which were members of clusters. However, in the last 2–3 years a series of observations have challenged this value and suggest a considerably steeper faint end slope.

The universality of the free parameters of the GLF have come under fire primarily because the GLF seems to be a function of morphological type. In their detailed study of the Virgo cluster, Sandage, Binggeli, and Tammann (1985) were able to convincingly demonstrate that the GLF for spiral galaxies was better fitted by a Gaussian function than the Schechter function. This actually had been known years earlier by H I observers, as plotting the distribution of rotation velocities in spiral galaxies – correlated with galaxy luminosity through the Tully–Fisher relation – generally yielded a Gaussian distribution. A detailed study of the CFA redshift survey by Marzke *et al.* (1994) has produced the important result that the faint end slope α also seems to depend on morphological type. For low mass irregular galaxies, Marzke *et al.* (1994) find a rather steep slope of $\alpha = -1.9$, close to the maximum allowable slope of $\alpha = -2.0$ (where the integral over the GLF becomes divergent). Further modifications of the GLF have occurred as a result of the discovery of LSB galaxies, discussed shortly.

6.2.2 Deep galaxy counts: the mystery of the faint blue galaxies

The second estimate for the space density of galaxies can be made by doing deep galaxy counts and plotting the surface density of galaxies as a function of their

Figure 6.3 (a) Schematic representation of the Schechter formulation that describes the galaxy luminosity function. This function is defined by a characteristic luminosity L^*, and normalized space density at that luminosity N^*, along with the slope of the faint end. The solid line indicates a relatively flat faint end slope, which represents what is usually obtained in samples that are not corrected for surface brightness selection effects. The dashed line shows the case of a steep faint end slope in which most of the galaxies in the Universe are faint, low mass objects. (b) Examples of various luminosity functions that have been obtained. The dashed line plots the Marzke *et al.* (1994) data for all morphological types in the CFA redshift survey. The dotted line plots the subset of low luminosity, irregular galaxies in that survey; note that the faint end slope is significantly steeper. The solid line is a maximum likelihood fit to the data for field LSB galaxies; it also shows a fairly steep faint end slope. Reproduced from Sprayberry *et al.* (1997).

apparent flux. In principle, if all galaxies had the same luminosity as a function of redshift (or if we had an independent way of picking galaxies of the same luminosity, say through some kind of filter selection), this diagram would be very sensitive to the geometry of the Universe, hence Ω. Realistically, the stellar populations of galaxies do evolve, hence we expect moderate to strong luminosity evolution for galaxies. The history of using deep galaxy counts as a cosmological probe is that of inconsistent, ambiguous and confusing results. The naive expectations are that (1) faint galaxies should generally be red (due to the effects of redshift) and (2) they should have number counts that decline significantly more rapidly than $(1 + z)^3$ since the volume per unit area decreases with distance in cosmologies based on the Robertson–Walker metric. When the first deep counts came in, based on long-exposure 4 m photographic plates, it quickly became apparent that neither of these two effects was seen in the data (e.g., Koo and Kron 1982). Most noticeable was a significant population of rather faint but fairly blue galaxies. These galaxies are known as faint blue galaxies (FBGs). Several other surveys verified the existence of the FBG population. This led to a widely accepted model in which these galaxies were experiencing a phase of significantly enhanced star formation. The high number density of FBGs could best be explained if these galaxies were located at redshifts $z = 1$–3, so that a very large volume was being sampled.

However, by the late 1980s the advent of multiobject fiber spectroscopy meant that a deep field containing several FBGs could be exposed for several hours, thus returning spectra of sufficient quality for redshift measurements. These redshift surveys generally showed that the FBGs were primarily a low redshift population ($z \leq 0.7$) (Broadhurst, Ellis, and Shanks 1988; Colless *et al.* 1993; Glazebrook *et al.* 1995). In general, blue galaxies imply significant star fromation. The basic CDM scenario allows relatively late formation of galaxies, hence the FBGs may represent galaxies undergoing their initial bursts of star formation, several billion years after the Universe was formed. The exact conditions that would cause this delayed star formation are not well understood. One possibility is that the gas in these potentials was ionized early on and took a few billion years to cool. But this scenario means that (1) we have to identify the sources of ionization and (2) we have to explain why only this population was affected. We obtain a partial explanation if the FBGs are relatively low mass and low density galaxies.

Still, high redshift FBGs have been detected. Cowie *et al.* (1988) were among the first to identify a high redshift ($z \geq 2.5$) population of FBGs. This high redshift population has been confirmed by a number of others, most recently Steidel *et al.* (1996), who present indisputable evidence that star-forming galaxies exist out to at least $z \approx 3.5$. These observations demonstrate that FBGs can be found over a wide redshift range, hence they are a very heterogeneous population.

The nature of the FBGs seen in the deep count data therefore continues to be elusive. However, if the FBGs are a significant population at moderate redshifts, they must somehow disappear by $z = 0$. This indicates either strong luminosity or strong density evolution of the GLF. In particular, if galaxy merging occurs as a result of galaxy–galaxy interactions, then galaxy number is not conserved and number density of galaxies increases with redshift. This is a serious complication

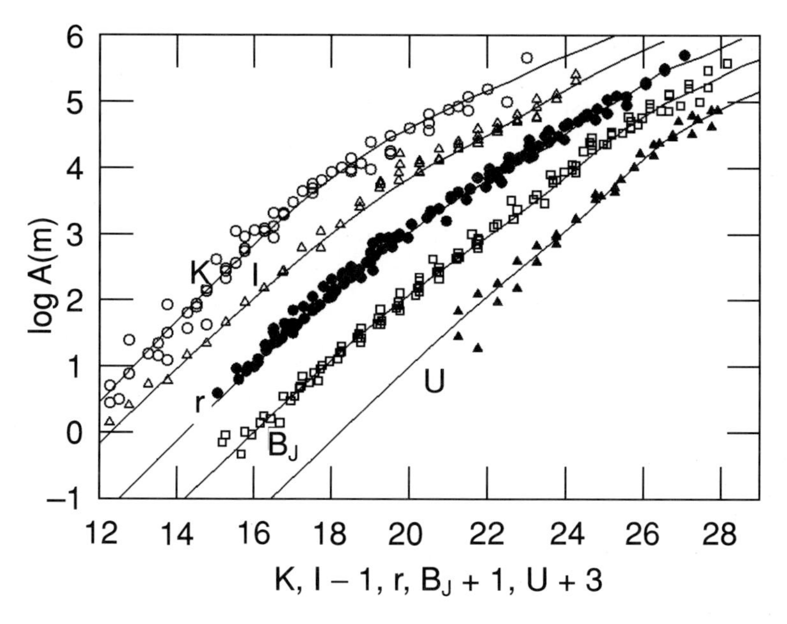

Figure 6.4 Summary of recent deep count data in various wavebands. Courtesy of Dave Koo and Carly Grownwall, University of California.

for structure formation theories that attempt to predict the GLF, since it implies that normalization of the GLF changes with redshift. For a sample of intermediate redshift galaxies, Patton *et al.* (1997) derive a merging rate of $(1+z)^{2.9\pm0.9}$. This exponent is close to the expected $(1+z)^3$ volume evolution, although the observed error bar is too large to confirm this. The other possibility of strong luminosity evolution suggests that the FBGs have a star formation history that allows them to fade rapidly, so that by $z=0$ they have extremely low surface brightness and are faint galaxies. This latter prediction is also testable to some extent (see below).

In the last four years, advances in detector technology have paved the way for performing deep galaxy counts at near-IR wavelengths. Figure 6.4 summarizes the current status of the deep count data. Note that in some cases there does not appear to be a significant turnover in the counts even down to very low flux levels. It is unclear whether this is a result of accessing a huge volume (i.e., the faintest galaxies are at the highest redshifts) or a reflection of a very steep faint end slope of the GLF (i.e., the faintest galaxies represent a very numerous population of low mass objects). And there appears to be a major inconsistency between the optical and the near-IR counts. Compared to the simple no-evolution (NE) models, the optical counts exceed the predicted counts by approximately a factor of 10 down to an apparent magnitude limit of $B=25$. In contrast, the near-IR counts are significantly lower and are fully consistent with the NE model. Recent redshift surveys have shown that the distribution in redshift space of faint galaxies is also quite consistent with the NE model. In particular, the mild luminosity evolution models, popular in the mid 1980s, would clearly predict a tail of $z \geq 1$ objects in the redshift distribution, which is not observed. And the FBGs generally have emission lines (indicative of star formation) and are more weakly clustered than redder galaxies of the same apparent flux.

There are several possible explanations for the FBGs, some of which are quite relevant to the question of where the baryons are at $z = 0$. Here are some of the most popular:

- The FBGs are a population of star-bursting dwarf galaxies located at modest redshift. This suggestion takes advantage of the fact that in any GLF with $\alpha \leq -1$, low mass dwarf galaxies dominate the space density. To produce the FBGs, however, these dwarf galaxies have to be at least an order of magnitude brighter at these modest redshifts, which requires a fairly significant star formation rate. Subsequent heating of the interstellar medium (ISM) by massive stars and supernovae should be sufficient to heat it beyond the escape velocity of these low mass systems (Wyse and Silk 1985). These galaxies would have a significant phase of baryonic blowout, after which they would fade to very low absolute luminosities and would become hard to detect at $z = 0$. This mechanism effectively gives the Universe a channel for making baryons "disappear" with time.
- The number density of galaxies is not conserved and the FBGs merge with other galaxies. It is difficult to support this hypothesis because (1) the FGBs are already weakly clustered and (2) the required merging rate is significantly higher than the rate measured at modest redshift by Patton *et al.* (1997). The merger idea works best if the FBGs are predominately at higher redshift, where the merger rate is higher, owing to the much smaller volume of the Universe.
- Over the redshift interval which contains most of the FBGs, the volume is larger due to a positive cosmological constant. Nonzero Λ Universes have larger volumes per unit redshift interval compared to $\Lambda = 0$ models. Like the fits to the power spectrum, nonzero Λ models also fit the deep count data rather well, although the volume effect is less pronounced if the FBGs are primarily at low redshift ($z \leq 0.7$).
- The FGBs represent an entirely new population of galaxies – a population defined by a star formation history and or initial mass function that allows only a limited window of visibility before the galaxies fade to extremely low surface brightness levels by $z = 0$. In general, it is dangerous to introduce a new population of objects into the Universe without strongly considering the possibilities of detecting the relic population (see below).
- The apparently high number density of the FBGs is an artifact of uncertainties in the determination of the local GLF (Gronwall and Koo 1995). In particular, the faint end slope of the GLF has been seriously underestimated from nearby samples (Sprayberry *et al.* 1997). This possibility remains highly viable (see below) and, in fact, incorporating a steeper faint end slope can remove much of the apparent excess.
- The local normalization $\Phi(0)$ of the GLF is too low. This could result if, for instance, deep surveys were more efficient at selecting LSB galaxies than nearby surveys. Although the evidence presented below strongly supports this idea, the effect of increasing the space density at $z = 0$ can only partially offset the excess FGB counts. A much larger lever arm is provided by increasing α.

Very recently, Lilly *et al.* (1995) have presented a redshift survey of ≈ 500 faint galaxies (see also Ellis *et al.* 1996). Their sample has excellent quality control, it is fairly free from selection effects and it is primarily aimed at determining the GLF up to $z \sim 1$. Their results have helped clear some of the confusion cited above. Their principal result is that, for blue galaxies, there is a change in the GLF by approximately one magnitude between $z \approx 0.38$ and $z \approx 0.62$, then another magnitude between $z \approx 0.62$ and $z \approx 0.85$. Moreover, many of these galaxies have been observed with the HST in order to measure characteristic surface brightnesses. Schade *et al.* (1995) find that the disks of these blue galaxies are ≈ 1 magnitude higher in surface brightness at $z = 0.8$ than at $z = 0.3$. Taken together this consitutes rather strong evidence for luminosity evolution in the FBGs. For a 15 Gyr Universe, there is approximately 3.3 billion years between $z = 0.85$ and $z = 0.38$. The data indicates that a typical FGB would decline in luminosity by a factor of 6 over this time period. This modest decline is quite consistent with standard population synthesis models involving a normal initial mass function (IMF). The decline in luminosity is primarily a reflection of the disappearance of the upper main sequence. At this rate, by $z = 0$, these galaxies will certainly not have faded to levels that preclude their detection, although many would be of LSB.

In contrast to the blue galaxies, the luminosity function for the red galaxies appears to show very little change back to $z \approx 1$. As it's these objects which should dominate the near-IR counts, the lack of evolution seen in those counts is not surprising. Still, caution should be exercised in the interpretation of "blue" vs. "red," as the color distinction is based on the colors of local galaxies (e.g., those of class Sbc) and it's unclear if the Lilly *et al.* division really allows one to be comparing the same galaxies at high redshift to those nearby. For instance, one could get evolution of the luminosity function with redshift for the "blue galaxies" because galaxy types that are in their "blue sample" at high redshift are in fact galaxies which would be in their definition of a "red sample" at low redshift due to natural evolution of their stellar populations. Adding to this confusion is the study of Im *et al.* (1996), whose morphologically selected sample of E and S0 galaxies does exhibit evolution of the luminosity function between $z = 0.5$ and $z = 1$ of ≈ 1 mag. The weakness is that, unlike the Lilly *et al.* sample, the Im *et al.* sample uses photometric redshifts, which are probably highly uncertain.

In sum, it seems likely there are two main populations of galaxies in the Universe: those that are evolving very slowly and those that are showing mild to perhaps rapid luminosity evolution. By $z = 0$ these two populations should evolve to a population of galaxies which exhibit a wide range of surface brightnesses. If this is the case, there could be a population of sufficiently diffuse galaxies that have escaped our detection. In turn, this would give rise to the "missing" baryon problem as well as providing the illusion that there are more galaxies at high redshift than at low redshift. For many years this wide range of surface brightness was not seen in the data. However, once the effects of surface brightness selection of galaxies became understood, these "missing" galaxies were found, and found in large numbers. What follows is the story of that particular scientific journey.

6.2.3 A new population of galaxies at $z = 0$

An important corollary to the cosmological principle asserts that all observers in the Universe should construct the same catalogs of galaxies. If this were not the case, then different observers might have biased views and information about (1) the nature of the general galaxy population in the Universe, (2) the three-dimensional distribution of galaxies, and (3) the amount of baryonic matter that is contained in a galactic potential. Thinking about the cosmological principle in terms of the homogeneity of observers' catalogs of galaxies raises an immediate and perhaps profound problem which can be stated very simply: If you see (detect) a galaxy, you can catalog and classify it; if you don't see it, you can't. Since galaxies are generally quite diffuse, low contrast objects with respect to the noisy background of the night sky, which has finite brightness, one can easily conceive of observing environments that would make galaxy detection difficult.

For instance, suppose that we lived on a planet that was located in the inner regions of an elliptical galaxy. The high stellar density would produce a night sky background that would be relatively bright and therefore not conducive to the discovery of galaxies. Similarly, if the Solar System, in its journey around the galaxy were unlucky enough to be located near or in a giant molecular cloud (GMC) at the same time that evolutionary processes produced telescopes on the Earth, then our observational horizon would be severely limited by the local dust associated with the GMC. Finally, consider the poor astronomer that lives on a planet which has two moderate-size moons in orbit about it, at least one of which is in the night sky at all times. Such a planet would have no "dark" time and observers would be hard-pressed to discover external galaxies. Although these are extreme situations, they illustrate the basic point that, for all observers, the finite brightness of their night sky acts as a visibility filter which, when convolved with the true population of galaxies, produces the population that appears in catalogs. Thus we have no guarantee that our location in the outer regions of a spiral galaxy, on a planet with 50% dark time per lunar orbital cycle, allows us to detect and catalog a representative sample of galaxies.

This potential bias in optical catalogs was first emphasized by Mike Disney in 1976 but was known at the time of Messier. Indeed, Hubble noted this problem as it pertains to galaxy classification. In his 1926 paper, Hubble wrote

> Subdivision of non-galactic nebulae is a much more difficult problem. At present and for many years to come, their classification must rest solely upon the simple inspection of photographic images, and will be confused, by the use of telescopes of widely differing scales and resolving powers. Whatever selection of types is made, longer exposures and higher resolving powers will surely cause a reclassification of many individual nebulae.

Hubble establishes that galaxy classification, and implicitly galaxy detection, is highly dependent upon observing equipment and resolution. There are two essential questions. How severe is the bias in terms of the potential component of the galaxy population that has been missed to date? And how would this affect our current understanding of galaxy formation and evolution?

Galaxies are detected on the basis of their surface brightness contrast with respect to the night sky background. Mild to strong luminosity evolution potentially produces galaxies with rather low luminosity per unit area at $z = 0$. Indeed, many have suggested that the fate of the FBGs is to fade to very low surface brightness levels by $z = 0$. Alternatively, there could be a population of intrinsically low surface mass density systems whose evolution is quite different from "normal" galaxies but which nevertheless are important repositories of baryonic matter. Disney suggested that such diffuse systems are hard to detect and therefore we could be missing an important constituent of the general galaxy population. Extragalactic astronomers lived for many years in a comfort zone, dismissing Disney's original hypothesis as, at best, applying to a limited and inconsequential population of objects. But let's examine this effect in greater detail.

6.3 Surface brightness selection effects

Within the idealized framework of azimuthally symmetric galaxy profiles, the luminosity profile of a disk galaxy is given by

$$\mu(r) = \mu_0 + 1.086(r/\alpha) \tag{6.10}$$

where μ_0 is the central surface brightness of the disk and α is its angular scale length, which corresponds to the physical scale length α_l at distance d. These two parameters characterize the light distribution of the idealized disk galaxy, and together they determine the integrated luminosity

$$L = 2\pi\alpha_l^2\Sigma_0 f(x) \tag{6.11}$$

Here Σ_0 is the central surface brightness in linear units, and

$$f(x) = 1 - (1 + x)e^{-x}$$

gives the fraction of the light contained within a finite number of scale lengths x, where

$$x = \frac{r}{\alpha} = 0.92[\mu(r) - \mu_0] \tag{6.12}$$

relative to that contained in an exponential profile extrapolated to infinity (Disney and Phillipps 1983). These simple formulas provide adequate fits to most spiral galaxies (de Vaucoulers 1959), and in particular to low surface brightness (LSB) galaxies (McGaugh and Bothun 1994; de Blok, van der Hulst, and Bothun 1995).

Freeman (1970) found that all spiral disks have essentially the same central surface brightness, $\mu_0 = (21.65 \pm 0.3)B$ mag arcsec^{-2}. This has become known as Freeman's law. If correct, the number of parameters relevant to galaxy selection reduces to one, as only variations in size modulate those in luminosity. Since μ_0 is a measure of the characteristic surface mass density of a disk, Freeman's law requires that all the physical processes of galaxy formation and evolution conspire to produce this very specific value for all galaxies. Either the surface mass density must be the same for all galaxies (in itself a peculiar result) with little variation in the mass-to-light ratio, or variations in the star formation history, collapse epoch and initial angular momentum content must all conspire to balance at this arbitrary value. It is thus important to rigorously test the reality of Freeman's law as the distribution of μ_0 may be directly related to the conditions of galaxy formation (Freeman 1970; McGaugh 1992; Mo, McGaugh, and Bothun 1994).

The Freeman value is about one magnitude brighter than the surface brightness of the darkest night sky. That the number of galaxies with faint central surface brightnesses appears to decline rapidly as $\mu_0 \rightarrow \mu_{sky}$ is suspicious, and if true of the real galaxy population, it implies that our observational viewpoint is privileged in that we are capable of detecting most of the galaxies that exist, at least when the Moon is down. This again is the essence of the argument voiced by Disney (1976) in characterizing the Freeman law as a selection effect. To assess the validity of these arguments, it is necessary to construct a "visibility" function which basically determines the volume sampled by galaxy surveys as a function of central surface brightness. Existing galaxy catalogs are the result of this visibility function convolved with the intrinsic bivariate galaxy distribution $\Phi(M, \mu_0)$. Here we the derive visibility as a function of the two disk parameters μ_0 and α_l by considering galaxy selection based on isophotal diameter limits or magnitude limits.

6.3.1 Diameter-limited catalogs

The two selection parameters which need to be specified for a galaxy catalog selected by diameter are the diameter limit θ_ℓ and the isophotal level μ_ℓ at which the diameter is measured. To show this, it is only necessary to calculate the relative volume sampled as a function of the disk parameters, so the explicit value of θ_ℓ is irrelevant. Disney and Phillipps (DP) derive the visibility as a function of surface brightness at fixed luminosity, then state it can be scaled by luminosity (i.e., $V \propto L^{3/2}$). For a diameter-limited catalog, they find that the volume sampled goes as

$$V(\mu_0) \propto (\mu_\ell - \mu_0)^3 10^{-0.6(\mu_\ell - \mu_0)} \tag{6.13}$$

at fixed luminosity (see their equation 31). In order to show the full functional dependence on two parameters which describe the disk, we scale by luminosity as they suggest, arbitrarily normalized at $M^* = -21$ as they chose to do. Their full visibility, including the effects of both central surface brightness and absolute magnitude, is thus

$$V(M, \mu_0) \propto (\mu_\ell - \mu_0)^3 \ 10^{-0.6[(\mu_\ell - \mu_0) + (M - M^*)]} \tag{6.14}$$

This is shown in Figure 6.5 for the case $\mu_\ell = 25$ mag arcsec^{-2}.

The DP visibility function has a broad peak which always occurs at $\mu_0 = \mu_\ell - 2.17$. For $\mu_\ell = 24.0$ this peak would occur at $\mu_0 = 21.8$. If there is an intrinsic distribution $\Phi(M, \mu_0)$ which populates a range of μ_0 as well as M, then the visibility function will cause catalogs to preferentially sample galaxies with μ_0 near this peak. That this is very close to the Freeman value is the root of Disney's argument. However, a robust prediction of the DP visibility formalism is that the central surface brightness typically found in diameter-limited surveys will grow fainter as the surveys are pushed deeper. Contrary to this expectation, the peak in the apparent distribution is observed *not* to vary with μ_ℓ (Phillipps *et al.* 1987), leading some to conclude that surface brightness selection effects are not important (van der Kruit 1987). Also, as noted by DP, the peak is too broad to explain the narrow observed distribution.

An alternative derivation of these effects can be made as follows. For diameter-limited surveys, the requirement is that $\theta = 2r \geq \theta_\ell$ when $\mu(r) = \mu_\ell$. From equation (6.5) it follows that

$$\theta = 1.84\alpha(\mu_\ell - \mu_0) \propto \frac{\alpha_l}{d}(\mu_\ell - \mu_0) \tag{6.15}$$

The maximum distance at which a galaxy can lie and meet the selection criteria occurs when $\theta = \theta_\ell$, so

$$d_{max} \propto \frac{\alpha_l}{\theta_\ell}(\mu_\ell - \mu_0) \tag{6.16a}$$

For arbitrary θ_ℓ the relative volume sampled, $V \propto d^3_{max}$, is

$$V(\alpha_l, \mu_0) \propto \alpha_l^3 (\mu_\ell - \mu_0)^3. \tag{6.16b}$$

This is plotted in Figure 6.6.

There is a very significant difference between Figures 6.5 and 6.6 (and between Figures 6.7 and 6.8). The "visibility" in Figure 6.5 does *not* have a peak at some preferred surface brightness, but increases without bound as μ_0 becomes brighter. In contrast, Figure 6.6 shows that the variation of V with μ_0 is extremely rapid, so we *expect* that the apparent surface brightness distribution should always be very

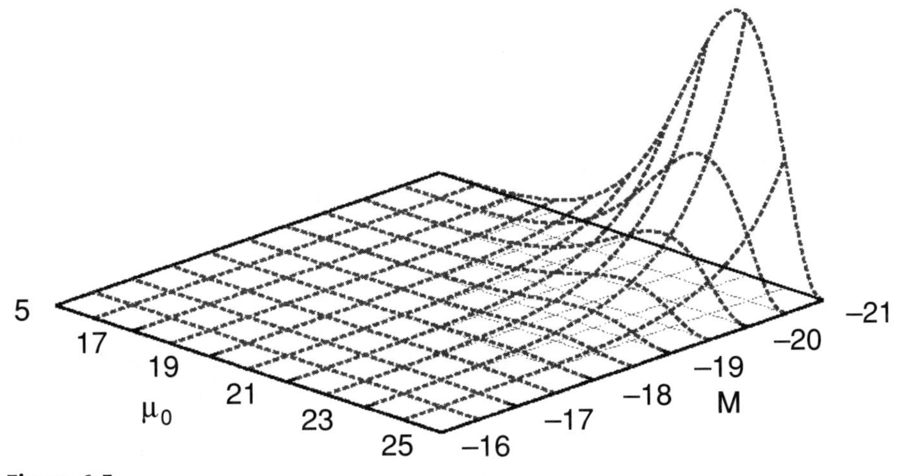

Figure 6.5

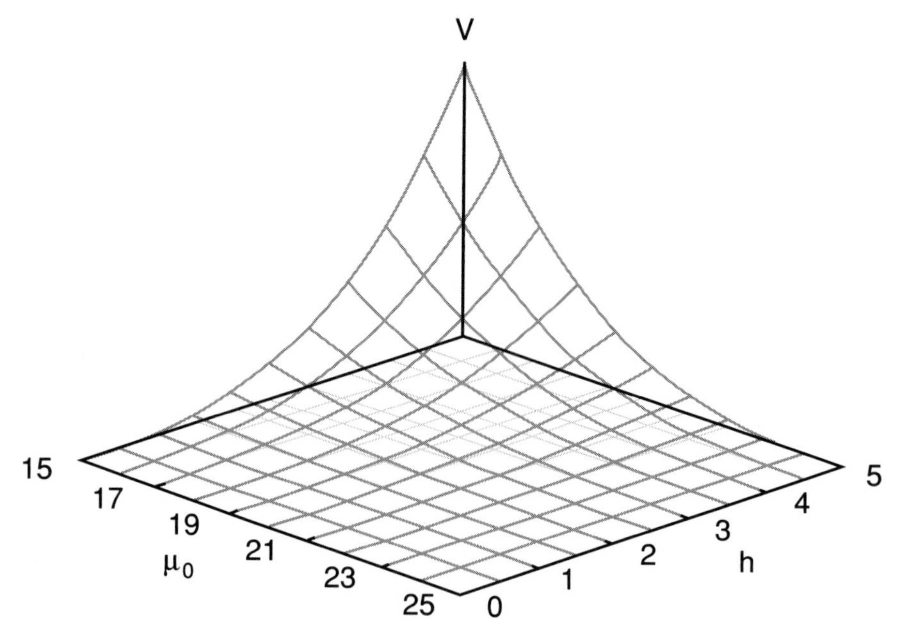

Figure 6.6

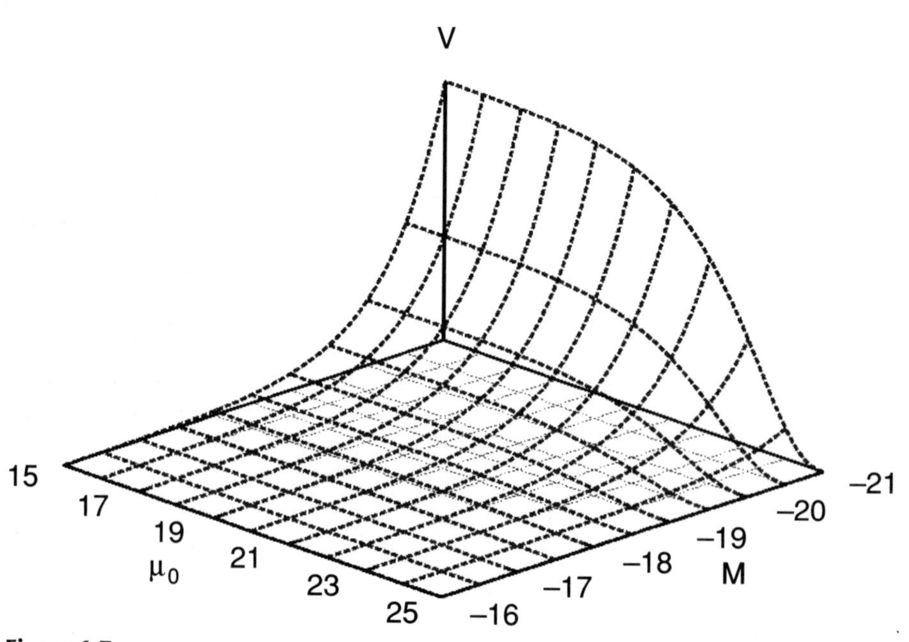

Figure 6.7

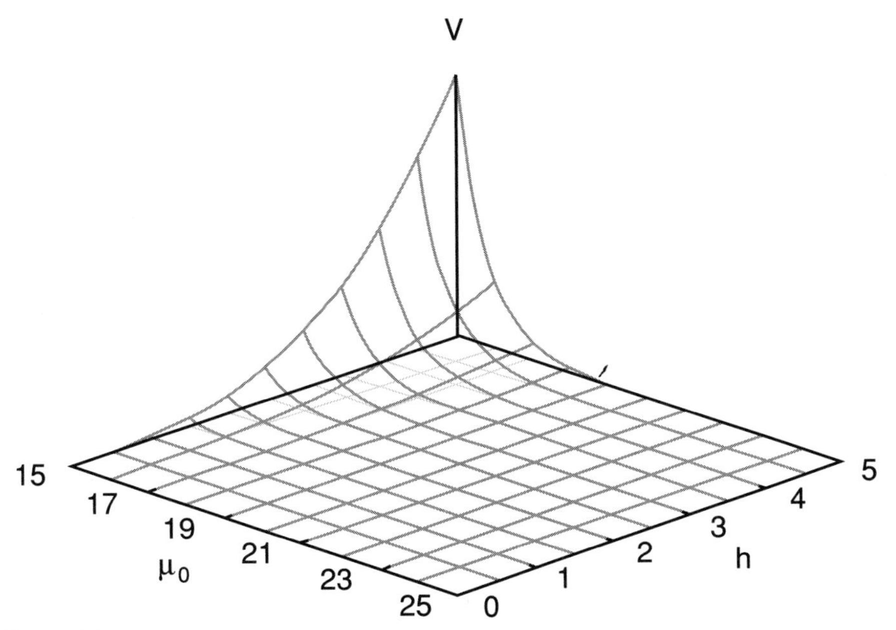

Figure 6.8

Figures 6.5 to 6.8 Representations of the visibility function of galaxies: using scale length and central surface brightness produces a substantially different representation than using absolute magnitude and central surface brightness. Adapted from McGaugh, Bothun, and Schombert (1995) and reproduced courtesy of Stacy McGaugh.

strongly peaked around the brightest value which exists in the intrinsic distribution, *regardless* of the value of μ_ℓ. In this treatment, galaxy size and surface brightness are orthogonal properties of a galaxy for purposes of selection, and it is really necessary to consider their separate effects fully. For instance, very luminous galaxies can be missed if their central surface brightness is very low. So far the available data strongly shows that μ_0 and α_l are uncorrelated (Figure 6.14).

Mathematically, one can transform α_l in equation (6.16b) into M to recover equation (6.14). Although algebraically correct, this makes little physical sense. Absolute magnitude and central surface brightness are not independent quantities, so the axes of Figure 6.5 have a high degree of covariance. This is the reason why representation of the selection effect in this manner is so misleading. In order to keep M fixed as μ_0 varies, it is necessary also to change α_l; this is unphysical. At issue is how any given galaxy will be selected. Imagine a galaxy of scale length α_l. As μ_0 varies, M varies with it but α_l remains fixed. And in order to hold M fixed, we must alter the size of the object so it is no longer the same object. In this sense, the characteristic size and surface brightness of a galaxy are more fundamental properties that its luminosity, a quantity which loses information by lumping together these two distinct pieces of information in a degenerate manner (through equation 6.11).

6.3.2 Flux-limited catalogs

For flux-limited samples where isophotal magnitudes are employed, the selection parameters are the magnitude limit m_ℓ and the isophotal limit μ_ℓ above which the flux is measured. Samples limited by total flux do not exist, since survey material always has an effective isophotal limit below which very diffuse galaxies cannot be identified, regardless of their total flux. Such galaxies do exist, e.g., Malin 1 and its cousins (Bothun *et al.* 1987; Impey and Bothun 1989); and Malin 2 (Bothun *et al.* 1990) has an apparent magnitude of $B = 14.2$ but is contained in neither the NGC, which contains many fainter galaxies, nor the UGC (Nilson 1973), which besides the limit $\theta_\ell = 1$ arcminute is also supposedly complete to the usually less demanding limit $B_\ell = 14.5$.

For magnitude-limited samples, DP (see their equation 31) find

$$V(M, \mu_0) \propto 10^{-0.6(M-M^*)}[f(x)]^{3/2} \tag{6.17}$$

which is plotted in Figure 6.7 ($f(x)$ is equivalent to their L_{ap}/L_T). $V \propto L^{3/2}$ can be decomposed into Σ_0 and α_l using equation (6.11):

$$V \propto [L(r < \theta_\ell/2)]^{3/2} \propto [\Sigma_0 \alpha_l^2 f(x)]^{3/2} \tag{6.17a}$$

so

$$V \propto \alpha_l^3 [\Sigma_0 f(x)]^{3/2} \propto \alpha_l^3 10^{-0.6(\mu_0 - \mu_0^*)}[f(x)]^{3/2} \tag{6.18}$$

Equation (6.18) is shown in Figure 6.8.

The essential feature to note is that flux-selected samples will have apparent distributions of μ_0 even more strongly peaked around the brightest extant value μ_0^* than diameter-limited samples, because the factor $10^{-0.6(\mu_0 - \mu_0^*)}$ varies more rapidly than $(\mu_0 - \mu_0^*)^3$. At any given (plausible) value of μ_ℓ, magnitude-limited samples will detect more galaxies in total than diameter-limited samples. However, they will be strongly dominated by α_l^*, μ_0^* (and hence L^*) galaxies. Diameter selection yields samples which are less strongly biased, hence more representative of the general field population. This is clearly seen from the significant number of LSB galaxies contained in the UGC (McGaugh and Bothun 1994; de Blok, van der Hulst, and Bothun 1995) compared with virtually none in the pseudo flux-selected Zwicky catalog (Bothun and Cornell 1990).

In either case the volume over which LSB galaxies can be detected is very small. It goes to zero if the central surface brightness happens to be fainter than the selection isophote, even if the galaxy in question is intrinsically luminous. Examples of luminous galaxies with such faint central surface brightnesses are now known to exist. Proper selection of galaxies requires a two-dimensional description of the completeness limit of a survey, not just a magnitude limit. At the end of this discussion we will use these volume corrections, in conjunction with surveys for LSB galaxies, to show that the Freeman law is an abysmal representation of the surface brightness distribution of disk galaxies and that at least one-half of the general galaxy population has been missed to date. Twenty years after Disney's conjecture, this is the major quantitative result that confirms his early wisdom.

6.4 Discovering LSB galaxies

Bearing in mind the diameter or apparent magnitude selection, we can now engage in the following thought experiment. Consider the five hypothetical galaxies listed below. The first four have pure exponential light distributions and similar total luminosity ($M_B \approx -21.1$); the fifth galaxy has the same scale length as galaxy B but its total luminosity is lower by a factor of 10. This adheres to the McGaugh, Schombert, and Bothun (1995) assumption that scale length and total luminosity are uncorrelated in a representative sample of disk galaxies. Central surface brightness (μ_0 above) in the blue is now denoted by $B(0)$.

- *Galaxy A*: $\alpha_l = 0.5$ kpc, $B(0) = 16.0$ mag arcsec^{-2}
- *Galaxy B*: $\alpha_l = 5.0$ kpc, $B(0) = 21.0$ mag arcsec^{-2}
- *Galaxy C*: $\alpha_l = 25.0$ kpc, $B(0) = 24.5$ mag arcsec^{-2}
- *Galaxy D*: $\alpha_l = 50.0$ kpc, $B(0) = 26.0$ mag arcsec^{-2}
- *Galaxy E*: $\alpha_l = 5.0$ kpc, $B(0) = 23.5$ mag arcsec^{-2}

Let's assume that the intrinsic space densities of these five galaxies are equal and let's suppose that we conduct a survey to catalog galaxies which have diameters measured at the $B = 25.0$ mag arcsec^{-2} level of greater than 1 arcminute. Under these conditions, we are interested in determining the maximum distance at which each galaxy can be detected:

- *Galaxy A*: This galaxy is quite compact (ratio of 1/2 light diameter to $D_{25} = 0.33$) and would fall below the catalog limit beyond $D = 60$ Mpc.
- *Galaxy B*: This is a typical large spiral (like M31); D_{25} corresponds to $3.74\alpha_l$, which is 18.7 kpc or a diameter of 37 kpc. This projects to an angular size of 1 arcminute at $D = 125$ Mpc.
- *Galaxy C*: D_{25} corresponds to $0.45\alpha_l$ or 11.5 kpc. This projects to an angular diameter of 1 arcminute at $D = 76$ Mpc.
- *Galaxy D*: D_{25} doesn't exist and this galaxy would *never* be discovered in such a survey.
- *Galaxy E*: D_{25} corresponds to $0.92\alpha_l$ or 4.6 kpc. This projects to an angular diameter of 1 arcminute at $D = 30$ Mpc.

The total survey volume is defined by galaxy B, as it can be seen to the largest distance. The ratio of sampled volumes for each of the other galaxy types is considerably smaller. For instance, the volume ratio of galaxy B to galaxy E is a factor of 70! Hence a survey like this would take the real space density distribution (which is equal) and, through the survey selection effect, produce a catalog which would contain 72% type B galaxies, 18% type C galaxies and 9% type A galaxies. This is a severe bias which would lead us to erroneously conclude that there is predominately one type of disk galaxy in the Universe and this leads to Freeman's law. Now in the real world of LSB galaxies, type C and D galaxies are quite rare but type E galaxies are common. Hence their local detection automatically means the space density is relatively large because they are so heavily selected against.

For the rest of the discussion, define an LSB disk galaxy as one which has $B(0)$ fainter than 23.0 mag arcsec^{-2}. This value is more than 4σ removed from the value Freeman found and, conveniently, is equal to the sky background in the blue at a dark site such as Mauna Kea or Las Campanas, Chile.

The story of the discovery and characterization of LSB galaxies as important members of the general galaxy population really begins in 1963 with the publication of the David Dominion Observatory (DDO) catalog of galaxies by Sydney van den Bergh. This catalog consists of galaxies which exhibit a diffuse appearance and that have angular sizes larger than 3 arcminutes. Although the DDO objects are the first bona fide collection of a sample of LSB galaxies, they are not at all representative of the phenomena. The galaxies contained in the DDO catalog are exclusively of low mass (some are members of the Local Group of galaxies). This has fostered an erroneous perception, which continues to this day, that all LSB galaxies are dwarf galaxies. This is not correct, as today we know that all masses of galaxies have representation in the LSB class.

The first serious contribution to our understanding and recognition that LSB galaxies do exist was made by William Romanishin and his collaborators, Steve and Karen Strom, in 1983. They derived their sample from the UGC. Although few of the Romanishin *et al.* galaxies actually meet our defined criteria for LSB, their results nevertheless pointed out a class of galaxies that had somewhat unusual properties, in particular, they had relatively large amounts of gas for their luminosity. By 1984 Allan Sandage published some of the first results on his Las Campanas photographic survey of the Virgo cluster. Contained in those papers were some vivid examples of dwarf galaxies in the Virgo cluster, galaxies which were quite diffuse and an extension towards lower surface brightness of the DDO objects discovered 20 years earlier.

With these two studies, it became clear that some quite diffuse galaxies do exist. Their existence raises five fundamental questions:

1. Since optical redshift surveys are biased against the detection of LSB galaxies, has our view of large-scale structure been biased because of the failure to include this population?
2. How can such diffuse and apparently fragile systems in the Virgo cluster be stable against tidal disruption by other cluster members or by the mean tidal field of the cluster?
3. Has failure to account for LSB galaxies significantly biased the determination of the GLF?
4. Could large numbers of hitherto unknown LSB systems be responsible for the plethora of observed QSO absorption lines?
5. Do LSB galaxies represent a different form of disk galaxy evolution compared with the Hubble sequence of spirals?

To set about answering these questions requires a large, systematic survey to detect and measure the properties of LSB galaxies. In 1985 this survey was begun by Chris Impey and Greg Bothun and is still ongoing.

The initial aspects of the survey concentrated on obtaining 21 cm redshifts of all LSB galaxies that were in the UGC and investigating the Virgo cluster to see if there were still more diffuse galaxies that had been missed in the Las Campanas survey. To improve the sensitivity of the search in Virgo, David Malin agreed to use his photographic amplification technique (Malin and Carter 1980) on 10 selected regions of the Virgo cluster using UK Schmidt plates. The Malin survey was going well and we were identifying very diffuse objects as candidates for LSB galaxies missed in the Las Campanas survey. Verification of these candidates was through follow-up CCD imaging. Several candidates were found and verified, but among them was a galaxy which was to become known as Malin 1, eventually taking its place as the largest disk galaxy every discovered. The fact that such a galaxy could remain undiscovered for so long is powerful testimony to the severity of surface brightness selection effects.

Malin 1 first stood out as unusual in the CCD survey on the basis of its morphology. In contrast to the other objects, which were structureless and extremely diffuse, Malin 1 appeared to have very faint spiral structure which was connected to a point-like nuclear region. On the Palomar sky survey this nuclear region is visible as a faint star but no associated nebulosity is apparent. Figure 6.9 shows the CCD image of Malin 1. The brightness of the stellar-like nucleus of Malin 1 was sufficient to obtain its spectrum using the Palomar 200 in telescope. In May 1986 a spectrum was obtained which showed an obvious emission line due to [O II] λ 3727. This line was redshifted to $z = 0.083$, indicating the object was well in the background to the Virgo cluster ($z = 0.004$). The problem was that the total angular size of the object on our CCD frame (Figure 6.9) was approximately 2.5 arcminutes yet it was apparently 20 times farther away than Virgo. A quick scaling then indicates that if a galaxy like this were indeed in Virgo, its angular size would be 1°. If it were as close as the Andromeda galaxy, its angular size would be about 20°; we would look right through it and never know it was there. This seemed an absurd situation for the structure of any galaxy. If real, this galaxy would then turn out to be the largest known disk galaxy, yet it was found in a survey that was looking for extremely LSB dwarf galaxies in the Virgo cluster. The expectation, then, was that Malin 1 was a composite system consisting of a background, nucleated emission line galaxy that just happened to be located behind an LSB dwarf galaxy in Virgo. This supposition could be verified through a 21 cm detection of the system at the Virgo cluster redshift.

In October 1986 several 21 cm observations of Malin 1 were made with the Arecibo radio telescope. The system was not detected in H I emission at the Virgo cluster redshift, but many hours of integration did produce a superb upper limit on any flux. After failing to detect the system at the Virgo cluster redshift, Arecibo was tuned to the emission line velocity as measured off the 200 in spectrum. After 10 min of integration a very large signal at 21 cm was detected at the emission line velocity. The signature of the 21 cm profile was exactly that of a rotating-disk galaxy (Figure 6.10). This provided confirmation that Malin 1 was indeed a huge disk galaxy that was very, very diffuse. The point-like structure in its center turns out to be the bulge component of Malin 1, which hosts a weak AGN. Surface photometry from

Figure 6.9 Malin 1: the original CCD image taken in blue light; the field size is approximately 3.5 arcmin × 2.5 arcmin. Reproduced from Bothun *et al.* (1987).

the CCD frame combined with the now known distance to Malin 1 revealed a scale length of 55 kpc (this is 20 times larger than our galaxy) and a central surface brightness of $B(0) = 26.0$ mag arcsec^{-2}. This is $\approx 15\sigma$ fainter than Freeman's law, and the galaxy had been discovered accidentally. Malin 1 is Disney's crouching giant; and having found one, there may be many more. The issue now became one of engaging in a larger survey to better determine the space density of these LSB galaxies.

Three new surveys were then initiated, all with the goal of discovering the extent and nature of this new population of galaxies which had low contrast with respect to the sky background and which had therefore remained undetected and uncataloged.

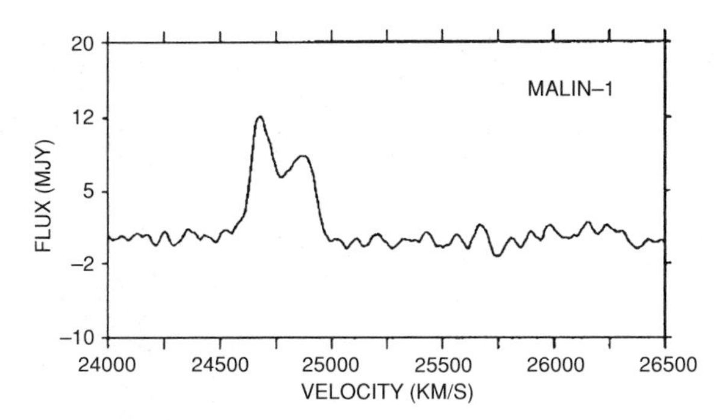

Figure 6.10 Malin 1: the 21 cm profile confirms it as being well behind the Virgo cluster. Reproduced from Bothun *et al.* (1987).

These surveys were: (1) a visual search of the new Palomar Observatory sky survey plates in a declination strip centered on 20° (so the objects are accessible from Arecibo); (2) applying the Malinization technique to UK Schmidt plates of the Fornax cluster to see if a similar population of very diffuse dwarfs to that in Virgo could be identified; and (3) using the automatic plate machine (APM) to scan UK Schmidt plates using an algorithm optimized to find galaxies of low contrast.

These surveys have largely been completed (and newer wide-field CCD surveys are now underway). Between them they have discovered approximately 1000 new nearby LSB galaxies, the subjects of extensive follow-up. A gallery of some selected members of the LSB class of galaxies is shown in Figure 6.11. The establishment that LSB galaxies are real and occur in the Universe in significant numbers (see below) has opened up a new field of inquiry in extragalactic astronomy, and many others are now involved with learning more about this population. Thus, in just over a decade, a whole new population of galaxies has been discovered. These LSB galaxies are of cosmological significance and have different properties than their HSB counterparts, which currently dominate the extant galaxy catalogs.

6.4.1 The space density of galaxies as a function of μ_0

With the results of these new surveys in hand, we can use our surface brightness selection formulism to make a preliminary estimate of the space density of galaxies as a function of surface brightness. To determine this, a correction for volume sampling effects must therefore be applied to the survey data (McGaugh, Bothun, and Schombert 1995). Ideally, we should determine the bivariate galaxy distribution $\Phi(\alpha_l, \mu_0)$ from complete catalogs for which the selection parameters μ_ℓ and θ_ℓ or m_ℓ are carefully specified and rigorously applied. Large catalogs which obey these strict criteria do not yet exist. Another requirement for the measurement of the bivariate distribution is that surface photometry should be performed on all objects (i.e., μ_0 and α_l must actually be measured). This has not often been done, and satisfying these criteria is really the best reason for performing a uniform digital sky survey for galaxy selection. It is therefore anticipated that the Sloan Digital Sky Survey (Chapter 7) will confirm much of what we present below.

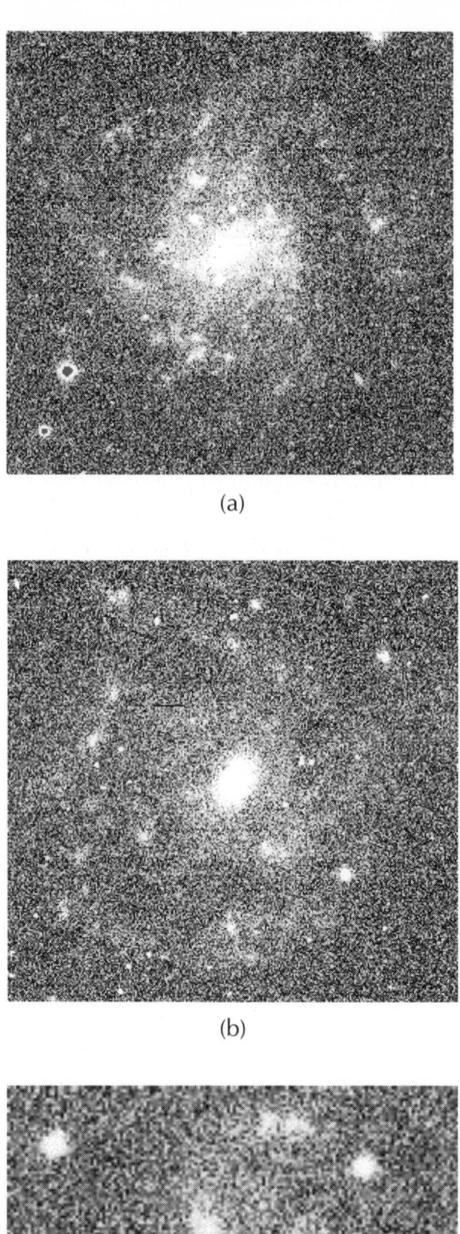

(a)

(b)

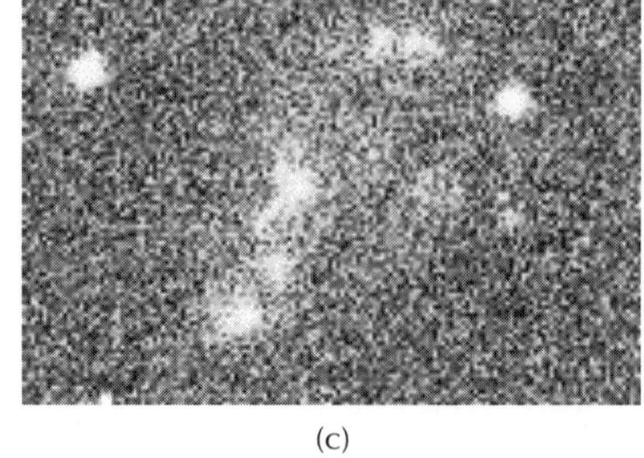

(c)

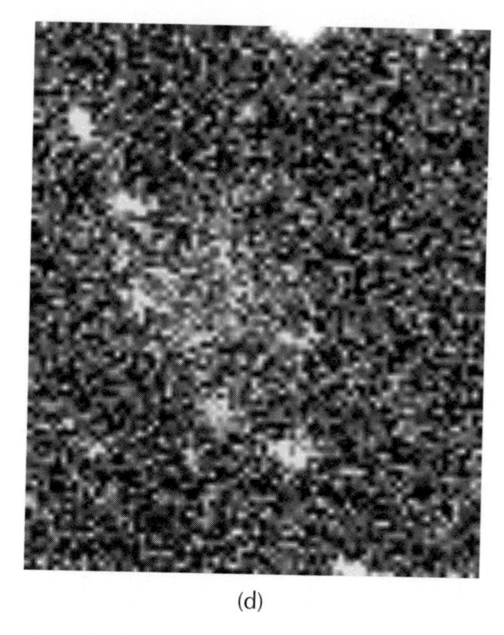

(d)

Figure 6.11 A selection of LSB galaxies from (a) the semiobvious to (d) the barely detectable.

A final requirement for measuring $\Phi(\alpha_l, \mu_0)$ is that redshifts should be measured in order to extract absolute information (i.e., α_l instead of α). No samples exist which meet all these requirements, though some come close (de Jong and van der Kruit 1994). Phillipps *et al.* (1987) and Davies (1990) do present data which meet the requirements for rigorous selection and measurement of μ_0, lacking only redshifts. These data sets consist of complete samples of several hundred galaxies selected by isophotal magnitude in the case of Phillipps *et al.* (1987) and both isophotal magnitude and diameter in the case of Davies (1990). The survey of Phillipps *et al.* (1987) is in the field of Fornax, but all higher surface brightness galaxies (those with $\mu_0 < 23$) are expected to be in the background field (Ferguson 1989; Irwin *et al.* 1990). Davies (1990) surveyed both Fornax and the adjacent field; we are concerned only with the field data. The isophotal level of selection for the Phillipps *et al.* (1987) data is $\mu_\ell = 25.5$ mag arcsec^{-2} and for the Davies (1990) data it is $\mu_\ell = 25.3$; both levels are in the B_J band. The relevant input data here is the number of galaxies detected at each central surface brightness, $N(\mu_0)$, i.e., the apparent distribution.

The most crucial data set in this investigation is provided by Schombert *et al.* (1992). In the sense described here, this catalog does not measure μ_0 for all galaxies in complete samples, but it does contain a large number of LSB galaxies (~ 200) with $\theta \geq 1$ arcminute measured at $\mu_\ell \approx 26$ B mag arcsec^{-2}. This catalog is characterized by disk galaxies of typical size (α_l^*) but low surface brightness, having a distribution very sharply peaked at $\mu_0 = 23.4$ (McGaugh and Bothun 1994;

de Blok, van der Hulst, and Bothun 1995). Since surface brightness is distance independent, we do not need redshifts to derive the surface brightness projection of the bivariate distribution. The relative number density of disk galaxies as a function of central surface brightness

$$\phi(\mu_0) = \frac{N(\mu_0)}{N(\mu_0^*)} \frac{V(\alpha_l^*, \mu_0^*)}{V(\alpha_l, \mu_0)} \tag{6.19}$$

follows from the apparent distribution corrected for volume sampling.

The surface brightness distribution follows directly from the observations, $N(\mu_0)$ and μ_ℓ, and equation (6.19), given one assumption. The volume correction factor depends on distance-dependent (scale length in physical units, α_l) as well as independent quantities (μ_0), so it is necessary to make an assumption about α_l. As before, we assume α_l is independent of μ_0. Thus, at any μ_0, the effects of volume sampling due to variations in α_l cancel out on average, and the only terms that matter are those due directly to μ_0. Hence we make the approximation

$$\frac{V(\alpha_l^*, \mu_0^*)}{V(\alpha_l, \mu_0)} = \frac{V(\alpha_l^*)}{V(\alpha_l)} \frac{V(\mu_0^*)}{V(\mu_0)} \sim \frac{V(\mu_0^*)}{V(\mu_0)}$$

Our assumption that scale length is uncorrelated with μ_0 is borne out by a wealth of observational data (Romanishin, Strom, and Strom 1983; Davies *et al.* 1988; Irwin *et al.* 1990; McGaugh and Bothun 1994; de Blok, van der Hulst, and Bothun 1995; McGaugh, Schombert, and Bothun 1995). Even if this assumption were incorrect, it does not alter the basic conclusion that there must be a relatively large space density of LSB galaxies, simply because $V^{-1} \to \infty$ as $\mu_0 \to \mu_\ell$. The detection of any galaxy with a central surface brightness within a few magnitudes of the selection isophote immediately implies a large density of such objects.

The surface brightness distribution determined from our assumption that α_l and μ_0 are uncorrelated is shown in Figure 6.12. The distribution has a long, approximately flat tail towards lower surface brightness. That is, approximately equal numbers of disk galaxies exist at each central surface brightness, as suspected by Disney (1976). This is only true faintwards of the Freeman value, which we have chosen as the fiducial μ_0^*. Brighter than this, there is an exponential cutoff. Though the extant results are not in perfect agreement as to the steepness of this cutoff, there is a clear turndown. The major revelation of Figure 6.12 is clear and far-reaching: diffuse galaxies do exist and they exist in substantial numbers; they are just harder to detect.

The Gaussian surface brightness distribution advocated by Freeman (1970) fails seriously to describe the true intrinsic distribution. It underestimates the number of galaxies with $\mu_0^* > 23$ mag arcsec^{-2} by over five orders of magnitude. No adjustment to the assumption about the scale length distribution made here can reconcile the data with a Freeman law. The monumental difference between what Freeman's law predicts and what is actually observed is a strong testimony to the fact that a proper survey for galaxies has yet to be done and that much of the baryonic material tied up in disks is contained in a fairly diffuse, hard-to-detect population.

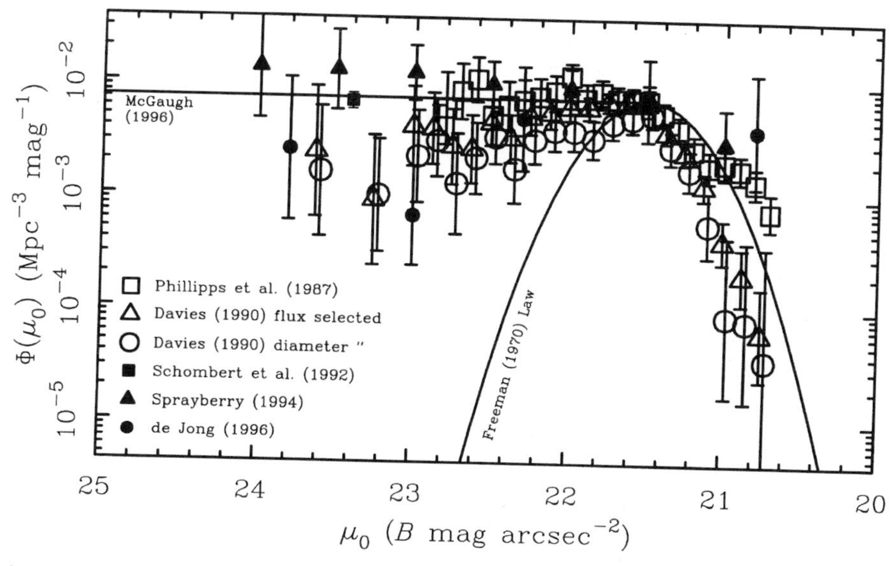

Figure 6.12 Space density as a function of central surface brightness resulting from the volumetric corrections discussed in this chapter. Adapted from McGaugh, Bothun, and Schombert (1995) and reproduced courtesy of Stacy McGaugh.

This constancy of disk galaxy number density as a function of μ_0 is the most significant result obtained to date on LSB galaxies and strongly alters the conventional view of the galaxy population, which suggests that they are at most a few percent of galaxies by number.

Figure 6.12 illustrates two points: (1) the large departure of real data from Freeman's law indicates that we do not yet have a representative sample of galaxies at $z = 0$, so it is very dangerous to use deep galaxy counts to proclaim there is some excess in the space density of galaxies relative to the density in the nearby Universe; (2) there is a substantial population of galaxies with low μ_0 that contain a significant amount of baryonic material, and a conservative estimate based on Figure 6.12 suggests that at least one-half the mass in disk galaxies is contained in systems with $\mu_0 \geq 22.0$ mag arcsec^{-2}.

6.4.2 The properties of LSB galaxies

LSB disk galaxies are not exclusively of low mass

This is clearly seen in the distribution of 21 cm velocity widths, which is indistinguishable from any sample of Hubble sequence spirals (Figure 6.13). LSB disks have similar sizes and masses as those high surface brightness (HSB) galaxies that define the Hubble sequence. However, despite this evidence to the contrary, the notion that LSB galaxies are exclusively low mass (or dwarf) galaxies still permeates the community. Indeed, a small percentage of the LSB population has truly

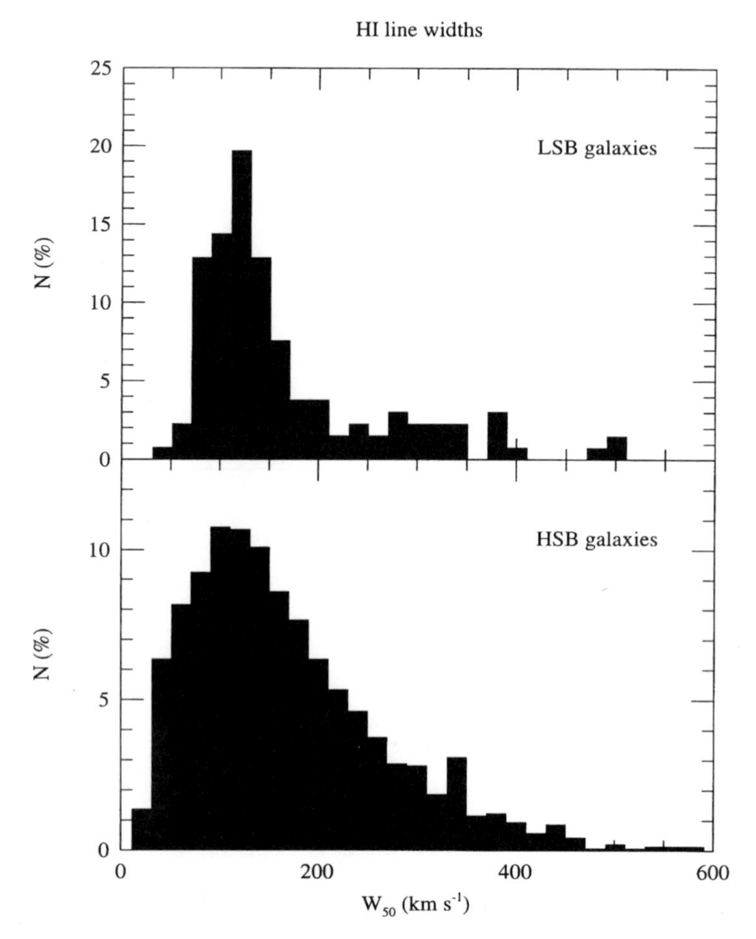

Figure 6.13 Comparison of 21 cm line widths: LSB samples versus HSB samples. The same range of rotational velocities is sampled with the same frequency of occurrence. It confirms the object as being well behind the Virgo cluster.

impressive overall sizes with scale lengths that exceed 15 kpc. Figure 6.14 shows that, when one closes in on a representative sample of galaxies, the entire disk structural plane defined by α_l and μ_0 is populated.

Very few LSB galaxies show evidence for nuclear activity

This is in marked contrast to HSB disk galaxies, where the percentage which have active galactic nuclei can be as high as 50%, depending on the mean luminosity of the sample. The most probable explanation for this is that LSB disks, in general, are lacking two structural features that facilitate gas flows and/or the formation of a compact object in the nucleus. These structural features are bulges and bars. In this context it is interesting to note that, of those large α LSBs (e.g., Malin 1 and cousins), approximately 50% do show some nuclear activity and all have a normal

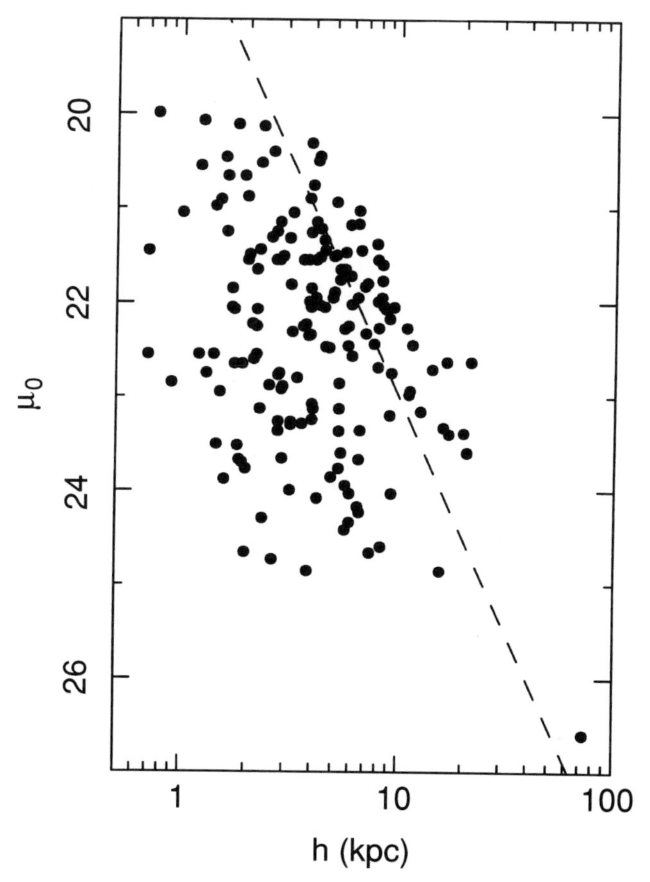

Figure 6.14 Parameter space defined by central surface brightness and scale length after LSBs are included in the overall sample. The dashed line is a line of constant disk luminosity. To appear in Bothun, Impey, and McGaugh (1997).

bulge component with luminosity $M_B = -18$ to 20. These galaxies are truly enigmatic because it appears that "normal" formation processes were at work to create the bulge component but no conspicuous stellar disk ever formed around this bulge.

LSB disks and HSB spirals contain similar amounts of neutral hydrogen

The amount of neutral hydrogen in LSB disks is very similar to the amount in HSB spirals, except for their lower than average surface densities. A typical case (Figure 6.15) shows that the gas density is generally below the threshold for molecular cloud formation, ruling out widespread disk star formation and metal production. Indeed, LSB spirals are significantly deficient in molecular gas compared to HSB spirals of the same mass. The observations of Schombert *et al.* (1990) failed to detect a single LSB disk in carbon monoxide, and virtually none of the LSB galaxies detected by Schombert *et al.* (1992), or Impey *et al.* (1996), are *IRAS* sources.

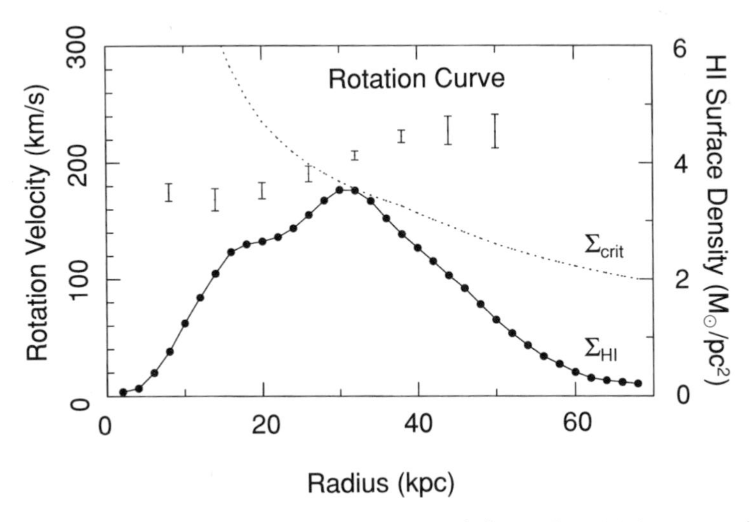

Figure 6.15 The 21 cm profile of the large LSB disk UGC 6614 in comparison with the critical density profile and the rotation curve. The surface density of H I is below the critical density at all points, hence star formation is not expected to occur in this disk. Reproduced from Pickering *et al.* (1997).

Although a typical LSB disk contains a handful of bright H II regions, indicating some current star formation, the average star formation rate is at least an order of magnitude lower than in galaxies of similar mass. Star formation in our galaxy appears to occur only in GMCs (e.g., Young and Scoville 1991). LSB galaxies may well be devoid of molecular clouds in their interstellar medium but may nevertheless have formed some stars. In fact, it is likely that the pressure–temperature–density manifold in LSB disks typically precludes molecular cloud formation but this has not been confirmed. The expectation of low metallicity is confirmed by the observations of McGaugh (1992). As derived from H II region spectroscopy, the abundance of heavy elements such as oxygen, nitrogen, neon and sulfur can be an order of magnitude lower than HSB disks of the same mass. On average, LSB disk galaxies with L^* luminosities have abundances of about one-third solar (McGaugh 1994). Thus, LSB galaxies strongly violate the mass–metallicity relation as defined by Hubble sequence spirals. This straightforwardly shows that multiple paths of chemical evolution exist for disk galaxies, which demands multiple star formation histories. In the case of LSB disks, that star formation history has not been robust; hence they represent some of the most unevolved objects in the Universe.

Rotational dynamics of LSB disks comes from very recent results

LSB disks galaxies rarely show the optical symmetry that is usually exhibited by Hubble sequence spirals, but see Zaritsky and Rix (1997). To first order, strongly symmetric optical disks are a signature of rotation-dominated kinematics within relatively cold disks. So the "chaotic" optical appearance of many LSB disks might suggest peculiar kinematics, but this is not consistent with the global 21 cm profiles, which generally show the two-horned signature of a rotating disk. The gas-rich nature of some LSB disks makes it possible to determine their rotation curves from aperture synthesis data. Two groups (de Blok, McGaugh, and van der Hulst 1996; Pickering *et al.* 1997) have used the VLA and WSRT radio arrays to determine the two-dimensional distribution of H I in about a dozen LSB galaxies. The results of these efforts can be summarized as follows:

- LSB galaxies have rotation curves that have a significantly shallower rise compared to HSB galaxies of the same rotation velocity and mass.
- The gaseous component of LSBs is dynamically significant at virtually all radii.
- LSB disks are highly dark matter dominated at virtually all radii. Compared to HSB galaxies of the same rotational velocity, LSB disks have higher global and local values of M/L (de Blok and McGaugh 1996).
- In contrast to HSB galaxies, no "maximum disk" mass model fits the rotation curve. This means there is no region in an LSB disk where the luminous (baryonic) matter dominates the potential, determining the form of the rotation curve.
- Mass models derived from the rotation curves of LSB and HSB galaxies show that LSB galaxies inhabit less dense and more extended dark matter halos. However, they have dynamical masses comparable to those of HSB galaxies.
- The most extreme examples give very hard upper limits for the ratio of disk mass to halo mass. These ratios produce baryon fractions of $\leq 10\%$ at the most, and more likely $\leq 3\%$. The value of f_b for LSB disks is significantly lower than for clusters of galaxies (e.g., White *et al.* 1993). So which are the more representative baryonic repositories, clusters of galaxies or LSB disks? Arguments given in Impey and Bothun (1997) show that LSB galaxies make a substantial contribution to the total baryonic mass density in the Universe, providing up to half of it depending on the faint end slope of the luminosity function (see below).

This dynamical information now allows LSB disks to be described in a manner that is fundamentally different than HSB disk galaxies, with the profound implication that the density profile of the dark matter halo ultimately determines the surface density of the galaxy which forms in that potential. Thus LSB galaxies may well have a fundamentally different dark matter distribution in them compared to HSB galaxies. This makes LSB disks physically distinct from HSB disks, even though they may have similar global properties. The paradoxical observation that LSB disks lie on the same Tully–Fisher relation as HSB disks (albeit with significantly more scatter) can then be understood only if the ratio of dark matter to light matter in disk galaxies is independent of the dark matter halo density. In this case, galaxies of similar circular velocity but dissimilar surface brightness can have the same luminosity because the LSB galaxy is defined by a larger radius, and the radius governs the mass that determines the circular velocity. This means that LSB disks also have lower surface mass density.

6.4.3 Cosmological relevance of LSB galaxies

The discovery of a substantial population of LSB galaxies at $z = 0$ is relevant to a number of cosmological issues. These issues are enumerated below and are more fully discussed in Impey and Bothun (1997) and Bothun, Impey, and McGaugh (1997).

QSO absorption lines

Increasing the number density of galaxies is a very good first step towards understanding the origin of QSO absorption lines. And the discovery of Malin 1, which has 10^5 kpc^2 of gaseous cross section at a level of $N_h \geq 3 \times 10^{19}$ cm^{-2} is a helpful addition to the zoology of objects that might produce damped $Lyman_\alpha$ systems.

Large-scale structure

To date, most studies of large-scale structure (LSS) are based on redshift surveys of optically selected samples which, by definition, do not contain many LSB galaxies. If LSB galaxies, for some reason, are better tracers of the mass distribution, then the biasing factor between mass and light has a surface brightness dependence. This would be an ugly complication to the various sophisticated attempts to determine the parameter $b\Omega^{0.6}$ (Strauss and Willick 1995). In an attempt to shed light on this, Bothun *et al.* (1985, 1986) performed a redshift survey of $\approx$ 400 LSB disk galaxies, primarily selected from the UGC. Few of these galaxies were in the CFA redshift survey at that time. That data, coupled with a thorough analysis of the clustering properties of LSB galaxies by Bothun *et al.* (1993) and Mo, McGaugh, and Bothun (1994) provides the following insights:

- On scales $\geq 5h^{-1}$ Mpc, LSB galaxies trace the identical structure as HSB galaxies.
- LSB galaxies generally avoid virialized regions.
- On scales $\leq 2h^{-1}$ Mpc, LSB galaxies are significantly less clustered. In fact, there is a significant deficit of companions within a projected distance of $0.5h^{-1}$ Mpc, indicating that LSB disks are generally isolated.
- LSB galaxies are not preferentially found in large-scale voids. For instance, there are none in either the CFA bubble (Bothun *et al.* 1992) or the Boötes void (Aldering *et al.* 1997).
- When found in the group or cluster environment, LSBs are usually near the edge of the galaxy distribution. The results of O'Neil, Bothun, and Cornell (1997) show that, in the cluster environment, there is a limiting value of μ_0 which depends on local galaxy density. If one considers that elliptical galaxies, which have the highest surface brightness of all, tend to occur in virialized regions, then we have the surface brightness extension of the morphology–density relation of Dressler (1980). That is, the local galaxy density determines the surface mass or luminosity density of galaxies.

This allows the following scenario to be advanced. If the initial spectrum of density perturbations which eventually form galaxies is Gaussian in nature, then there should be many more low density perturbations than high density perturbations. Many of these low density perturbations are subject to disruption or assimilation into other perturbations; hence they will not produce individual galaxies. However, if a substantial percentage survive to produce individual galaxies, we would expect them to dominate the galaxy population. We tentatively identify our discovery of LSB galaxies with these 1σ–2σ peaks in the initial Gaussian perturbation spectrum. If that is the case, we expect LSB galaxies (1) to have formed in isolation and (2) to be fair tracers of the mass distribution on large scales. This isolation on small scales must clearly affect their evolution because, compared to HSB galaxies, LSB galaxies have experienced fewer tidal encounters with nearby galaxies over the last Hubble time. Tidal encounters are effective at clumping gas and driving global star formation. Without this external hammer, LSB galaxies would continue to evolve slowly and passively.

Steeper end slope

After correcting for surface brightness selection effects, the faint end slope of the galaxy luminosity function is significantly steeper than when only HSB galaxies are selected. This is best seen in the Virgo and Fornax clusters, where the kinds of LSB dwarfs that were missed in earlier surveys but detected using the Malin method were very diffuse but fairly large. Thus, the integrated luminosity of these galaxies, although they are diffuse, was brighter then many of the Sandage diffuse dwarfs, owing to their larger scale length. Luminosity functions are traditionally corrected for incompleteness on the basis of apparent magnitude only. This is conceptually incorrect for galaxies since they are selected on the basis of surface brightness, not total luminosity. The surveys of Virgo and Fornax Found galaxies that were up to three magnitudes brighter than the nominal magnitude limit of the Sandage survey but which were not included in that survey. Figure 6.16 shows that, when properly accounting for these galaxies, there is a significant increase in the faint end slope of the GLF. LSB galaxies dominate numerically in clusters and probably in the field (although this has not yet been established); if M/L increases with decreasing surface brightness, they might form a significant component of baryonic mass in the Universe. It is interesting to speculate in the case of cluster LSBs that they might represent galaxy evolution following a phase of intense baryonic blowout at higher redshifts. So these relic galaxies might be the source of the gas seen in clusters as well as the enrichment of that gas. In the Virgo cluster there are approximately twice as many dwarfs as brighter galaxies.

LSB galaxies have many FBG properties

LSB galaxies have many of the same properties – color, luminosity, mean surface brightness, clustering amplitude – as the enormous number of FBGs seen in deep CCD surveys. Indeed, it may well be these LSB galaxies, in their initial phase of

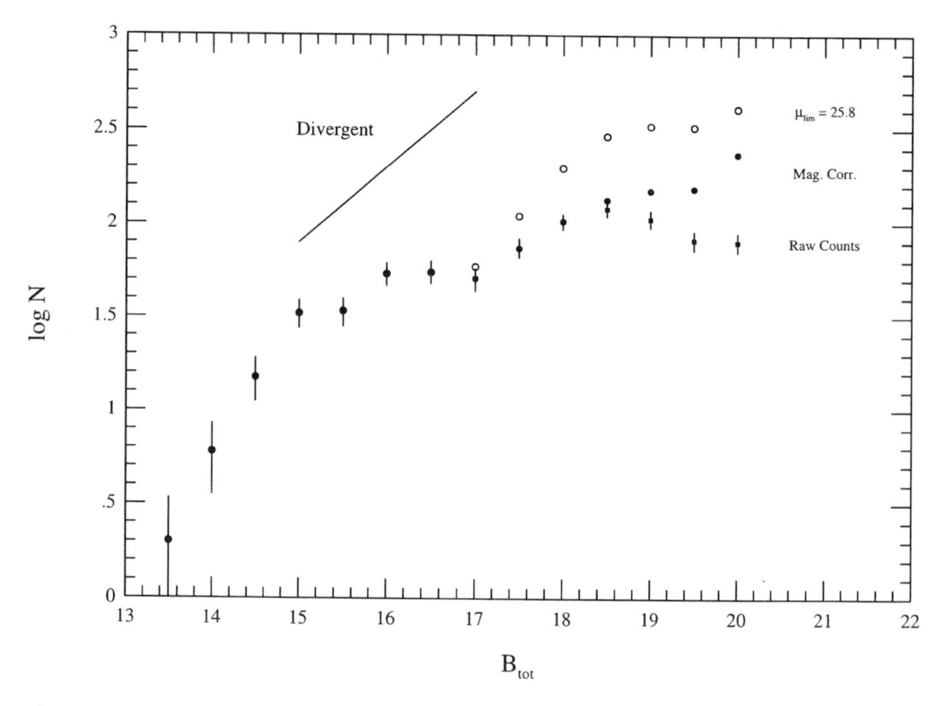

Figure 6.16 The corrected luminosity function that results from applying surface brightness selection effects as opposed to erroneously applying only corrections for magnitude incompleteness. Virgo cluster data from Impey, Bothun, and Malin (1988).

active star formation, that are being discovered in deep CCD surveys. However, the current space density of known LSBs remains too low to fully explain the FBGs. This might imply that we are only seeing the tip of the iceberg and that a much larger population of faded LSBs lurks below the current sensitivity limits. This possibility remains viable in the face of the rather inexplicable color distribution of LSBs. Practically all of the LSB galaxies discovered to date are blue to very blue, despite the lack of star formation. Furthermore, there is no correlation between surface brightness level and overall color, as would be expected in any scenario which suggests that LSB galaxies are the faded remnants of HSB galaxies after star formation has subsided. These blue colors remain difficult to understand and, in many cases, they demand that the galaxy has recently formed. This is consistent with the observed low densities of these objects, as the dynamical timescale is almost an order of magnitude larger than for HSB galaxies. Thus we expect many LSB galaxies to have late collapse times, hence delayed formation of their first stars. Perhaps our surveys for LSB galaxies have recovered this population but we are still missing a component that should be present. The still outstanding question is, Where is the red LSB galaxy population?

A new CCD survey for LSB galaxies (O'Neil, Bothun, and Cornell 1997) has very recently found some examples of red LSB objects, though they do not dominate the

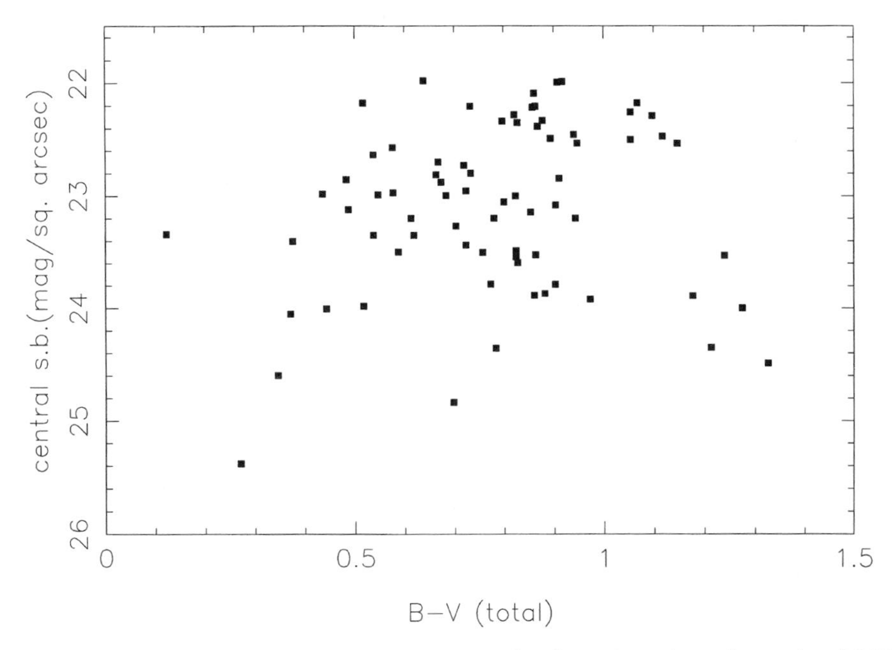

Figure 6.17 Color–surface brightness relation for the CCD-selected sample of LSB galaxies from O'Neil *et al.* (1997) where genuinely red LSB galaxies have been detected for the first time. Filled symbols are the CCD-selected LSB galaxies and the crosses are LSBs found in previous photographic surveys. To appear in Bothun, Impey, and McGaugh (1997).

kinds of LSBs in that survey. Figure 6.17 shows the data which indicates this new population. On average, the red LSBs are no different in scale length or μ_0 than the blue LSBs. The discovery of this component is heartening in the sense that we know galaxies will begin to fade once their star formation stops due to gas depletion. At some time in the future, the typical disk galaxy in the Universe will be red and of low μ_0. Since the evolutionary timescale of a disk galaxy is set both by its surface gas density and its environment, it is logical to assume that some galaxies have already evolved to this endpoint. Surveys have now recovered this population.

6.5 Summary

This chapter has considered the issue of the distribution of baryons in the Universe. The following major conclusions can be drawn from the data:

1. Although there is some evidence for the presence of an intergalactic medium (IGM), the total population of baryons that are in the IGM is at least an order of magnitude less than the baryons which are located in bound potentials. This indicates that dark matter potentials in the early Universe were fairly efficient

at sweeping up and collecting baryonic material. At high redshift, many of the smaller potentials could be easily ionized by QSOs, thus creating the seed population of "clouds" that are observed to produce QSO absorption lines. The high redshift Universe may well have most of the baryons contained in these "clouds," which constitute a warm IGM composed of discrete sources (Weinberg, Hernquist, and Katz 1997).

2. Exotic scenarios that appeal to the production of ionizing photons from the late decay of massive particles do not appear to be viable, making it unlikely that such massive particles exist and dominate the mass density of the Universe.
3. The isotropic distribution of gamma ray bursters remains the most astrophysically puzzling of all cosmological observations. Although it seems likely that the production originates in some compact object (center of a galaxy, dense globular cluster), the overall space density of these sources is unknown. Potentially, such objects can provide a significant source of energy feedback to the IGM at high redshifts.
4. The most recent nucleosynthesis constraints (Copi, Schramm, and Turner 1995) combined with the most probable range for H_0 (e.g., 70–90) lead to

   $$0.018 \leq \Omega_b \leq 0.042$$

   The baryonic inventory by Persic and Salucci (1992) yields a total $\Omega_b = (2.2 \pm 0.06) \times 10^{-3} h^{-3/2}$ (Chapter 3). For our range of H_0 this means that most baryons must be dark. LSB galaxies are not included in the Persic and Salucci census. The discovery of LSB galaxies has increased the faint end slope of the GLF. And LSBs contain substantial amounts of dark matter. Under the assumption that baryons scale with dark matter, needed to explain the Tully–Fisher relation, the contribution of LSB galaxies to Ω_b can be as high as 0.013 (Impey and Bothun 1997) to 0.025 (Bristow and Phillips 1994). Hence LSB galaxies can easily be the sites of much of the "missing" baryonic material in the Universe.
5. Deep galaxy surveys have uncovered an enigmatic population of faint blue galaxies (FBGs). Redshift surveys (e.g., Lilly *et al.* 1995) now provide evidence of luminosity evolution in this population. Thus, many of the FBGs could have faded by $z = 0$ to produce the red population of LSBs now detected in the survey of O'Neil *et al.* (1997). However, most of the LSB population so far detected is quite blue. This is best understood if these galaxies have long dynamical timescales (owing to low matter density), hence delayed formation times. As such, LSB galaxies may represent the more smoothly distributed and more numerous 1–2σ peaks in the initial Gaussian density distribution. Therefore, they are fair tracers of the mass distribution on a large scale. At small scales, LSBs are fairly isolated, which likely reflects the requirement on low density potentials that allows them to collapse only when they are isolated. Those that are not isolated are assimiated into nearby, denser potentials.
6. LSB galaxies offer a window into galaxy evolution that is different from the traditional approach. Surveys have so far detected galaxies with μ_0 as low as

27.0 mag arcsec^{-2} in the blue, implying a volume density 100 times lower than a typical spiral galaxy. LSBs are very much a manifestation of a different sequence of galaxy formation and evolution. They may be predestined to take this different evolutionary route due to lower density, dark matter halos. These halos account for the extended rotation curves recently measured in some LSB systems.

7. The space density of galaxies as a function of surface brightness appears to be flat for μ_0 fainter than 22.0 mag arcsec^{-2}. The space density of LSB disks is several orders of magnitude higher than would be predicted by extrapolation of Freeman's law. This is why they are significant contributors to the baryon density. These LSB disks are relatively inefficient at converting gas into stars, probably because their interstellar medium conditions do not permit the formation of molecular clouds which subsequently foster widespread star formation. As a result, LSB galaxies remain relatively unevolved. Since there is no hint of a drop-off in the space density, we expect future surveys to find yet more galaxies of even lower surface brigthtness. The formation of these very diffuse systems, and the existence of any stars in them at all, will provide a severe challenge to galaxy formation that, including ellipticals, must now explain why there is approximately a factor of 1000 range in the volume mass density of galaxies.

References

ALDERING, G., BOTHUN, G., MARZKE, R., and KIRSHNER, R. 1997 preprint

ANDERSON, S. and MARGON, B. 1987 *Astrophysical Journal* **314**, 111

BERNSTEIN, G., FISCHER, M., RICHARDS, P., PETERSON, J., and TIMUSK, T. 1989 *Astrophysical Journal Letters* **337**, L1

BERTSCH, D. *et al.* 1996 *Astrophysical Journal* **416**, 587

BOTHUN, G. and CORNELL, M. 1990 *Astronomical Journal* **99**, 1004

BOTHUN, G., IMPEY, C., and MCGAUGH, S. 1997 *Publications of the Astronomical Society of the Pacific* **109**, 745

BOTHUN, G., LONSDALE, C., and RICE, W. 1989 *Astrophysical Journal* **341**, 29

BOTHUN, G., BEERS, T., MOULD, J., and HUCHRA, J. 1985 *Astronomical Journal* **90**, 2487

BOTHUN, G., BEERS, T., MOULD, J., and HUCHRA, J. 1986 *Astrophysical Journal* **308**, 510

BOTHUN, G., IMPEY, C., MALIN, D., and MOULD, J. 1987 *Astronomical Journal* **94**, 23

BOTHUN, G., SCHOMBERT, J., SCHNEIDER, S., and IMPEY, C. 1990 *Astrophysical Journal* **360**, 427

BOTHUN, G., GELLER, M., KURTZ, M., HUCHRA, J., and SCHILD, R. 1992 *Astrophysical Journal* **395**, 349

BOTHUN, G., SCHOMBERT, J., IMPEY, C., SPRAYBERRY, D., and MCGAUGH, S. 1993 *Astronomical Journal* **106**, 530

BRISTOW, P. and PHILLIPS, S. 1994 *Monthly Notices of the Royal Astronomical Society* **267**, 13

BROADHURST, T., ELLIS, R., and SHANKS, T. 1988 *Monthly Notices of the Royal Astronomical Society* **235**, 827

CHENGALUR, J., GIOVANELLI, R., and HAYNES, M. 1995 *Astronomical Journal* **109**, 2415
CLARK, G. 1972 Gamma ray astronomy, *Scientific American* **May**, 52–61
COLLESS, M. *et al.* 1993 *Monthly Notices of the Royal Astronomical Society* **261**, 19
COMASTRI, A., SETTI, G., ZAMORANI, G., and HASINGER, G. 1995 *Astronomy and Astrophysics* **296**, 1
COPI, C., SCHRAMM, D., and TURNER, M. 1995 *Astrophysical Journal Letters* **455**, L95
COWIE, L., LILLY, S., GARDNER, J., and MCLEAN, I. 1988 *Astrophysical Journal Letters* **321**, L29
DAVIDSEN, A., KRISS, G., and ZHENG, W. 1996 *Nature* **380**, 47
DAVIDSEN, A. *et al.* 1991 *Nature* **351**, 128
DAVIES, J. 1990 *Monthly Notices of the Royal Astronomical Society* **244**, 8
DAVIES, J. *et al.* 1988 *Monthly Notices of the Royal Astronomical Society* **232**, 239
DE BLOK, E. and MCGAUGH, S. 1996 *Astrophysical Journal Letters* **469**, L89
DE BLOK, E., MCGAUGH, S., and VAN DER HULST, T. 1996 *Monthly Notices of the Royal Astronomical Society* **283**, 18
DE BLOK, E., VAN DER HULST, T., and BOTHUN, G. 1995 *Monthly Notices of the Royal Astronomical Society* **274**, 235
DE JONG, R. and VAN DER KRUIT, P. 1994 *Astronomy and Astrophysics Supplements* **106**, 451
DE VAUCOULEURS, G. 1959 in *Handbuch der Physik* **53**, 275
DERMER, C. and SCHLICKEISER, R. 1992 *Science* **257**, 1642
DESSENNE, C. *et al.* 1996 *Monthly Notices of the Royal Astronomical Society* **281**, 977
DISNEY, M. 1976 *Nature* **263**, 573
DISNEY, M. and PHILLIPS, S. 1983 *Monthly Notices of the Royal Astronomical Society* **205**, 1253
DRESSLER, A. 1980 *Astrophysical Journal* **236**, 351
ELLIS, R. *et al.* 1996 *Monthly Notices of the Royal Astronomical Society* **280**, 235
FANG, Y. and CROTTS, A. 1995 *Astrophysical Journal* **440**, 69
FERGUSON, H. 1989 *Astronomical Journal* **98**, 367
FRAIL, D. *et al.* 1994 *Astrophysical Journal Letters* **437**, L43
FREEMAN, K. 1970 *Astrophysical Journal* **160**, 811
GIACCONI, R. 1987 *Astrophysical Letters* **26**, 7
GIALLONGO, E. *et al.* 1994 *Astrophysical Journal Letters* **425**, L1
GIOVANELLI, R. and HAYNES, M. 1989 *Astrophysical Journal Letters* **346**, L5
GLAZEBROOK, K., ELLIS, R., SANTIAGO, B., and GRIFFITHS, R. 1995 *Monthly Notices of the Royal Astronomical Society* **275**, 19
GRONWALL, C. and KOO, D. 1995 *Astrophysical Journal Letters* **440**, L1
GUDEHUS, D. 1989 *Astrophysical Journal* **342**, 617
HASINGER, G. *et al.* 1993 *Astronomy and Astrophysics* **275**, 1
HEISLER, J. and OSTRIKER, J. 1988 *Astrophysical Journal* **332**, 543
HENRY, R. and MURTHY, J. 1993 *Astrophysical Journal Letters* **418**, L17
IM, M., GRIFFITHS, R., RATHATUNGA, K., and SARAJEDINI, V. 1996 *Astrophysical Journal Letters* **461**, L79
IMPEY, C. 1996 *Astronomical Journal* **112**, 2667
IMPEY, C. and BOTHUN, G. 1989 *Astrophysical Journal* **341**, 89
IMPEY, C. and BOTHUN, G. 1997 *Annual Reviews of Astronomy and Astrophysics* **35**, 267
IMPEY, C., BOTHUN, G., and MALIN, D., 1988 *Astrophysical Journal* **330**, 634
IMPEY, C., BOTHUN, G., MALIN, D., and STAVELEY-SMITH, L. 1990 *Astrophysical Journal Letters* **351**, L33

IMPEY, C. *et al.* 1996 *Astrophysical Journal Letters* **351**, L33
IRWIN, M. *et al.* 1990 *Monthly Notices of the Royal Astronomical Society* **245**, 289
JANSKY, K. 1933 *Proceedings of the Institute of Radio Engineers* **21**, 1387
KOO, D. and KRON, R. 1982 *Astronomy and Astrophysics* **105**, 107
KORANYI, D. *et al.* 1995 *Monthly Notices of the Royal Astronomical Society* **276**, L13
LAHAV, O., LOEB, A., and MCKEE, C. 1990 *Astrophysical Journal Letters* **394**, L9
LARSON, S., MCCLEAN, I., and BECKLIN, E. 1996 *Astrophysical Journal Letters* **460**, L95
LILLY, S., TRESSE, L., HAMMER, F., CRAMPTON, D., and LEFEVRE, O. 1995 *Astrophysical Journal* **455**, 108
LYNE, A. *et al.* 1990 *Nature* **347**, 650
MCGAUGH, S. 1992 PhD Thesis, University of Michigan
MCGAUGH, S. 1994 *Nature* **367**, 538
MCGAUGH, S. and BOTHUN, G. 1994 *Astronomical Journal* **107**, 530
MCGAUGH, S., BOTHUN, G., and SCHOMBERT, J. 1995 *Astronomical Journal* **110**, 573
MCGAUGH, S., SCHOMBERT, J., and BOTHUN, G. 1995 *Astronomical Journal* **109**, 2019
MALIN, D. and CARTER, D. 1990 *Nature* **285**, 643
MARZKE, R., GELLER, M., HUCHRA, J., and CORWIN, H. 1994 *Astronomical Journal* **108**, 437
MATHER, J. *et al.* 1990 *Astrophysical Journal Letters* **354**, L37
MATSUMOTO, T. *et al.* 1988 *Astrophysical Journal* **329**, 567
MCGAUGH, S. 1992 PhD Thesis, University of Michigan
MO, H., MCGAUGH, S., and BOTHUN, G. 1994 *Monthly Notices of the Royal Astronomical Society* **267**, 129
NILSON, P. 1973 *The Uppsala General Catalog of Galaxies*, Uppsala: Societatis Scientarium Upsaliensis
OEGERLE, W. and HOESSEL, J. 1989 *Astronomical Journal* **98**, 1523
O'NEIL, K., BOTHUN, G., and CORNELL, M. 1997 *Astronomical Journal* **113**, 1212
O'NEIL, K., BOTHUN, G., CORNELL, M., SCHOMBERT, J., and IMPEY, C. 1997 *Astronomical Journal*, in press
OSTRIKER, J. and STRASSLER, M. 1989 *Astrophysical Journal* **338**, 579
PATTON, D., PRITCHET, C., YEE, H., ELLINGSON, E., and CARLBERG, R. 1997 *Astrophysical Journal* **475**, 29
PERSIC, M. and SALUCCI, P. 1992 *Monthly Notices of the Royal Astronomical Society* **258**, 14P
PHILLIPS, S., DISNEY, M., KIBBLEWHITE, E., and CAWSON, M. 1987 *Monthly Notices of the Royal Astronomical Society* **229**, 505
PICKERING, T., IMPEY, C., VAN GORKOM J., and BOTHUN, G. 1997 preprint
REBER, G. 1940 *Astrophysical Journal* **91**, 621
ROMANISHIN, W., STROM, S., and STROM, K. 1983 *Astrophysical Journal Supplements* **53**, 105
SALZER, J. *et al.* 1991 *Astronomical Journal* **101**, 1258
SANDAGE, A., BINGGELI, B., and TAMMANN, G., 1985 *Astronomical Journal* **90**, 1759
SARGENT, W., YOUNG, P., BOKSENBERG, A., and TYTLER, D. 1980 *Astrophysical Journal Supplements* **42**, 41
SCHADE, D. *et al.* 1995 *Astrophysical Journal Letters* **451**, L1
SCHECHTER, P. 1976 *Astrophysical Journal* **203**, 297
SCHOMBERT, J., BOTHUN, G., IMPEY, C., and MUNDY, L. 1990 *Astronomical Journal* **100**, 1523
SCHOMBERT, J., BOTHUN, G., SCHNEIDER, S., and MCGAUGH, S. 1992 *Astronomical Journal* **103**, 1107

SCIAMA, D. 1990 *Nature* **348**, 617

SCIAMA, D. 1995 *Monthly Notices of the Royal Astronomical Society* **276**, 1

SPITZER, L. and HART, M. 1971 *Astrophysical Journal* **164**, 399

SPRAYBERRY, D., IMPEY, C., IRWIN, M., and BOTHUN, G. 1997 *Astrophysical Journal* **482**, 104

STECKER, F. and SALAMON, M. 1996 *Astrophysical Journal* **464**, 600

STEIDEL, C., GIAVALISCO, M., PETTINI, M., DICKINSON, M., and ADELBURGER, K. 1996 *Astrophysical Journal Letters* **462**, L17

STRAUSS, M. and WILLICK, J. 1995 *Physics Reports* **261**, 271

SUBRAHMANYAN, R. and COWSKI, R. 1989 *Astrophysical Journal* **347**, 1

TAYLOR, G. and WRIGHT, E. 1989 *Astrophysical Journal* **339**, 619

TREVESE, D., CIRIMELE, G., and APPODIA, B. 1996 *Astronomy and Astrophysics* **315**, 365

VAN DER KRUIT, P. 1987 *Astronomy and Astrophysics* **173**, 59

VOGEL, S., WEYMANN, R., RAUCH, M., and HAMILTON, T. 1995 *Astrophysical Journal* **441**, 162

VON MONTIGNY, C. *et al.* 1995 *Astronomy and Astrophysics* **299**, 680

WALTERBOS, R. and SCHWERING, P. 1987 *Astronomy and Astrophysics* **180**, 27

WEINBERG, D., HERNQUIST, L., and KATZ, N. 1997 *Astrophysical Journal* **477**, 8

WHITE, S., NAVARRO, J., EVRARD, A., and FRENK, C. 1993 *Nature* **366**, 429

WRIGHT, E. *et al.* 1995 *Astrophysical Journal* **420**, 450

WYSE, R. and SILK, J. 1985 *Astrophysical Journal Letters* **296**, L1

YOUNG, J. and SCOVILLE, N. 1991 *Annual Reviews of Astronomy and Astrophysics* **29**, 581

ZARITSKY, D. and RIX, H. 1997 *Astrophysical Journal* **477** 118

CHAPTER SEVEN

Concluding Remarks and Future Prospects

7.1 Revisiting the questions of Chapter 1

It is time to revisit the questions we asked in Chapter 1. Throughout this book we have attempted to confront each question with the best available observational data. Later we will consider a series of promising new cosmological observations that can be made with improving instrumentation and telescope aperture. These observations have the potential to offer even more constraining power on competing cosmological models. For now we assess our cosmological knowledge, as of mid 1997, which is anchored firmly in observations and a realistic assessment of possible systematic errors in them. In evaluating this assessment, we should keep in mind the two themes articulated in Chapter 1: (1) at any given time in history, everyone believes their cosmological models are correct; (2) history shows that cosmological models are subject to great change once better observations become available. In fact, falsification is the scientific history of cosmological model making and our current cosmology will certainly be subject to heavy revision in the years to come. The following summary positions are consistent with the data for each of the key questions we posed earlier on.

> *What is the age of the Universe as determined from the observed expansion rate and cosmological distance scale? Is there a need to invoke the cosmological constant to reconcile the ages of the oldest stars with the value of the Hubble constant?*

This was the subject of Chapter 2. At two recent meetings on the distance scale, one in May 1996 at the Space Telescope Science Institute and the other in June 1996 at Princeton University, many of the participants felt that a convergence on the estimation of H_0 had finally occurred. However, this convergence is still a manifestation of overlapping error bars and probably small systematic effects associated with choice of sample. There is still no reason to be particularly optimistic that H_0 has been derived to an accuracy of 10%. Although it does seem true that the decades old $H_0 = 50$ vs. $H_0 = 100$ battle has converged to $H_0 = 60$ vs. $H_0 = 85$, the entire range remains open, given the error bars. The major source of uncertainty in the distance scale continues to be disagreement over distances to nearby galaxies, particularly the LMC. This disagreement notably manifests itself by a significant difference in zero point between the Cepheid distance scale and the RR Lyrae distance scale. Probably the most encouraging development in the last few years is the convergence on the distance to the Virgo cluster. Three somewhat independent methods, Cepheids, surface brightness fluctuations, and the planetary nebula luminosity function all give a consistent distance to Virgo. The consistency of this distance is now even more firmly anchored in the latest Cepheid results from the HST Key Program (Graham *et al.* 1997). However, the distance to Virgo is a significant step removed from the determination of H_0 due to the tremendous difficulty of determining the true cosmic velocity of Virgo. This difficulty is caused by the structure of the Virgo cluster itself and the unknown infall velocity of the Local Group. Using a consistent set of values for infall and the cosmic velocity of Virgo does yield an estimate of H_0 which is 80 ± 10, with a possible systematic effect of +10% if the distance to the LMC is $(m - M) = 18.35$, as suggested by several indicators.

At odds with these estimations is the determination of the extragalactic distance scale using type Ia supernovae as standard candles. After applying corrections based on the luminosity vs. decline rate relation, SN Ia appear to give very good relative distances between galaxies. The most recent absolute calibration of that data set, based on Cepheid distances to the calibrators, yields $H_0 = 65 \pm 5$ (Riess, Press, and Kirshner 1996). However, this does not mean that cosmologists should be content in believing that H_0 is now pinned down to occupy the range 65–80. As Sandage (1996) is quick to point out, there remain several possible sources of systematic error that can push the data down to $H_0 = 50$. As this debate is now almost 35 years old, it seems that the best resolution will come when distance estimates from more direct means become available. Very recently an opportunity has arisen.

In December 1994 one of the components of the lensed QSO system 0957 + 061 brightened by 10%. Previous timing observations in the radio and optical bands have failed to distinguish between a delay of 536 days and a delay of 415 days. Analysis of the light curve since the December 1994 event by Kundic *et al.* (1997) seems to have resolved this ambiguity and produced a delay time of ≈ 410 days. Although there is still considerable uncertainty in the mass distribution within the lensing system (Fischer *et al.* 1997), a tentative value of $H_0 \approx 82$ can be derived using the Grogin and Narayan (1996) model of the mass distribution. Until more such systems are measured, this value again should be regarded as consistent with other determinations but in no way is it definitive.

If a consistent extragalactic distance scale can ever be obtained from observations which indicate that H_0 is greater than 70 and if we believe that stellar evolutionary theory is sufficiently understood to yield fairly precise ages for globular cluster stars, then we have little choice but to appeal to nonzero Λ as the principal means of resolving this conflict. Throughout this book we have presented other lines of observational evidence that favor nonzero Λ. In particular, if H_0 is greater than 70 then all $\Lambda = 0$ CDM structure formation models can be ruled out, leaving behind various nonzero Λ models which have already been shown to fit the data on large-scale structure rather well. Thus the determination of the extragalactic distance scale and the measure of the power spectrum of galaxies on large scales can be made consistent in a nonzero Λ cosmology. This is either a coincidence or observations are telling us something fundamental about the nature of the Universe.

The major effect of a positive value of Λ is that the volume element per unit redshift is significantly higher than in the $\Lambda = 0$ case. In principle the space density of QSO absorption lines (Malhotra and Turner 1996) and gravitational lens systems are sensitive probes of the volume per unit redshift (Schneider 1995). At the moment, the data set is not large enough to adequately constrain Λ – but see Kochanek (1996) for an optimistic counterargument – leaving very much open the possibility that the current data is best reconciled with nonzero Λ. In a possibly prescient piece of work, Pen (1998) shows that it's not possible to reconcile CDM with both standard Big Bang nucleosynthesis and the luminosity–temperature relation for clusters of galaxies. Pen concludes that the luminosity–temperature relation, in conjunction with the measurment of a possibly low deuterium abundance by Tytler, Fan, and Burles (1996), greatly favors the Λ-dominated Universe.

Hence, several lines of observational evidence now suggest nonzero Λ. As a consequence, models which now incorporate Λ are being seriously considered. Indeed, the best evidence of this is the current plethora of papers which deal specifically with nonzero Λ models and/or attempts to verify or refute it with observations of gravitational lenses (e.g., Klypin, Nolthenius, and Primack, 1997; Kashlinsky and Jimenez 1997; Horack *et al.* 1996; Mo and Fukugita 1996; Kochanek 1996; Amendola 1996; Lin *et al.* 1996; Wu and Shude 1996; Efstathiou 1995; Bunn and Sugiyama 1995; Malhotra and Turner 1995; Sivaram 1994; Ratra and Peebles 1994; Peebles 1994). These models were virtually absent from the literature just a few years ago. Whether this is just a temporary condition or a manifestation of real learning will only be discerned when future data is available to confront these new models. In any event, the historical comments made in Chapter 1 are relevant here. Cosmological models change with time – should future data demonstrate that Λ is indeed 0 and that H_0 is greater than 70, and that the Universe is at least 15 billion years old, then very little of what has been presented in this book can be correct.

What is the nature of the large-scale distribution of matter in the Universe as traced by the three-dimensional galaxy distribution?

This was the subject of Chapter 3. There is no doubt from the observations that the distribution of light (galaxies) is clustered on a variety of size scales (Bahcall 1995). Redshift surveys such as the CFA slice survey, or the Southern Sky Strip survey have revealed a remarkably complex galaxy distribution. Attempts to push deeper into redshift space have produced power spectra which have features on very large scales. This high degree of clustering makes it very difficult to determine whether a fair volume of the Universe has yet been sampled. If light traces the distribution of mass fairly on large scales (another open question), then the high degree of galaxy clustering suggests inhomogeneities in the mass distribution. These inhomogeneities will manifest themselves by exerting a gravitational influence on other galaxies (which now are test particles probing the velocity field) causing a deviation from pure expansion velocity. These peculiar velocities which arise from variations in the density field need to be accounted for in any full kinematical description of the local velocity field. As a result, determining H_0 from local observations (e.g., the distance to the Virgo cluster) is severely compromised by uncertainties in our model of the local velocity field. This is why a determination of the distance to Coma, a very massive cluster which ought to have very little peculiar velocity, is much better for determining H_0.

Although we are far from having a complete picture of large-scale structure, the following ideas seem fairly well-established:

- On small scales both the galaxy–galaxy correlation length and the pairwise velocity dispersion are known fairly accurately. These values are $\approx (8 \pm 1)h^{-1}$ Mpc and 300 ± 100 km s^{-1} respectively. These values could reflect the conditions of structure formation or they could be modifications as a result of significant small-scale merging.
- There is a definitive relation between surface brightness (or morphological type) and local galaxy density. Ellipticals (very high surface brightness systems) are prevalent in the virialized cores of clusters whereas spiral galaxies (and LSB galaxies) occur in regions of lower galaxy density. This suggests that sparse sampling of spiral galaxies might be the best tracers of the large-scale density field.
- Clusters of galaxies themselves seem to be strongly correlated (more so than individual galaxies) and form larger superclusters. In turn there is some evidence from orientation measures that even superclusters are correlated. The most intriguing observations in this regard are those of West (1994), who demonstrates that the major axis of many powerful radio galaxies in the cores of rich clusters is aligned to the axis of the host supercluster. This remarkable observation suggests that structure formation maintains its memory of initial conditions over many length scales.
- It's clear that we live in a void-filled Universe with the void filling factor approaching 75%. The characteristic size of a void is ≈ 3500 km s^{-1} in diameter and the larger ones are ≈ 6000 km s^{-1}. The once astounding Boötes void now appears to be a typical large void of diameter 5000–6000 km s^{-1} (Aldering *et al.* 1997). Although distortions in redshift space caused by peculiar velocities can render the appearance of physical structure in redshift surveys, measurements of relative distances to galaxies on the near and far side of voids (bubbles) indicate that the voids are very nearly spherical structures which are massless (Bothun *et al.* 1992). The intersections of these spherical voids produce surfaces inhabited by galaxies, giving rise to a filamentary component in the galaxy distribution. The most striking example of this is the Great Wall (Geller and Huchra 1989), which is now known to be a real structure in physical space. Distance measurements by Dell'Antonio, Geller, and Bothun (1996) show that the wall is a cold, thin structure with little peculiar or shear velocity.

- The peculiar velocity field of the nearby Universe is complex. Although the signature of a dipole moment is unmistakable, the source of that perturbation remains unidentified. The existence of quadrupole moments in the nearby velocity field now seem well established. However, there is no clear resolution between bulk flow models in which virialized structures themselves are participating in the flow, and infall models (e.g., the Great Attractor) generated by virialized structures which are at rest. Attempts to infer the large-scale density field from well-defined redshift surveys (e.g., Fisher *et al.* 1995) to predict the all-sky peculiar velocity vectors have had only marginal success in matching the observations. The scale over which the peculiar velocity field converges remains unknown. The Lauer and Postman (1994) result suggests that volumes with diameters as large as 15 000 km s^{-1} have a bulk peculiar velocity of a few hundred kilometers per second. Such a flow cannot be accounted for in any structure formation model given the observed degree of anisotropy in the CMB as measured by *COBE*. Recent attempts at recovering the LP vector from other data sets (e.g., Riess, Press, and Kirshner 1996; Giovanelli *et al.* 1996; Borgani *et al.* 1995; Dell'Antonio, Geller, and Bothun 1996) are marginally inconsistent with the original result but cannot rule it out.
- Attempts to measure Ω directly from the peculiar velocity field have yielded ambiguous results. Sample- and scale-dependent results permeate the literature. Moreover, under the linear biasing model it is only the parameter

$$\beta = \frac{f(\Omega_0, \Lambda)}{b} \sim \frac{\Omega_0^{0.6}}{b}$$

  which is measured. In their excellent and comprehensive review of the current situation, Strauss and Willick (1995) use 24 different methods to estimate β. Those estimates can be characterized by a quasi-Gaussian distribution with mean 0.75 and dispersion 0.30. It is thus clear that β is unknown to at least a factor of 2.5. Furthermore, since the value of b is unknown (although most recent results suggest $b \approx 1$), this introduces more uncertainty into the determination of Ω_0.
- Clusters of galaxies are very ambiguous tracers of Ω. On the one hand, the high degree of substructure observed in most clusters suggests a long formation process and high Ω as structure freezes out (e.g., stops growing) roughly at $z = 1/\Omega$. On the other hand, recent observations of the baryonic mass fraction in clusters strongly argues for $\Omega \leq 0.3$. Clearly, clusters of galaxies are not a clean laboratory for determinations of Ω.

What is the evidence for the existence of dark matter and what is its overall contribution to the total mass density of the Universe?

This was the subject of Chapter 4 and the nature of the dark matter (DM) remains a profound cosmological mystery. The evidence of some amount of DM, based on galaxy rotation curves, the velocity dispersion of clusters of galaxies, and gravitational lens systems is overwhelming and unambiguous. In general, these systems have total M/L ratios which suggest $\Omega \approx 0.2$. On the other hand, if $\Omega = 1$ then it is clear that the great majority of the DM must be nonbaryonic. The existence of nonbaryonic gravitating DM may be key to the formation of galaxies in the Universe, as such material will be much less affected by radiation drag in the early Universe; hence density perturbations can begin to grow at very early times. Fluctuations in purely baryonic matter are easily damped out by the radiation when baryonic matter is coupled to the photon field. The presence of significant amounts of nonbaryonic DM means that an initially very smooth Universe (thought to characterize its early state) can give rise to structure at later times.

The elegant inflationary paradigm, which accurately predicts most of the large-sale properties of the Universe, strongly predicts a spatially flat Universe. For $\Lambda = 0$ this means $\Omega = 1$ and dominance by nonbaryonic matter. Although the inflationary paradigm is deeply rooted in the (unknown) physics of the very early Universe, and hence one is predisposed to favor it as the proper paradigm, there is rather little observational evidence to support its $\Omega = 1$ prediction. Thus spatial flatness is achieved through some combination of Ω and Λ, or the paradigm is incorrect, or the Universe is still more complex than standard inflation can account for. Clearly we will be unable to distinguish between these possibilities until we actually know what the DM is. Although the list of candidates appears to grow each year, the current status can be summarized as follows:

- *Baryonic DM*: The constraints on the total baryonic content of the Universe from considerations of primordial nucleysynthesis and the observed abundances of light elements are fairly severe. At most, they allow Ω_b to be 0.1. However, since the luminous parts of galaxies only contribute ≈ 0.005 to Ω, we know that baryonic DM must exist and is most likely in the form of low mass stars or stellar remnants in normal galaxies and a significant population of LSB galaxies. Direct searches for baryonic DM are now underway and the microlensing experiments towards the LMC and the galactic bulge have produced a significant number of events. The line of sight toward the LMC is more sensitive to halo baryonic DM and to date, a total of eight events have been detected by the MACHO project. Analysis of these events is still ongoing, but the distribution of lensing masses is consistent with a population of low mass main sequences stars and/or white dwarfs. Filling up the halo with Jupiter mass objects can be ruled out. The total amount of baryonic DM in the halo inferred from these lensing events is consistent with that inferred from the rotation curve of our galaxy and observations of satellite companion galaxies to bright spiral hosts whose relative velocities suggest total halo masses of $(1\text{–}2) \times 10^{12} M_{\odot}$ and radial extents of ≈ 200 kpc (Zaritsky *et al.* 1997).

- *CDM*: Although CDM remains the preferred theory for structure formation (due to the intrinsic shape of the CDM power spectrum), there is still no identifiable CDM particle from a plethora of candidates. Supersymmetric theory in combination with R-parity conservation predicts that a stable particle must exist, and this is the best hope for the CDM particle. Candidate particles come and go each year in the particle physics world and it becomes a matter of picking the favorite "ino" (neutralino, higgsino, photino, etc.). The search for these particles is ongoing in terrestrial accelerators, and ultimately the fundamental nature of the DM, and hence our cosmology, may be found at the end of a beam pipe instead of a telescope.

- *HDM*: At least HDM has a known particle which may have a mass, the neutrino. Furthermore, its cosmological density is known. A plausible resolution to the observed deficit of solar neutrinos appeals to oscillations between muon and electron neutrinos as each has a small mass. Positive combined neutrino masses in the range of 1–10 eV have high cosmological significance as they would supply much of the observed power on large scales. A neutrino-dominated Universe, however, is ruled out by the observation that the ages of the galaxies are approximately H_0^{-1}.

How did structure form in the Universe and what formation scenarios are consistent with the current observational data?

This was the subject of Chapter 5. A quick summary is trivial: we don't have a structure formation scenario that works. More definitively, no single structure formation scenario can satisfy the simultaneous large- and small-scale constraints. The traditional CDM, $\Omega = 1$ model overproduces small-scale structure when normalized to the *COBE* data. Purely baryonic fluctuations (the PBI model) don't appear likely as (1) they require Ω_b to be in excess of the primordial nucleosynthesis constraint and (2) they are inconsistent with the (admittedly uncertain) degree-scale anisotropy observed in the CMB. To suppress the production of small-scale structure, CDM must be augmented in a manner that essentially extends the time it takes for matter and radiation to equilibrate. This can be accomplished by lowering the overall matter density (low H_0), increasing the radiation density by adding extra relativistic particles, tilting the primordial spectrum to suppress small-scale fluctuations, adding gravity waves as a source of the observed CBR anisotropy, giving neutrinos a small mass (the MDM model), or having Λ dominate over Ω in establishing spatial flatness.

Each of these adjustments to the standard CDM theory is somewhat exotic and renders structure formation all the more complex. Many of these augmentations are nearly ruled out via observations of H_0 and the highly Gaussian nature of the CMB fluctuations. Given this, it seems highly significant that several investigators' characterizations of the power spectrum of galaxies appear to converge and they strongly favor either open Universe or nonzero Λ models. Finally, although the gravitational instability paradigm remains key, the overall topology of the Universe strongly suggests that feedback (from explosions?) and other nongravitational forces have played a role in structure formation. This has led to the incorporation of hydrodynamics into N-body simulations with often spectacular results (see animations on the Web page). Now, if we could only understand the role that magnetic fields play in collapsing structures, we would have all the physics input necessary to construct a more rigorous set of models. As a consequence of this, models of galaxy formation are necessarily incomplete.

> *Where do the baryons reside? Are they predominately inside or outside of galaxies? How efficient was the process of galaxy formation? What is the true nature of the galaxy population? Do we have a representative survey of galaxies in the nearby Universe from which coherent arguments about galaxy formation and evolution can be made?*

This was the subject of Chapter 6. In general, the various extragalactic backgrounds (aside from the CMB) that have been discovered can be well explained by an aggregate discrete source population. There is very little evidence for a diffuse intergalactic medium (IGM) and certainly no evidence that most of the baryons are in the IGM. The nature of the IGM itself and the evolution of its ionization state with redshift is strongly constrained by the density and ionization states of various QSO absorption line systems. At high redshift, these pressure-confined IGM clouds, which won't rapidly collapse to form galaxies, probably contain much of the baryonic material (Weinberg, Hernquist, and Katz 1997). The evolution of these clouds is not well known but they could cool to form dwarf galaxies at some later epoch. Thus baryons seem to be strongly confined to galactic-size potential wells, which argues that the capture of baryonic gas by these potentials was fairly efficient. Despite this efficiency, we have never actually observed a galaxy in the process of formation, despite many dedicated searches (Thompson, Djorgovski, and Trauger 1995; Pahre and Djorgovski 1995). This implies that either galaxy formation is hidden from us – i.e., it's masked by the presence of QSO activity or early star formation is shrouded in dust (Zepf and Silk 1996) – or that the bulk of stars even in spheroid-dominated galaxies form on a timescale much longer than the dynamical timescale of the system. The morphology of galaxies in the Hubble deep field (Figure 5.3) has been used as evidence that galaxy formation is an extended process that involves assembling and accreting subgalaxy-size regions (Pascarelle *et al.* 1996).

At much lower redshift, the recent discovery of LSB galaxies and their now inferred large space density is a strong indication that surface brightness selection effects have been severe and quite effective at preventing a representative sample of $z = 0$ galaxies to be obtained. We have thus predominately observed only one mode of galaxy evolution, evolution which leads to a conspicuous HSB object. But if Figure 6.12 is an accurate representation, it tells us that we are only just beginning the kinds of observations we will need when constructing a representative sample of galaxies. Thus great caution must be exercised in trying to compare the properties and space density of galaxies at modest and distant redshifts to those nearby. As is often the case in extragalactic astronomy, we seem to have more knowledge about distant samples of galaxies than nearby samples. The discovery of LSB galaxies has significantly increased the baryonic census of the nearby Universe (bringing it better into accord with the primordial nucleosynthesis constraints) and has fundamentally altered the faint end slope of the galaxy luminosity function. These galaxies have also given us a window into a different avenue of galaxy evolution, which was closed to us for many years when we failed to detect these diffuse objects.

7.2 Significant observations as a function of redshift

Here is a list of significant observations as a function of redshift, observations which have been made or are waiting to be made. Each one bears directly on many of the issues discussed in this book:

- $z = 1100$: The *COBE*-measured quadrupole anistotropy provides excellent confirmation that gravitational instability and the associated Sachs–Wolfe effect, at some level, must have provided the seeds for structure formation. It is particularly reassuring that the observed level of anisotropy is close to the value predicted from very simple arguments.
- $z \approx 100$: The Universe is 10^{-3} of its present age at this redshift. This time interval corresponds to the dynamical timescale of globular clusters. It is thus possible that the first star formation and metal production occurs at this redshift. The photon flux from this putative formation event would be redshifted to wavelengths of 10–100 μm, where it would be largely undetectable due to a very weak signal in the presence of a noisy background.
- $z \approx 20$: At this redshift the Universe is approximately 1% of its expansion age, which coincides with the dynamical timescale of a typical elliptical galaxy. We thus expect spheroid formation to occur at this redshift, but so far there is no observational evidence for this.
- $z \approx 5$: The first luminous signposts appear in the form of QSOs. Since the energy budget of QSOs requires infall of gas into supermassive black holes, the engines for the QSOs had to form at higher redshifts. If these supermassive black holes have evolved from the gravitational coalescence of stellar remnants in a massive stellar cluster, the precursor to the engine formation could have been the first era of metal production in the Universe. This could have occurred at $z \approx 100$.
- $z \approx 4.5$: The highest redshift, damped Lyman$_\alpha$ system is now observed along with the presence of Lyman$_\alpha$ emitting galaxies. This indicates that (1) the Universe is not completely ionized at this redshift, (2) large column density clouds of H I exist, and (3) star formation in some galaxies is occurring. If the IGM is completely ionized at some postrecombination redshift, it would take some time for components to cool and collapse into high column density, neutral hydrogen clouds. Although it is difficult to estimate the cooling times in the absence of many metals (one has to appeal to complex models of cooling involving molecular hydrogen line emission) the cooling time is not likely to be less than 10^8 years. This suggests a lower limit for the epoch of complete reionization by QSOs to be $z \approx 10$; see also the arguments given in Peebles (1993).
- $z \approx 3.5$: Deep imaging and follow-up spectroscopy has now revealed the presence of apparently normal star-forming galaxies at this redshift. The inferred star formation and metal production rates are modest. In one case, a galaxy at $z \approx 3$ appears to have a spectrum consistent with a stellar population that is a few gigayears old. If true, this is further evidence of a significant conflict between the

expansion age of the Universe and the ages of the oldest stars and galaxies that inhabit it.

- $z \approx 3.0$: Observations of metal-line QSO absorption systems suggest a mean metallicity at this redshift of ≈ 0.01 solar.
- $z \approx 3.0$: The amount of gas which is locked up in damped Lyman$_\alpha$ systems appears to reach a maximum at this redshift. This probably marks the beginning of the formation of the galaxies' disk components, likely to be a slow process that continues over many epochs of z.
- $z \approx 1.5$: Observations of metal-line QSO absorption systems now suggest a mean metallicity at this redshift of ≈ 0.1 solar. This factor of 10 increase in mean metallicity since $z \approx 3.0$ can be attributed to the first generations of stars which form in galactic disks. Age dating of our own galaxy via use of the white dwarf cooling curve (Winget *et al.* 1987) strongly suggests that the oldest stars in our disk formed at $z = 1$–1.5.
- $z \approx 1.0$: The highest redshift X-ray cluster yet detected (Hattori *et al.* 1997) indicates that the virialization of rich clusters did occur by $z \approx 1$. Observations of high redshift radio galaxies and QSOs are also consistent with some clustering occurring up to $z \approx 2$.
- $z = 0.083$: The discovery of Malin 1 at this redshift indicates that very large gaseous disks can remain cold and relatively devoid of stars even at very recent epochs. This shows that disk galaxy evolution can indeed be very, very slow and that some baryons are fairly effective at hiding.

7.3 Future observations

What will the future bring? Will new and better data resolve most of the problems addressed in this book or will it open up a whole new set of cosmological challenges that indicate a more complex Universe than we would care to deal with? This is what makes the study of cosmology invigorating. The influx of good data can't be ignored and often times good data illuminates things that we don't understand instead of confirming things we thought we knew. Such is the balance between observations and theory, and the following set of planned observations will most certainly upset that balance in an unpredictable manner:

- *The Sloan Digital Sky Survey*: This five-color survey for galaxies and follow-up spectroscopy to reach a goal of 10^6 redshifts should produce a uniform set of data that can be corrected for surface brightness selection effects to finally produce a real catalog of galaxies. This will be the acid test of the preponderance of LSB systems in the Universe and should provide a definitive measure of the local luminosity function of galaxies, hence a proper census of baryonic material. A million redshifts should also provide us with a "fair" sample from which the large-scale peculiar velocity field can be determined to greater accuracy.

- *CMB anisotropy measurements on the 0.5–1.0° scale*: Improvements in ground-based and balloon-borne instruments are now detecting anisotropy on this scale but with some uncertainty in the overall amplitude. Measures of the power spectrum reveal large-scale power on the 100–200h^{-1} Mpc scale, which corresponds to an angular scale of 0.5–1.0° at $z \approx 1000$. This data should effectively discriminate between the various exotic fixes to CDM theory as well as the alternative theories such as explosions, PBI or topological defects.
- *Keck 10 m observations of QSO absorption line systems*: These observations reflect the metal production history of the Universe over the range $z = 1$–5. The use of a large aperture allows fainter systems to be probed at higher resolution, resulting in much more secure line identification (Lu *et al.* 1996).
- *The Hubble deep field (Figure 5.3)*: This deep survey has produced 2000 interesting galaxies that are at various stages of formation at various redshifts. Already it has been claimed on the basis of color analysis of objects in the HDF that the very reddest galaxies likely are at $z \geq 4$ (Clements and Couch 1996). A candidate very red object is shown in Figure 7.1. This object is purported to be at $z \geq 6$ (Lanzetta, Yahil, and Fernandez-Soto 1996). Obviously this awaits spectroscopic confirmation, but in general the HDF is the first true extragalactic gold mine of information produced by the HST and it's the subject of much new analysis (see the suggested further reading). The redshift distribution of some of these HDF galaxies obtained recently with the Keck telescope is shown in Figure 7.2. Most of these are at low redshift (a manifestation of the FBG problem discussed in Chapter 6) with a handful between redshifts 2 and 3.5.
- *Surveys for weak gravitational lensing*: In essence this is a survey to paint a gravity map of the Universe via lensing distortions. The space density of lensed systems as a function of redshift is a direct reflection of the volume per unit redshift. It is surveys like these that will ultimately produce a positive detection of Λ or a strong upper limit; hence they will rank as some of the most important cosmological observations ever made.
- *The continued quest for H_0 and Ω*: If distance estimation techniques continue to improve, we may one day look forward to direct measures of these parameters to an accuracy of a few percent. Although past history suggests this level of accuracy isn't going to be achieved any time soon in a manner that is accepted by most (unless some new physical technique emerges), it is clear that our techniques are getting better. If and when these parameters are measured at this level, it will slam the door on a number of cosmological models. The most interesting aspect of all of this would be a reliable determination that Ω is low, meaning either Λ dominates or the Universe is very open. This would send theorists scrambling for new structure formation models.
- *Particle physics and cosmology*: The detection of a supersymmetric particle and/or a definitive measurement of neutrino mass would elevate experimental particle physics to observational cosmology. Direct detection of the DM, by any means, is probably the single most important new cosmological observation that can be made.

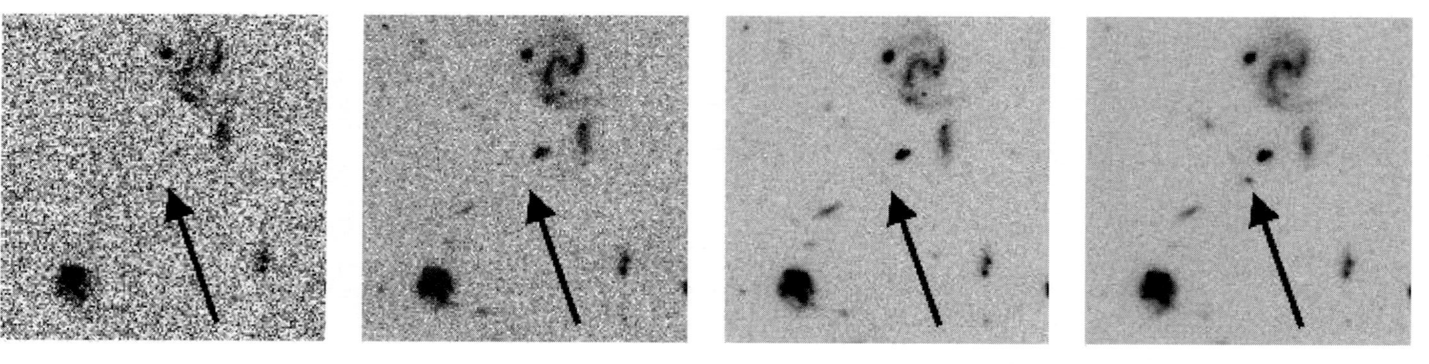

Figure 7.1 Filter photometry of a very red galaxy in the Hubble deep field. From left to right, filters are 3000, 4500, 6000 and 8100 Å. The object appears strongest in the 8100 Å filter. Under the assumption that this increase in brightness is due to redshift Lyman$_\alpha$ radiation, the redshift of this object would be $z > 6$. Courtesy of Ken Lanzetta, State University of New York.

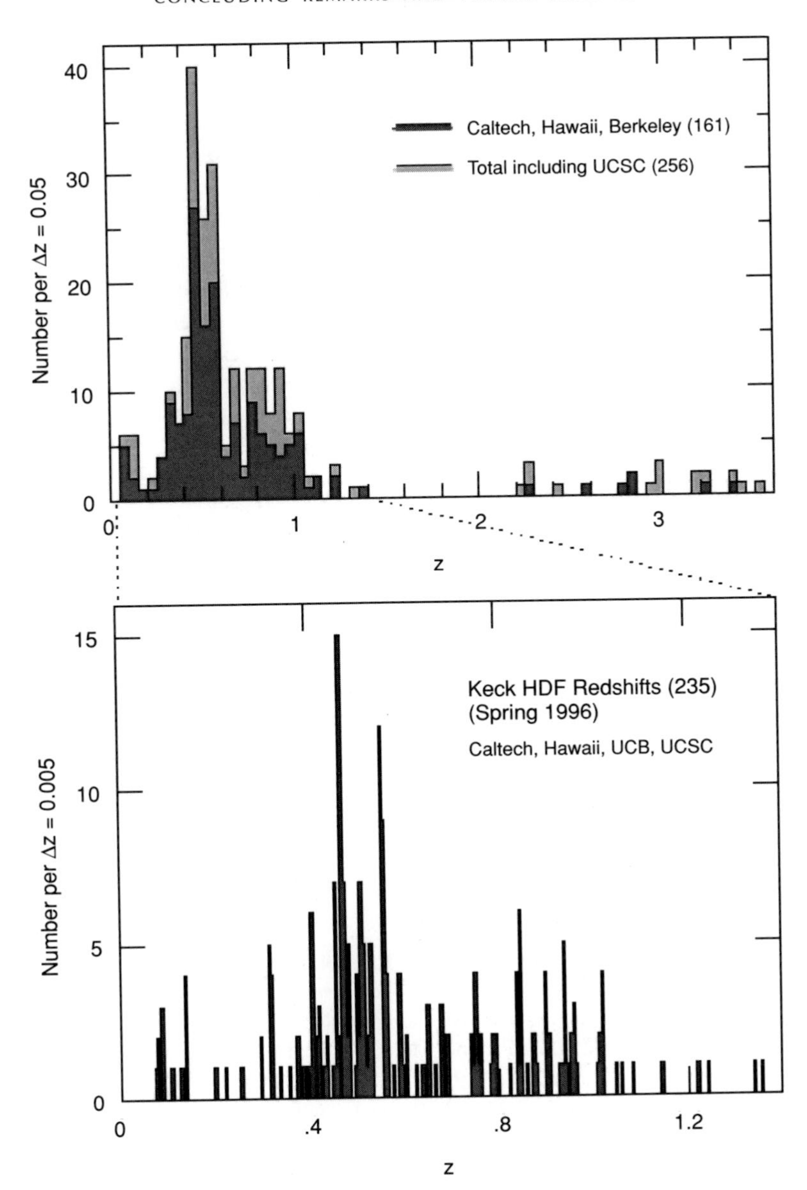

Figure 7.2 Distribution of redshifts of objects in the Hubble deep field obtained with the Keck 10 m telescope. The majority of objects with measured redshifts to date have relatively low redshift. This is consistent with the hypothesis that galaxy formation occurs over a long period and there are rather few very high redshift galaxies in the HDF (Cohen *et al.* 1996). Courtesy of Dr Andrew Phillips, University of California.

- *The Advanced X-ray Astronomical Facility (AXAF)*: Like HST and CGRO, AXAF is the third of the great observatory programs of NASA. AXAF will be significantly more sensitive than any previous X-ray mission and should provide the observations necessary to construct the X-ray luminosity function of galaxies and its evolution with redshift. The detection of X-ray clusters with $z \geq 2$ would be rather unexpected in CDM structure formation scenarios.
- *Shuttle Infrared Telescope Facility (SIRTF)*: This is the last of the great observatories. SIRTF will have the sensitivity to make a better determination of the character of the diffuse extragalactic infrared light and determine, once and for all, whether dusty shrouds around protogalaxies have so far prevented their detection. Peering through dusty curtains into the heart of a forming galaxy is akin to witnessing a primal birth and may have spiritual ramifications that transcend our cosmological model.

This array of new observations will provide a clear test of the validity and usefulness of this book. How much of what is described here will be valid when the new data comes in? Do we have the basic framework correct but just lack the details, or have we missed something fundamental? Will our fervor for a nonzero cosmological constant survive the test of these new observations? Whichever is the case, it seems that our grand and insatiable curiosity about the nature of the Universe will continue to grow, and as our observational and theoretical knowledge base expands, we can look forward to new and more complex challenges in the near future. The quest for knowledge seems to have sufficient momentum to strongly propel us into the next millennium in search of better cosmological models that ultimately forge the connection between human beings and the Cosmos.

References

ALDERING, G., BOTHUN, G., MARZKE, R., and KIRSHNER, R. 1997 preprint

AMENDOLA, L. 1996 *Monthly Notices of the Royal Astronomical Society* **283**, 983

BAHCALL, N. 1995 *Publications of the Astronomical Society of the Pacific* **107**, 790

BORGANI, S., PLIONIS, M., COLES, P., and MOSCARDINI, L. 1995 *Monthly Notices of the Royal Astronomical Society* **277**, 1191

BOTHUN, G., GELLER, M., KURTZ, M., HUCHRA, J., and SCHILD, R. 1992 *Astrophysical Journal* **395**, 349

BUNN, E. and SUGIYAMA, N. 1995 *Astrophysical Journal* **446**, 49

CLEMENTS, D. and COUCH, W. 1996 *Monthly Notices of the Royal Astronomical Society* **280**, L43

COHEN, J. *et al.* 1996 *Astrophysical Journal Letters* **471**, L5

DELL'ANTONIO, I., GELLER, M., and BOTHUN, G. 1996 *Astronomical Journal* **112**, 1780

EFSTATHIOU, G. 1995 *Monthly Notices of the Royal Astronomical Society* **274**, L73

FISCHER, P., BERNSTEIN, G., RHEE, G., and TYSON, A. 1997 *Astronomical Journal* **113**, 521

FISHER, K., LAHAV, O., HOFFMAN, Y., LYNDEN-BELL, D., and ZAROUBI, S. 1995 *Monthly Notices of the Royal Astronomical Society* **272**, 885

GELLER, M. and HUCHRA, J. 1989 *Science* **246**, 897

GIOVANELLI, R. *et al.* 1996 *Astrophysical Journal Letters* **464**, L99

GRAHAM, J. *et al.* 1997 *Astrophysical Journal* **477**, 535
GROGIN, N. and NARAYAN, R. 1996 *Astrophysical Journal* **464**, 92
HATTORI, M. *et al.* 1997 preprint
HORACK, J. *et al.* 1996 *Astrophysical Journal* **472**, 25
KASHLINSKY, A. and JIMENEZ, R. 1997 *Astrophysical Journal Letters* **474**, L81
KLYPIN, A., NOLTHENIUS, R., and PRIMACK, J. 1997 *Astrophysical Journal* **474**, 533
KOCHANEK, C. 1996 *Astrophysical Journal* **466**, 638
KUNDIC, T. *et al.* 1997 *Astrophysical Journal* **482**, 75
LANZETTA, K., YAHIL, A., and FERNANDEZ-SOTO, A. 1996 *Nature* **381**, 759
LAUER, T. and POSTMAN, R. 1994 *Astrophysical Journal* **425**, 418
LIN, H. *et al.* 1996 *Astrophysical Journal* **471**, 617
LU, L., SARGENT, W., WOMBLE, D., and TAKADA-HIDA, M. 1996 *Astrophysical Journal* **472**, 509
MALHOTRA, S. and TURNER, E. 1995 *Astrophysical Journal* **445**, 553
MO, H. and FUKUGITA, M. 1996 *Astrophysical Journal Letters* **467**, L9
PAHRE, M. and DJORGOVSKI, S. 1995 *Astrophysical Journal Letters* **449**, L1
PASCARELLE, S., WINDHORST, R., KEEL, W., and ODEWAHN, S. 1996 *Nature* **383**, 45
PEEBLES, P.J.E. 1993 *Principles of Physical Cosmology*, Princeton NJ: Princeton University Press
PEEBLES, P.J.E. 1994 *Astrophysical Journal Letters* **432**, L1
PEN, U. 1998 *Astrophysical Journal*, in press
RATRA, B. and PEEBLES, P.J.E. 1994 *Astrophysical Journal Letters* **432**, L5
RIESS, A., PRESS, W., and KIRSHNER, R. 1996 *Astrophysical Journal* **473**, 88
SANDAGE, A. 1996 Lecture given at May 1996 Meeting on the Distance Scale, Space Telescope Sciences Institute
SCHNEIDER, P. 1995 *Astronomy and Astrophysics* **302**, 639
SIVARAM, C. 1994 *Astrophysics and Space Science* **219**, 135
STRAUSS, M. and WILLICK, J. 1995 *Physics Reports* **261**, 271
THOMPSON, D., DJORGOVSKI, G., and TRAUGER, J. 1995 *Astronomical Journal* **110**, 963
TYTLER, D., FAN, X., and BURLES, S. 1996 *Nature* **381**, 207
WEINBERG, D., HERNQUIST, L., and KATZ, N. 1997 *Astrophysical Journal* **477**, 8
WEST, M. 1994 *Monthly Notices of the Royal Astronomical Society* **268**, 79
WINGET, D., HANSEN, C., LIBEBERT, J., VAN HORN, H., FONTAINE, G., NATHER, R., KEPLER, O., and LAMB, D. 1987 *Astrophysical Journal Letters* **315**, L77
WU, X. and SHUDE, M. 1996 *Astrophysical Journal* **463**, 404
ZARITSKY, D., SMITH, R., FRENK, C., and WHITE, S. 1997 *Astrophysical Journal* **478**, 39
ZEPF, S. and SILK, J. 1996 *Astrophysical Journal* **466**, 114

Suggestions for Further Reading

General cosmology books

- *Masks of the Universe*
  E. Harrison 1985
  Focuses on historical cosmologies; very interesting treatment

- *The Hidden Universe*
  Mike Disney 1984
  Focuses on the missing mass problem and properties of galaxies

- *Principles of Cosmology and Gravitation*
  M.V. Berry 1989
  Good description of Robertson–Walker metric and spacetime geometries

- *Cosmology*
  E. Harrison 1981
  General overview of cosmology at nontechnical level

- *Introduction to Cosmology*
  M. Roos 1994
  Modern discussion of general cosmological principles

- *Modern Cosmology and the Dark Matter Problem*
  D. Sciama 1993
  Theoretical overview of the dark matter issue

- *The Cosmological Distance Ladder: Distance and Time in the Universe*
  M. Rowan-Robinson 1985
  A bit dated but a generally good overview of the distance scale ladder discussed extensively in Chapter 2

General cosmology articles

- Observational tests of world models. *Annual Reviews of Astronomy and Astrophysics* **26** (1988) 561 (A. Sandage)
- Alternatives to the Big Bang. *Annual Reviews of Astronomy and Astrophysics* **22** (1987) 157 (G. Ellis)

Recent articles relevant to individual chapters

Chapter 2: Distance scale

- The age of the oldest globular clusters. *Ap.J.* **479** (1997) 665 (Salaris *et al.*)
- Time delay of QSO 0957 + 561 and cosmological implications. *Ap.J.* **479** (1997) L89 (Oscoz *et al.*)
- Testing the Hubble law with the IRAS 1.2 Jy redshift survey. *Ap.J.* **477** (1997) 36 (Koryani and Strauss)
- The Tully–Fisher relation and H_0. *Ap.J.* **477** (1997) L1 (Giovanelli *et al.*)
- Implications for the Hubble constant from the first seven supernovae at $z = 0.35$. *Ap.J.* **476** (1997) L63 (Kim *et al.*)
- H_0: the incredible shrinking constant, 1925–1975. *PASP* **198** (1996) 1073 (Trimble)

Chapter 3: Large-scale structure

- Galaxy groups, CDM/CHDM cosmologies, and the value of Ω_0. *Ap.J.* **480** (1997) 43 (Nolthenius *et al.*)
- The dynamical equilibrium of galaxy clusters. *Ap.J.* **476** (1997) L7 (Carlberg *et al.*)
- Probing large-scale structure using percolation and genus curves. *Ap.J.* **476** (1997) L1 (Sahni *et al.*)
- Estimating Ω from galaxy redshifts: linear flow distortions and nonlinear clustering. *Ap.J.* **475** (1997) 414 (Bromley *et al.*)
- X-ray clusters: towards a new determination of the density parameter of the universe. *A&A* **317** (1997) 1 (Oukbir and Blanchard)
- The creation of large-scale voids by explosions of primordial supernovae. *MNRAS* **283** (1996) 912 (Miranda *et al.*)
- The primeval mass fluctuation spectrum and the distribution of the nearby galaxies. *Ap.J.* **473** (1996) 57 (Peebles)

Chapter 4: Dark matter and inflation

- The MACHO project: 45 candidate microlensing events from the first-year galactic bulge data. *Ap.J.* **479** (1997) 119 (Alcock *et al.*)

- Detection of (dark) matter concentrations via weak gravitational lensing. *MNRAS* **283** (1996) 837 (Schneider)
- Galaxy dark matter: galaxy–galaxy lensing in the Hubble deep field. *Ap.J.* **473** (1996) L17 (Dell'Antonio and Tyson)
- Light neutralinos as dark matter in the unconstrained minimal supersymmetric standard model. *Ap.H.* **6** (1996) 1 (Gabutti *et al.*)
- A comparison of direct and indirect mass estimates for distant clusters of galaxies. *Ap.J.* **479** (1997) 70 (Smail *et al.*)
- On the origin of cusps in dark matter halos. *Ap.J.* **477** (1997) L9 (Fukushige and Makino)

Chapter 5: Structure formation and the cosmological constant

- Old galaxies at high redshift and the cosmological constant. *Ap.J.* **480** (1997) 466 (Krauss)
- A measurement of the cosmological constant using elliptical galaxies as strong gravitational lenses. *Ap.J.* **475** (1997) 457 (Im *et al.*)
- Molecules at high redshift: the evolution of the cool phase of protogalactic disks. *Ap.J.* **480** (1997) 145 (Norman and Spaans)
- New cosmological structures on medium angular scales detected with the Tenerife experiments. *Ap.J.* **480** (1997) L83 (Guitterez *et al.*)
- The 4 year COBE normalization and large-scale structure. *Ap.J.* **480** (1997) 6 (Bunn and Martin)
- Statistical tests for CHDM and λ CDM cosmologies. *Ap.J.* **479** (1997) 580 (Ghigna *et al.*)

Chapter 6: Distribution of baryons and formation of galaxies

- High-redshift supernovae and the metal-poor halo stars: signatures of the first generation of galaxies. *Ap.J.* **478** (1997) L57 (Miralde-Escude and Rees)
- The effects of a photoionizing ultraviolet background on the formation of disk galaxies. *Ap.J.* **478** (1997) 13 (Navarro and Steinmetz)
- The baryon fraction in the Perseus cluster. *Ap.J.* **476** (1997) 479 (Cruddace *et al.*)
- High-redshift galaxies in the Hubble deep field: colour selection and star formation history to $z\sim4$. *MNRAS* **283** (1996) 1388 (Madau *et al.*)
- Baryonic dark halos: a cold gas component? *Ap.J.* **472** (1996) 34 (Gerhard and Silk)

Visit the www site: `http://zebu.uoregon.edu/tandf.html`

Index